Seismic Analysis and Design of Ocean and Underground Structures

Seismic Analysis and Design of Ocean and Underground Structures

Guest Editors

Xin Bao
Jingqi Huang
Piguang Wang

MDPI

Basel • Beijing • Wuhan • Barcelona • Belgrade • Novi Sad • Cluj • Manchester

Guest Editors
Xin Bao
Key Laboratory of Coastal and Offshore Engineering
Dalian University of Technology
Dalian
China

Jingqi Huang
Beijing Key Laboratory of Urban Underground Space Engineering
University of Science and Technology Beijing
Beijing
China

Piguang Wang
Key Laboratory of Urban Security and Disaster Engineering of Ministry of Education
Beijing University of Technology
Beijing
China

Editorial Office
MDPI AG
Grosspeteranlage 5
4052 Basel, Switzerland

This is a reprint of the Special Issue, published open access by the journal *Applied Sciences* (ISSN 2076-3417), freely accessible at: https://www.mdpi.com/journal/applsci/special_issues/4CYIFZNKLU.

For citation purposes, cite each article independently as indicated on the article page online and as indicated below:

Lastname, A.A.; Lastname, B.B. Article Title. *Journal Name* **Year**, *Volume Number*, Page Range.

ISBN 978-3-7258-7118-6 (Hbk)
ISBN 978-3-7258-7119-3 (PDF)
https://doi.org/10.3390/books978-3-7258-7119-3

Contents

About the Editors

Xin Bao

Xin Bao is a researcher at the Faculty of Infrastructure Engineering at Dalian University of Technology, China. His research interests include structural seismic analysis, fluid–structure interaction, soil–structure interaction, and impact engineering. Against the strategic backdrop of "deep sea and deep earth" construction, his work focuses on the development of equipment for marine energy exploitation and the seismic and blast resistance of major marine and underground structures. He has led several research projects involving the safe development of marine resources and the design of offshore artificial islands.

Jingqi Huang

Jingqi Huang is a faculty member at the School of Civil and Resource Engineering, University of Science and Technology Beijing, China. His research expertise lies in the seismic response of underground structures and tunnels, utilizing advanced numerical simulation and analytical methods to enhance structural safety. He has participated in several research projects aimed at improving the seismic resilience of urban underground transportation systems. His scholarly achievements primarily focus on developing theoretical and computational frameworks for analyzing the stability of subterranean infrastructure under dynamic seismic loading.

Piguang Wang

Piguang Wang is a faculty member at the School of Architecture, Civil and Transportation Engineering, Beijing University of Technology, China. His main research interests include the seismic performance of marine structures, fluid–structure interaction, and pile–soil interaction. He has participated in numerous academic projects exploring the dynamic stability of marine foundations and the behavior of coastal engineering structures under extreme environmental loads. His professional achievements include advancing the scientific understanding of the complex interactions between marine structures and their surrounding soil–fluid environments during seismic events.

Preface

The rapid advancement of global maritime strategies and the intensive development of urban subsurface spaces have ushered in a new era for major infrastructure projects. As civil and ocean engineering projects transition into deeper and more hostile environments, the necessity for specialized seismic analysis and design has become a critical priority for ensuring structural integrity and public safety. This Reprint, titled "Seismic Analysis and Design of Ocean and Underground Structures", focuses on the intersection of advanced computational mechanics, innovative material science, and resilient engineering design.

The motivation for this work stems from the inherent limitations of traditional seismic design methods when applied to the extreme conditions of the deep sea and the deep earth. Conventional approaches often overlook the complex fluid–structure interactions (FSI) and soil–structure interactions (SSI). These challenges are further compounded by environmental factors such as high water pressure, corrosion, and non-uniform soil sites. This Reprint aims to address these design gaps by presenting cutting-edge research that emphasizes structural resilience and disaster mitigation.

The scope of this Reprint covers a wide range of topics, including the validation of the coupled Eulerian–Lagrangian (CEL) method for complex underwater simulations, the analysis of modular artificial reefs, and the assessment of water-intake tunnels for nuclear power plants. A primary focus is placed on structural innovation, such as the application of split column technology to subway stations in weak interlayer sites, which offers a robust method for shifting failure modes from brittle shear to ductile bending. Furthermore, the work extends into intelligent monitoring and maintenance, covering compound-reinforced pipelines and deep-learning-based subsurface imaging.

This Reprint is addressed to researchers, academic scholars, and professional engineers within the fields of civil, ocean, and geotechnical engineering. It is also intended as a reference for graduate students seeking to understand the latest developments in seismic resilience for major infrastructure. By synthesizing these diverse research efforts, we hope to provide a comprehensive technical foundation that supports the safe and sustainable exploitation of marine and underground resources.

Xin Bao, Jingqi Huang, and Piguang Wang
Guest Editors

applied sciences

MDPI

Editorial

Seismic Analysis and Design of Ocean and Underground Structures: State-of-the-Art and Future Perspectives

Xin Bao [1,*], Kuichen Li [1], Jingqi Huang [2] and Piguang Wang [3]

1 Key Laboratory of Coastal and Offshore Engineering, Dalian University of Technology, Dalian 116024, China; kcl@mail.dlut.edu.cn
2 Beijing Key Laboratory of Urban Underground Space Engineering, School of Future Cities, University of Science and Technology Beijing, Beijing 100083, China; huangjingqi11@163.com
3 Key Laboratory of Urban Security and Disaster Engineering of Ministry of Education, Beijing University of Technology, Beijing 100124, China; wangpiguang1985@126.com
* Correspondence: baoxin@dlut.edu.cn

1. Introduction

Driven by the advancement of the global blue economy strategy and the rapid expansion of urbanization into deep underground spaces, the scale of critical infrastructure, ranging from cross-sea bridges and undersea tunnels to offshore wind farms and deep-buried utility tunnels, has reached unprecedented levels [1,2]. However, these massive structures must operate for extended periods in extreme environments characterized by high hydrostatic pressure, severe corrosion, and complex geotechnical conditions. Consequently, their seismic safety assessment faces significant challenges arising from the highly nonlinear behaviors of fluid–structure interaction (FSI), soil–structure interaction (SSI), and multi-physics coupling [3].

Traditional seismic analysis and design theories, primarily rooted in onshore engineering practices, often neglect the effects of FSI, SSI, and multi-physics coupling. This oversight can lead to critical design deficiencies when structures are subjected to earthquakes or cascading disasters. However, in recent years, with the rise of artificial intelligence (AI), high-performance materials, new structural systems, and the concept of resilient cities, research methods in this field have undergone profound changes. Consequently, analytical methods are transitioning from simplified physical modeling to refined simulations driven by physics–data fusion. Similarly, structural health monitoring (SHM) is evolving from traditional sensing to intelligent monitoring and predictive analytics. Furthermore, risk assessment paradigms are moving beyond isolated hazard scenarios toward systematic, comprehensive management strategies that integrate environmental evolution and multi-hazard cascading effects.

In light of these developments, recent progress in the seismic analysis and design of marine and underground structures is summarized herein. Key topics include refined modeling strategies for complex environments, intelligent SHM, high-performance materials, and multi-hazard risk assessment. These discussions offer essential theoretical support and technical guidance for ensuring the seismic safety and resilience of major infrastructure projects in the future.

2. An Overview of Published Articles

2.1. New Methods for Refined Modeling and Simulation in Complex Environments

Given the unique physical environments of marine and underground structures, recent seismic analysis research has prioritized improving simulation accuracy regarding complex boundaries and multi-physics coupling. For instance, Liu [4] proposed an FSI analysis method based on a curvilinear coordinate system and constructed a dual complex boundary model containing a rough sea surface and an irregular seabed. The equations of acoustic waves and elastic waves were mapped to the boundary coordination grid for solution. The numerical results showed that, compared with the ideal flat model, the irregular reflection interface induces strong ghost waves and multiple wave effects, which significantly change the propagation characteristics of seismic waves. Li et al. [5] reported the application of the coupled Eulerian–Lagrange (CEL) method in the seismic analysis of submerged floating tunnels. The study authors highlight that the traditional Morison equation, due to its assumption of a static structure, struggles to accurately simulate relative displacements in two-way fluid–structure interactions. In contrast, the CEL method, by discretizing fluids within an Eulerian grid and solids within a Lagrangian grid, provides a more realistic reproduction of the dynamic interaction between the structure and the water.

Refined modeling is often hindered by huge computational costs, however. Therefore, a new meshless simulation method that integrates labelled data and the physical information neural network (PINN) has shown great potential [6]. Although the traditional PINN has an inherent spectral bias problem when approximating high-frequency wave fields, Song and Wang [7] proposed an improved structure that embeds Fourier features to address this issue. This structure enables the network to effectively capture high-frequency wavefield details that are difficult to preserve using traditional methods by mapping the input data to a higher dimension. This method allows a single trained network to simultaneously output scattered wave fields from arbitrary source locations and with multiple frequency components, substantially improving the efficiency of multi-source, multi-frequency simulation.

2.2. Intelligent Structural Health Monitoring

As the scale of marine and underground engineering expands, SHM is shifting from traditional passive sensing to intelligent monitoring and prediction. To address the inefficiencies and bandwidth bottlenecks associated with transmitting massive data in large-scale networks, edge computing and decentralized architectures have emerged as critical solutions. AL-ZURIQAT et al. [8] proposed a decentralized diagnostic method for sensor faults based on the AI of Things (AIoT). This method breaks away from the traditional model that relies on a central server by directly embedding neural network models into wireless sensor nodes and portable master stations, enabling local fault detection and isolation. By transmitting only the diagnostic results rather than raw data, this design significantly reduces the data communication load and enhances the system's real-time performance and reliability.

Sensing methods are evolving from single-point monitoring to non-contact and distributed wide-area, full-scale coverage. For large-scale structures such as cross-sea bridges, Interferometric Synthetic Aperture Radar (InSAR) has become a key technology for capturing minute deformations [9]. Zhang et al. [10] combined empirical mode decomposition and machine learning to further develop the application of distributed acoustic sensing (DAS) in structure–environment integrated perception. They proved that DAS can not only monitor lining strain over long distances but also effectively capture and classify micro-earthquakes in external environments such as rockfalls and construction. DAS provides strong support for unmanned early warning of underground engineering disasters.

Furthermore, significant progress has been made in research on sensor environmental adaptability to the extreme service environments unique to deep-sea engineering, such as high hydrostatic pressure and strong corrosion. Compared to traditional electrical sensors, Fiber Bragg Grating (FBG) sensors have shown great potential in deep-sea exploration due

https://doi.org/10.3390/app16020919

to their inherent advantages of being lightweight, corrosion-resistant, and resistant to electromagnetic interference. Zhao et al. [11] systematically reviewed the latest achievements made in the field of FBG sensors and clarified the technical path of FBG as the preferred solution for stress and temperature monitoring in extreme deep-sea environments.

2.3. Innovation in Tough Materials and New Structural Systems

Faced with extreme operating conditions characterized by the convergence of deep-sea environments, deep-earth geotechnics, and strong seismic activity, conventional structural optimization alone is often insufficient to ensure the safety of ultra-large structures. Consequently, there is an imperative to incorporate high-performance materials and innovative structural systems capable of withstanding these extreme loading regimes. At the material level, Ultra-High-Performance Concrete (UHPC) has emerged as a pivotal solution. Its outstanding strength and superior interfacial bonding properties significantly enhance the resilience of infrastructure in complex environments. For instance, addressing the high hydrostatic pressure of deep-sea environments, Zhu et al. [12] developed a specialized UHPC formulation for subsea applications. This material effectively mitigates autogenous shrinkage microcracking, thereby overcoming durability challenges common in deep-sea structures. Regarding seismic design, Xie et al. [13] proposed a prefabricated connection scheme utilizing local UHPC reinforcement. By leveraging UHPC's superior bonding strength, this design successfully relocates seismic damage from the vulnerable joint core to non-critical regions. Furthermore, Xu et al. [14] systematically evaluated the seismic performance of split columns versus conventional monolithic columns in subway stations situated in soil profiles with weak interlayers. They observed that weak interlayers significantly amplify deformation and internal forces. The introduction of split columns effectively reduces lateral stiffness, shifting the failure mechanism from brittle shear failure to a more desirable ductile bending failure.

The development of novel smart metal materials has redefined the boundaries of structural energy dissipation and self-centering capabilities. For instance, a novel Fe-Mn-Si alloy damper exhibits a fatigue life and total energy dissipation capacity 10 times higher than those of traditional steel dampers, significantly enhancing the structural resistance against long-duration earthquakes [15]. In addition, significant progress has been made in the application of shape memory alloys (SMAs) in the field of engineering structures. In the design of flexible tunnel joints, in particular, SMAs can not only dissipate seismic energy but also achieve self-recovery [16].

At the macro-structural level, to overcome the problems of long construction cycles and severe sea condition constraints in traditional marine engineering, engineering-driven modular floating structures (MFSs) are gradually becoming the mainstream paradigm. Compared to traditional land reclamation, MFSs not only boast advantages such as environmental friendliness, flexible site selection, and ease of expansion but also effectively isolate the direct transmission of seabed seismic motion due to their unique floating characteristics, demonstrating excellent natural seismic isolation and earthquake resistance [17].

2.4. Risk Assessment and Defense Strategies in Complex Environments

Traditional land-based seismic models often suffer from applicability biases due to the unique geographical environments of marine and underground engineering projects. The authors of recent studies have focused on addressing this deficiency through refined modeling. For instance, in the seismic hazard analysis of the Gulf of Cádiz, research proved that ignoring the amplification effect of the soft seabed layer would lead to great design risks. Consequently, a modified seismic motion prediction model was proposed to account for these site-specific conditions [18]. With the implementation of new-generation

https://doi.org/10.3390/app16020919

standards such as ISO 19901-2, it is clearly stipulated that the variation in seismic motion input with depth must be considered. This factor marks a shift in seismic design from empirical estimation to refined analysis, effectively reducing the uncertainty in deep sea or deep earth engineering design [19]. In terms of physical disaster mitigation strategies, Bi et al. [20] studied the further development of a structure-tuned mass damper to significantly improve the toughness of submarine pipelines under dynamic loads.

In the face of increasingly complex environmental challenges, single earthquake hazard assessments are no longer sufficient to address systemic risks. Consequently, research focus is shifting toward multi-hazard coupling and cascading effects. Hain et al. [21] surveyed the post-disaster reconstruction of Puerto Rico and revealed that Hurricane Maria caused the accumulation of hidden damage to the structure. The structural corrosion and foundation scouring caused by the hurricane significantly weakened the load-bearing capacity of the infrastructure in the subsequent 2020 earthquake. Yu et al. [22] introduced the time dimension into the evaluation system, explaining that the environmental degradation caused by chloride ion corrosion and other factors is considered in conjunction with the dynamic disaster of seismicity, and the seismic reliability of marine structures will decrease nonlinearly throughout the entire life cycle. Ignoring this time-varying characteristic will lead to a serious underestimation of the risks in the later stages of service.

The macro-management model is also undergoing profound changes. Grandjean [23] highlighted in their latest analysis based on literature statistics that multi-hazard research is transforming from a single technical perspective to cross-departmental comprehensive risk management. In particular, for the planning of urban underground space and marine infrastructure, there is an urgent need to break down the barriers of separate disaster management and establish a cascaded defense standard covering floods, landslides, and earthquakes in order to achieve overall control of systemic risks in complex environments.

3. Conclusions

In summary, recent research on seismic resistance of marine and underground structures is moving towards high-precision simulation, intelligent monitoring and prediction, and the construction of resilient defense systems. This movement lays a solid theoretical and technological foundation for building the next generation of major infrastructure with self-adaptive, self-diagnostic, and highly resilient capabilities.

Author Contributions: Writing—original draft preparation, X.B. and K.L.; writing—review and editing, J.H. and P.W. All authors have read and agreed to the published version of the manuscript.

Funding: This research was funded by the National Natural Science Foundation of China [grant number 52478497] and the Fundamental Research Funds for the Central Universities [grant number DUT23RC(3)069].

Conflicts of Interest: The authors declare no conflicts of interest.

References

1. Amran, A.; Khaw, T.Y.; Harymawan, I. Navigating the blue economy: Challenges, governance, and pathways for enterprises. *J. Clean. Prod.* **2025**, *523*, 146407. [CrossRef]
2. Cui, J.; Broere, W.; Lin, D. Underground space utilisation for urban renewal. *Tunn. Undergr. Space Technol.* **2021**, *108*, 103726. [CrossRef]
3. Chen, G.; Ruan, B.; Zhao, K.; Chen, W.; Zhuang, H.; Du, X.; Khoshnevisan, S.; Juang, C.H. Nonlinear Response Characteristics of Undersea Shield Tunnel Subjected to Strong Earthquake Motions. *J. Earthq. Eng.* **2020**, *24*, 351–380. [CrossRef]
4. Liu, X. Modeling seismic waves in ocean with the presence of irregular seabed and rough sea surface. *J. Geophys. Eng.* **2023**, *20*, 49–66. [CrossRef]
5. Li, H.; Cheng, X.; Pan, H. Numerical Simulation of a Submerged Floating Tunnel: Validation and Analysis. *Appl. Sci.* **2024**, *14*, 3589. [CrossRef]

https://doi.org/10.3390/app16020919

6. Zhang, Z.; Li, Z.; Dias, D. Seismic stability of cracked rock slopes based on physics-informed neural networks. *Int. J. Rock Mech. Min. Sci.* **2025**, *192*, 106147. [CrossRef]
7. Song, C.; Wang, Y. Simulating seismic multifrequency wavefields with the Fourier feature physics-informed neural network. *Geophys. J. Int.* **2022**, *232*, 1503–1514. [CrossRef]
8. Al-Zuriqat, T.; Geck, C.C.; Dragos, K.; Manolis, G.D.; Smarsly, K. Decentralized sensor fault diagnosis for wireless structural health monitoring systems using artificial intelligence of things. In Proceedings of the Structural Health Monitoring 2025, Stanford, CA, USA, 9–11 September 2025.
9. Ren, L.; Liu, C.; Ou, J. Deformation Pattern Classification of Sea-Crossing Bridge InSAR Time Series Based on a Transfer Learning Framework. *Remote. Sens.* **2025**, *17*, 3567. [CrossRef]
10. Zhang, T.-Y.; Shi, B.; Zhang, C.-C.; Xie, T.; Yin, J.; Li, J.-P. Tunnel disturbance events monitoring and recognition with distributed acoustic sensing (DAS). In Proceedings of the ISRM Regional Symposium–11th Asian Rock Mechanics Symposium, Beijing, China, 21–25 October 2021.
11. Zhao, X.; Wei, C.; Zeng, L.; Sun, L.; Li, Z.; Chen, H.; Liu, G.; Qiao, Z.; Qu, Y.; Xu, D.; et al. Research Progress in Fiber Bragg Grating-Based Ocean Temperature and Depth Sensors. *Sensors* **2024**, *25*, 183. [CrossRef] [PubMed]
12. Zhu, J.; Chen, F.; Dai, X.; Tan, Y.; Duan, L.; Zhang, Z.; Leng, Y.; Wang, S.; Yin, T.; Yu, R. Development of a novel ultra-high performance concrete (UHPC) suitable for underwater operation: Design and performance evaluation. *J. Build. Eng.* **2023**, *75*, 107030. [CrossRef]
13. Xie, J.; Xu, Q.; Lu, L.; Sun, J.; Wu, C.; Han, S. Experimental and simulation investigation on seismic performance and failure mechanism of a new-type middle slab joints in prefabricated assembled metro stations. *Eng. Fail. Anal.* **2025**, *181*, 109955. [CrossRef]
14. Xu, Z.; Xia, Z. Seismic Mitigation Effect and Mechanism Analysis of Split Columns in Underground Structures in Sites with Weak Interlayers. *Appl. Sci.* **2025**, *15*, 798. [CrossRef]
15. Wang, W.; Fang, C.; Ji, Y.; Lu, Y.; Yam, M.C. Experimental and Numerical Studies on Fe–Mn–Si Alloy Dampers for Enhanced Low-Cycle Fatigue Resistance. *J. Struct. Eng.* **2022**, *148*, 04022170. [CrossRef]
16. Long, X.; Ma, Y.; Miao, Y.; Ruan, B. An innovative shape memory alloy flexible circumferential joint for tunnel seismic mitigation. *Tunn. Undergr. Space Technol.* **2023**, *131*, 104783. [CrossRef]
17. Loe, P.Y.A.; Kim, M.; Jin, C.; Bakti, F.P.; Kim, D.K. Recent advances in Discrete-Module-Beam-based hydroelasticity method as an efficient tool approach for continuous very large floating structures. *Ocean Eng.* **2025**, *340*, 122229. [CrossRef]
18. Rosario Beltré, A.J.; Paredes Bartolomé, C.; Llorente Isidro, M. Probabilistic and Deterministic Seismic Hazard Assessments of the Area Comprised between West Gulf of Cádiz and East Alboran Sea. *EGUsphere* **2025**, *2025*, 1–40. [CrossRef]
19. *ISO 19901-2:2022*; Petroleum and Natural Gas Industries—Specific Requirements for Offshore Structures—Part 2: Seismic Design Procedures and Criteria. International Organization for Standardization: Geneva, Switzerland, 2022.
20. Bi, K.; Hao, H. Using pipe-in-pipe systems for subsea pipeline vibration control. *Eng. Struct.* **2016**, *109*, 75–84. [CrossRef]
21. Hain, A.; Zaghi, A.E.; Padgett, J.E.; Tafur, A. Case studies of multihazard damage: Investigation of the interaction of Hurricane Maria and the January 2020 earthquake sequence in Puerto Rico. *Front. Built Environ.* **2023**, *9*, 1128573. [CrossRef]
22. Yu, X.; Li, Z.; Yang, A.; Li, Y.; Lu, D.; Dai, K. Time-Dependent Seismic Fragility of Coastal RC Frames Considering Effect of Distance from Coastline. *Buildings* **2025**, *15*, 737. [CrossRef]
23. Grandjean, G. A Bibliographic Analysis of Multi-Risk Assessment Methodologies for Natural Disaster Prevention. *Geohazards* **2025**, *6*, 41. [CrossRef]

https://doi.org/10.3390/app16020919

applied sciences

MDPI

Article

Numerical Simulation of a Submerged Floating Tunnel: Validation and Analysis

Hao Li [1,2], Xiaohui Cheng [2,*] and Hua Pan [1]

1 Institute of Geophysics, China Earthquake Administration, Beijing 100081, China; lihao@cea-igp.ac.cn (H.L.); panhua.mail@163.com (H.P.)
2 Department of Civil Engineering, Tsinghua University, Beijing 100084, China
* Correspondence: chengxh@tsinghua.edu.cn

Abstract: The dynamic response analysis of submerged floating tunnels (SFTs) under seismic action is a complex two-way fluid–structure coupling problem that requires expertise in structural dynamics, fluid mechanics, and advanced computational methods. The coupled Eulerian–Lagrangian (CEL) method is a promising method for solving fluid–structure interaction problems, but its application to SFTs is not well established. Therefore, it is crucial to verify the accuracy and reliability of the CEL method in fluid–structure coupling simulations. This study verified the applicability of the CEL method for simulating one-way and two-way fluid–structure coupling cylindrical flow problems, and then applied the CEL method for the analysis of a shaking table test of a model SFT. A comparison of results obtained with the CEL method with those obtained in a previous indoor model test of an SFT demonstrates the agreement between the results of the CEL method and the overall trend of the experimental results, indicating the reliability of the method for the seismic analysis of SFTs. Moreover, the analysis of the dynamic response characteristics of SFTs under seismic conditions provides data support and a technological means for the seismic design of SFTs.

Keywords: submerged floating tunnel (SFT); coupled Euler–Lagrange (CEL); circular cylinder flow; fluid–structure coupling interaction; seismic dynamic response

1. Introduction

Submerged floating tunnels (SFTs), which consist of a tubular structure supported by its buoyancy and anchored to two banks via anchor cables and foundations, offer outstanding environmental and economic prospects as transportation structures. However, despite the increasing research attention paid to SFTs, none have been built anywhere in the world. Moreover, SFTs were listed as one of the 60 major engineering technology problems at the 2018 Annual Meeting of the Chinese Association of Science and Technology [1].

Scholars in China and other countries have extensively studied the dynamic responses of SFT structures under seismic action, the vibration characteristics of tethered anchor structures, and the worst-affected response areas of SFT structures. For instance, Wu et al. [2] analyzed the dynamic response of SFTs under combined wave flow–earthquake action through scaled-down model tests and numerical simulations. They concluded that the dynamic response of SFTs under the action of waves and flow alone is relatively small, but that the effect of seismic loading dominates under the combined action of earthquakes and wave flow, and that the flow can attenuate the effects of high-frequency earthquake motions on SFTs. Li et al. [3] used the finite-element method to analyze the static and dynamic responses of a tension-leg SFT. They found that the seismic dynamic response of the SFT could be approximated as the superposition of the two following parts: the dynamic water pressure generated by a seismic wave propagating through seawater, and the additional external force induced by the offset of the submarine support. Their calculation results showed that the weak links were the two ends of the tube body, which were embedded in the rock layer and at the cable location, respectively. Therefore, special attention needs to be paid to these areas during the design process. Xiang et al. [4] believes that, due to the absence of comprehensive analysis methods and calculation procedures for SFTs under seismic and wave forces, it is crucial to establish a multi-scale theoretical model and conduct the corresponding experimental studies. This will facilitate obtaining more precise analysis results and verifying the effectiveness of various seismic protection measures. Wu et al. [5] believes that the ratio of parametric frequency to natural frequency, the direction and magnitude of earthquake excitation, the initial tension in the cable, and the damping ratio all have significant influences on the hydrodynamic and seismic responses of the cable. Therefore, these effects must be considered rigorously during the design of anchor systems for submerged floating tunnels. Fogazzi [6] devised a calculation method that considers the mutual coupling between a fluid, a structure, an anchor foundation, and soil, and an anchor's geometric nonlinearity when determining the seismic response of anchored SFTs. Martinelli et al. [7,8] have devised a new method for obtaining a response spectrum based on the median pseudo-acceleration spectrum and demonstrated its suitability for SFT engineering design calculations. Brancaleoni et al. [9] derived an equation for calculating the dynamic response of an SFT under seismic and wave effects that can be applied to both short-span SFTs with anchorages and long-span SFTs with underwater piers. Martire et al. [10] examined the seismic response characteristics of SFTs under multi-point support excitation, and they found that it excites more vibration modes in SFTs with longer spans than in SFTs with shorter spans. Xiao et al. [11] considered various simple boundary conditions (such as hinged, solid, and elastic supports) for the SFT with a channel barge structure and showed that a support with the appropriate elastic stiffness can significantly reduce seismic effects on an SFT structure. Chen et al. [12] found that the dynamic response of SFTs to seismic traveling waves is significant, with the most unfavorable response site being the connection between an SFT and its bank foundation. Chao and Xiang [13,14] proposed an anchor-displacement calculation method that integrates along the height of the SFT by parts, which demonstrated the advantages of simplicity and feasibility. Sun and Chen [15–18] have conducted the first experimental studies of a scaled-down model of an SFT under seismic conditions using an underwater shaking table, which revealed that horizontal seismic excitation could cause the horizontal rotation of an SFT tube. Additionally, the seismic response under an anchor cable inclination angle of 45° was less than that under an anchor cable inclination angle of 60°.

As described above, there have been detailed studies of SFTs under seismic conditions, based on the models and numerical calculations of coupling between SFTs' tubes and support systems, and the fluids in which they are submerged. These studies have focused on various aspects of SFTs, such as their dynamic responses, how they are affected by fluid forces, and the dynamic characteristics of anchor cables. The calculation methods and research results generated from these studies have significantly contributed to the successful implementation and completion of future SFT projects.

Scaled-down model tests of SFTs serve as a crucial means for understanding the force characteristics of SFTs' structural systems. However, due to the complex water environment surrounding SFTs, it is challenging to model prototypical structures with perfect fidelity, so satisfactory test results often cannot be obtained. There are three primary challenges. First, SFT model testing is performed on a model consisting of a tunnel body, an anchor cable, and a soil body, and therefore requires the use of multiple materials, and it is difficult to maintain physical similarity between model materials and a prototype. Second, it is challenging to accurately replicate a prototype structure's detailed characteristics in a large-scale model, which hinders large-scale model testing. Third, calculating the dynamic boundary effects of the scaled-down model system requires the refinement of the model test results via fast and efficient numerical simulation techniques.

The application of the CEL method in the calculation of hydrodynamic loads significantly overcomes some limitations of traditional methods. Conventional methods Huaqing Zhang and Zhiwen Yang [19] for calculating hydrodynamic loads typically assume that underwater structures are fixed to obtain external hydrodynamic loads. Then, using the known hydrodynamic loads, they calculate the structural response. For instance, the Morrison equation [20] assumes that the structure is stationary, while, in reality, the submerged floating tunnel undergoes certain relative displacements under the influence of hydrodynamic loads. This phenomenon is often challenging to accurately simulate using traditional methods.

In the case of employing the CEL method, we can more precisely consider the interaction between the submerged floating tunnels and the surrounding water, achieving bidirectional fluid–structure coupling calculations. Compared to traditional methods like the Morrison equation, the CEL method is more in line with reality because it thoroughly considers the potential relative displacements of underwater structures under hydrodynamic loads. This characteristic enables the CEL method to provide more accurate and realistic numerical results, offering a more reliable basis for engineering designs and structural response analysis.

Therefore, by adopting the CEL method, we not only address the limitations of traditional methods in considering structural displacements, but we also simulate hydrodynamic loads in conditions closer to real-world scenarios. This advantage highlights the significant innovation and practicality of the CEL method in underwater structure research.

In recent years, new numerical calculation methods (such as the smoothed particle hydrodynamics method, the arbitrary Lagrangian–Eulerian method, and the volume of fluid method) have provided powerful technical support for solving complex fluid–structure coupling problems relevant to SFTs. At present, few scholars have used the CEL method for the seismic response analysis of SFT. This paper reports the first attempt to adopt the coupled Eulerian–Lagrangian (CEL) method to conduct the seismic response analysis of an SFT.

2. CEL Method for One- and Two-Way Fluid–Structure Coupling Validation

2.1. CEL Method

Lagrangian methods are widely used for static analysis in geotechnical and structural engineering. Matsui and San [21] simulated slope stability problems, and Shiau et al. [22] analyzed the ultimate bearing capacity of sandy soil foundations based on the limit analysis method. In the Lagrangian method, the cell grid is unified with the object to be analyzed. Thus, during calculation, the shape of the analyzed object remains the same, while that of the finite element mesh changes, and the material does not flow from cell to cell.

In contrast, the Eulerian method is characterized by a fixed finite element mesh in terms of shape, size, and spatial location throughout the numerical computation. Material can flow between the mesh and the grid, so the method is ideal for dealing with material distortion and certain large deformation problems [23]. However, the Eulerian method struggles to accurately capture boundary information about the objects and materials, and so is typically used in the computational analysis of fluid dynamics.

Pure Lagrangian and pure Eulerian methods each have strengths and weaknesses, and by combining these methods, it is possible to take advantage of their strengths while overcoming their weaknesses. The CEL method was devised by Noh [24] for this purpose, and has since been used in fluid–solid coupling analysis, where the Lagrangian mesh is used to discretize the solid, whereas the Eulerian mesh is used to discretize the fluid domain. The Lagrangian mesh can pass through the fluid domain formed by the Eulerian mesh, thereby defining the contact between the fluid and solid materials. The CEL method has also been used to numerically simulate a variety of phenomena, including the scouring of vertical breakwaters [25] and a steel water-storage tank under blast loading [26], and has been employed for the development of a Reynolds-averaged Eulerian–Lagrangian model of vortex interaction that simulates the turbulent suspension of flow in a thin plate [27].

Specifically, the basic idea of the CEL method is to divide the computational domain into small Eulerian regions, with the physical quantities within each Eulerian region being described by a Lagrangian function. During the time-stepping process, the physical quantities within the Eulerian regions are first calculated, and then these quantities are transferred to the adjacent Eulerian regions through interpolation, forming a complete Lagrangian description of the computational domain. Therefore, the CEL method can control the accuracy of different spatial scales through multi-level grids, and can adapt to different physical processes via the use of different time steps.

The calculation process of the CEL method includes the following steps:

(1) Partitioning Eulerian regions: Divide the computational domain into several small Eulerian regions.
(2) Constructing Lagrangian functions: For each physical quantity within each Eulerian region, construct a corresponding Lagrangian function.
(3) Solving Euler equations: For each physical quantity within each Eulerian region, use the Euler method to solve its corresponding Euler equation.
(4) Interpolation transfer: Use interpolation methods to transfer physical quantities within Eulerian regions to adjacent Eulerian regions, forming a complete Lagrangian description of the computational domain.
(5) Time stepping: Repeat steps 3 and 4 to perform time-stepping calculations.

2.1.1. Fluid Motion Equations (Eulerian Equations)

The Eulerian equations describe the motion of the fluid, typically in the form of the Navier–Stokes equations. For incompressible fluids, the Navier–Stokes equations [24] can be written as

Momentum conservation equation:

$$\frac{\partial \mathbf{u}}{\partial t} + (\mathbf{u} \cdot \nabla)\mathbf{u} = -\frac{1}{\rho}\nabla\left(p_0 + \frac{1}{2}\rho|\mathbf{u}|^2\right) + \nu\nabla^2\mathbf{u} + \mathbf{g} \tag{1}$$

Here, ρ is the density, $\mathbf{u}$ is the velocity vector, ν is the dynamic viscosity coefficient, $\nabla\mathbf{u}$ is the velocity gradient, p_0 represents the static pressure, and $\mathbf{g}$ is the gravitational acceleration vector.

2.1.2. Solid Motion Equations (Lagrangian Equations)

In the Lagrangian coordinate system, the motion of the solid is typically described using elastic mechanics or other solid dynamics models (Noh, 1964). The basic equation for solid motion can be written as

Motion equation:

$$\rho_s \frac{d^2\mathbf{u}_s}{dt^2} = \nabla \cdot \left(E \cdot \left(\frac{1}{2}\left(\nabla\mathbf{u}_s + (\nabla\mathbf{u}_s)^T\right)\right)\right) + \rho_s\mathbf{g} \tag{2}$$

Here, ρ_s is the density of the solid, $\mathbf{u}_s$ is the displacement vector of the solid, E is the elastic modulus, and $\mathbf{g}$ is the gravitational vector.

2.1.3. Coupling

As mentioned above, in the CEL (coupled Eulerian-Lagrangian) method, there exist motion equations in two different coordinate systems (Eulerian and Lagrangian). A common strategy to tackle this issue in coupling methods involves introducing interface coupling conditions to coordinate the exchange of information between the two systems. These interface conditions may encompass characterizing the forces exerted by the fluid on the solid and the influence of the solid on the fluid in terms of forces and displacements. By using the CEL method, the contact between the Eulerian domain and Lagrangian domain is discretized using the general contact algorithm, which is based on the penalty contact method [28]. The penalty contact method is less stringent when compared to the kinematic contact method. Seeds are created on the Lagrangian element edges and faces, while the anchor points are created on the Eulerian material surface. The penalty method approximates hard pressure-overclosure behavior. This method allows the small penetration of the Eulerian material into the Lagrangian domain. The contact force F_P, which is enforced between seeds and anchor points, is proportional to the penetration distance d_P.

$$F_P = k_P d_P \quad (3)$$

The factor k_P is the penalty stiffness which depends on the Lagrangian and Eulerian material properties.

The detailed principles of the CEL method are described in Section S5 of Supplementary Materials. So far, the CEL method is mainly used to address significant object displacement problems, such as pile infill, landslides, collisions, and fluid dynamics, and has not been used in the field of SFTs. Therefore, careful verification is needed to determine whether the CEL method can be applied in fluid–solid coupling simulations. Accordingly, in this study, the CEL method was employed to simulate the classical hydrodynamic problem of circular cylinder flow to validate the accuracy of the CEL method in this context and provide data to support subsequent research.

2.2. Numerical Simulation of Circular Cylinder Flow

The circular cylinder flow problem is an obtuse-body winding problem and thus exhibits the unique phenomenon of vortex shedding into its wake. This phenomenon was first formally proposed by von Karman in 1912, and forms a vortex array known as the "Karman vortex street". Numerous researchers, including Roshko [29], Taneda [30], and Zhu [20], have studied this phenomenon experimentally, and Zhu [20] found that it is related to the Reynolds number, which is defined in Equation (4) as follows:

$$Re = \frac{\rho V D}{\eta} \quad (4)$$

where ρ, V, and η represent the density, velocity, and viscosity of the fluid, respectively; D is the diameter of the cylinder; and Re is the Reynolds number. As shown in Table 1, when $15 < Re < 40$, two Föppl vortices were produced in the flow field upstream and downstream of the rear sides of a cylinder, and, as Re increased, the tail vortex gradually became longer but remained attached to the rear side of the cylinder. When $150 < Re < 300{,}000$, the two tail vortices were shed alternately, influenced by the downstream fluid propagation, and propagated at a slightly lower speed than the fluid flow, thereby forming the Karman vortex street. As Re increased, the shedding frequency of the tail vortices increased.

Table 1. Relationship between vortex shedding and Reynolds number (Re) [31].

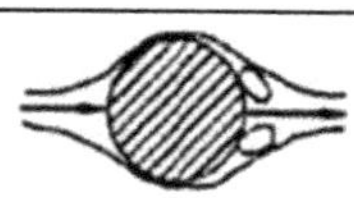	$15 < Re < 40$: A group of small, fixed vortices appeared after a cylinder
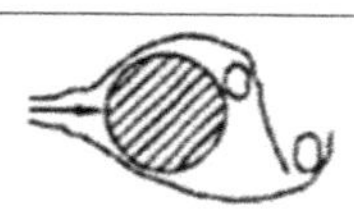	$150 < Re < 300{,}000$ Periodic alternating turbulent vortex discharge occurred

The calculation parameters of the hypothetical model [32] used in the current study are as follows. The circular rod diameter was 0.02 m and the length of the part of the rod submerged in the water body was 0.14 m. The water body was simulated using Eulerian elements, with a density of 1000 kg/m^3, a viscosity of 0.001 Pa·s, and a speed of sound of 1480 m/s^2. The upper surface of the water body was a free boundary, and the initial inflow velocity was set for the fluid. An Euler inflow boundary was applied on one side and an Euler outflow boundary on the other side. The inflow boundary had an initial velocity of V_x, while the outflow boundary allowed the material to flow freely (Figure 1). A smooth boundary was used at the bottom and on both sides, where V_z (the velocity in the z direction) and V_y (the velocity in the y direction) were set to zero. The contact between the cylinder and the fluid was defined as a smooth hard contact, where no mutual intrusion was allowed in the normal direction and the tangential friction was zero.

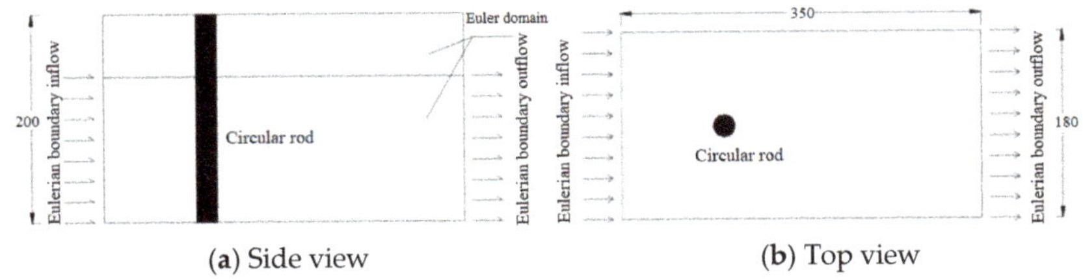

Figure 1. Simplified modeling diagram of cylindrical flow.

Figure 2 shows the results of a CEL-based simulation of the circular cylinder flow experiment. When $Re = 20$, a long vortex area was generated downstream of the cylinder. When Re was increased to 200, a Karman vortex street formed, and the volume of the detached vortex increased. When Re was increased to 2000, the Karman vortex street intensified, and the volume of the detached vortex decreases. When Re was again increased to 20,000, the Karman vortex street became even more intense and formed a periodic alternating turbulent vortex discharge. These results indicate that the CEL-based simulation method accurately models the Karman vortex street in the circular cylinder flow problem.

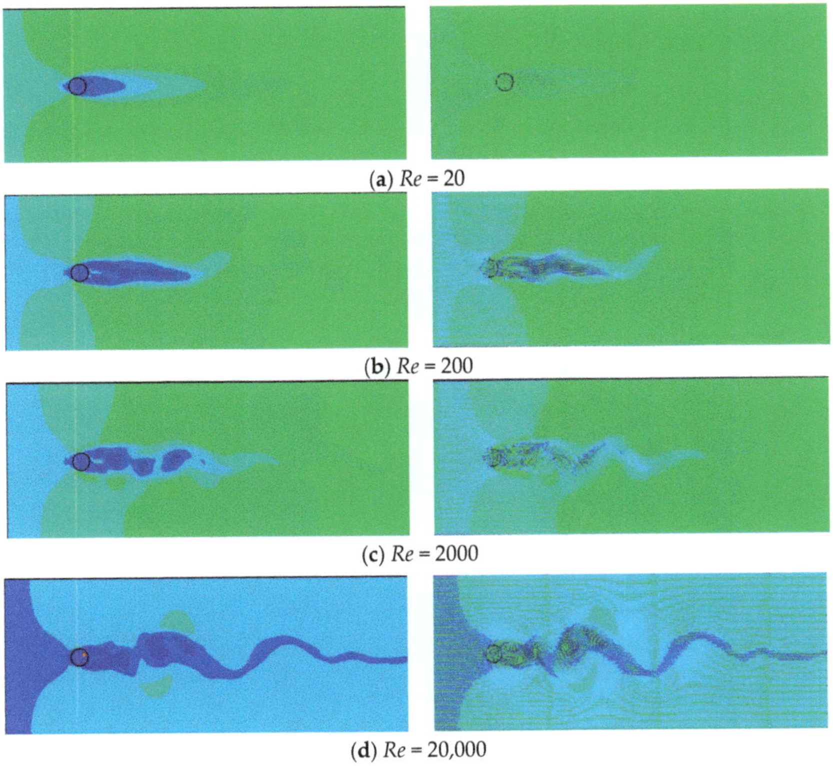

Figure 2. Velocity clouds and vector diagrams of simulated circular cylinder flow at various Reynolds numbers (Re) (the left and right figures correspond to the same time).

The above simulation of the Karman vortex street was based on a one-way fluid–structure coupling calculation. However, as SFTs involve two-way fluid–structure coupling, it is important to verify the accuracy of the CEL method for simulating two-way fluid–structure coupling.

2.3. Verification of Round-Rod Dragging Results

The CEL method was applied in a two-way fluid–structure coupling analysis to verify that it can be applied for studying SFTs. To achieve this, a circular rod was given an initial velocity in a body of water that remained stationary. To validate the CEL method in a two-way fluid–structure coupling calculation, the drag force obtained from the numerical simulation was compared with the drag force calculated using the Morison equation (Equation (5)) as follows:

$$F = C_D \frac{1}{2} \rho V^2 D \quad (5)$$

where C_D is the drag force coefficient and F is the drag force. In the hypothetical model [32], the circular rod had a diameter of 0.02 m, and 0.14 m of its length was submerged in the water body. The water body was simulated using Eulerian elements, with a density of 1000 kg/m^3, a viscosity of 0.001 Pa·s, and a speed of sound of 1480 m/s. The upper surface of the water body was a free boundary, while the bottom and the four sides were smooth boundaries, with the velocity set to zero. The horizontal degree of freedom was relaxed at the top of the circular rod, while the degrees of freedom in the other directions were constrained. The contact between the cylinder and the fluid was defined as a smooth hard contact, allowing no mutual intrusion in the normal direction and setting the tangential friction to zero. V_x was set to the top of the cylinder for calculation, with different horizontal velocities.

Figure 3 illustrates the velocity field clouds of the circular rod at different stages during its motion through the fluid. The results demonstrate that when the circular rod moved through the fluid at a certain speed, the velocity fields of both the rod and its surrounding fluid underwent significant changes. During the motion, a bulge formed in front of the circular rod, which led to the separation of the boundary layer at the maximum point of the column section width. This resulted in the formation of a free shear layer that extended to the back of both sides of the column. Within this shear layer, the outer part had a lower flow rate than the inner part, leading to a rotational tendency in the fluid and the formation of a vortex.

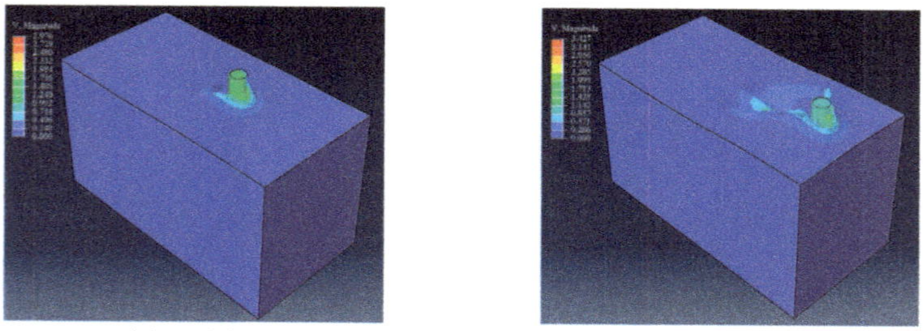

(**a**) t = 0.05s (**b**) t = 0.1s

Figure 3. Velocity field clouds at different stages in a model of two-way fluid–solid coupling.

The empirical curve of C_D vs. *Re* is shown in Figure 4.

In this study, the drag force on a circular rod was simulated at various *Re* values by changing the horizontal velocity of the rod. The empirical values of the drag force, F, were calculated using Equations (4) and (5), with the joint calculation of the motion velocity, V. The simulated drag force was calculated as the drag force in the horizontal direction between the contact surface of the circular rod and the fluid, consisting primarily of frictional resistance and differential pressure resistance. The drag force under each velocity condition was plotted as a fluctuating temporal curve, and the average value was taken as the simulated solution for the drag force on the circular rod. Table 2 shows the simulated and empirical solutions for the drag force of the cylinder at various *Re* values. At low *Re* values, the effect of fluid viscosity was prominent, resulting in vortex dissipation. The gradual dissipation of the momentum and energy of the fluid resulted in a relatively stable flow. However, at high *Re* values, complex turbulence phenomena occurred. Generally, *Re* values needed to be controlled within the range of 10^4 to 10^5, such that the fluid formed a complex flow field on the surface of the structure, generating nonlinear effects such as vortices and turbulence, which caused significant friction and drag. Due to the large inertial force of the fluid, its vibration response was also significant at high *Re* values, causing noticeable dynamic responses within the structure.

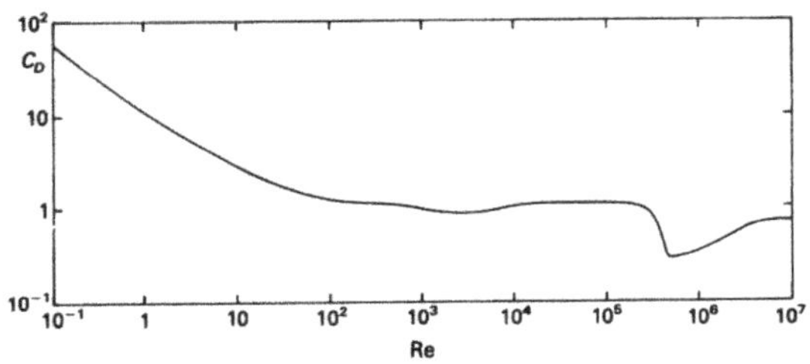

Figure 4. Empirical curve of the drag force coefficient of circular cylinder flow vs. Reynolds number (*Re*) [20].

Table 2. Comparison of empirically calculated and simulated values of drag force at various Reynolds numbers (*Re*).

V (m/s)	C_D	*Re*	Drag Force: Empirical Value (N)	Drag Force: Simulated Value (N)	Relative Error %
0.5	1.08	10,000	2.7	2.9	7.4
1	1.12	20,000	11.2	10.5	6.3
2	1.13	40,000	45.2	44.4	1.77
5	1.1	100,000	275	256.2	6.8

Relative error = ABS (|simulated solution − empirical solution|/empirical solution), C_D is the drag force coefficient and was taken from Figure 4, and *V* is the horizontal velocity of the column.

As shown in Table 2, when *Re* values were between 10^3 and 10^5, the simulated drag force using the CEL method was close to the empirical solution, with a relative error of less than 10%. This indicates the reliability of the CEL method for simulating two-way fluid–solid coupling problems.

The two examples presented above demonstrate the reliability of the CEL method in calculating both one-way and two-way fluid–structure coupling problems. Next, the CEL method was used to simulate the dynamic responses of SFTs under seismic action.

3. Numerical Study of Shaking Table Model for Scaled-Down SFT

The CEL method was utilized to simulate the dynamic response problem of an SFT under seismic action, and the accuracy of the CEL-calculated results was verified. The simplified model was established as follows: (1) the fluid was simulated by incompressible Eulerian elements; (2) the SFT tube body was simulated as a rigid body; and (3) the anchor cable was simulated using spring elements, which were subject to tension but not pressure, whereas the anchor cable mass was ignored. The accuracy of the model was verified by comparison with the results of underwater shaking table model tests reported by Sun and Chen (2008, 2012). Despite the referenced literature being from 2008, it represents a milestone, being the first study to conduct a shaking table test of a model for submerged floating tunnel vibrations, holding significant historical significance. On this basis, multiple sets of seismic wave conditions were simulated to analyze the coupled dynamic response of the SFT fluid–cable system under seismic action.

3.1. Experimental Overview

The entire SFT model had a length of 3 m, an outer diameter of 0.16 m, a wall thickness of 4 mm, and a modulus of elasticity of 8.7537 GPa. The model included a 1.5 mm-diameter anchor cable, which contained steel strands spaced 100 cm apart and had a modulus of elasticity of 210 GPa. The anchor cable had a tilt angle of 45°, and the model was placed at a water depth of 80 cm, with a tunnel placement depth of 35 cm. A schematic of the model test is presented in Figure 5.

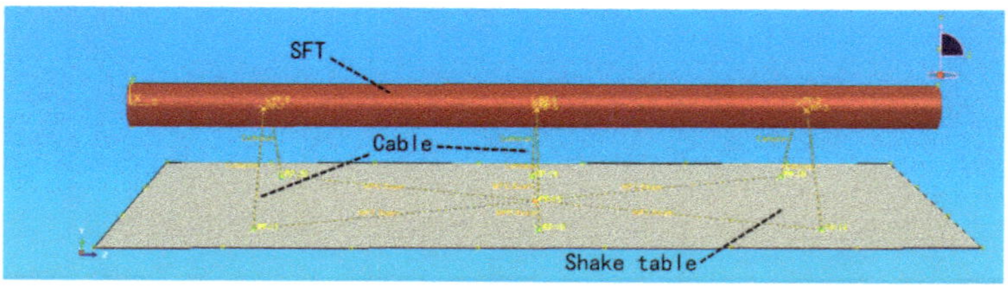

Figure 5. Experimental diagram of the submerged floating tunnel (SFT) model.

The seismic excitation waveform applied to the model was a horizontal sine wave with a frequency of 4 Hz, an acceleration amplitude of $0.01\times g$, and a vibration time of 1 s, as depicted in Figure 7d. Table 3 shows the relationship between the magnitudes of the main physical parameters involved in the actual tests and the similarity constants, i.e., the ratios of the model physical indicators to the prototype's physical indicators.

Table 3. Model test similarity constants.

	Physical Indicators	Dimensional System $[L][\rho][\varepsilon][g]$	Similarity Constant (N = 50)
Geometric features	Geometric dimension	$[L]$	$1/N$
	Area	$[L]^2$	$1/N^2$
	Inertia moment	$[L]^4$	$1/N^4$
Material behavior	Density	$[\rho]$	1
	Elasticity modulus	$[L][\rho][g][\varepsilon]^{-1}$	1
	Mass	$[\rho][L]^3$	$1/N^3$
Dynamic characteristics	Input vibration acceleration	$[g]$	1
	Field acceleration	$[g]$	1
	Force	$[\rho][L]^3[g]$	$1/N^3$
	Input vibration time	$[L]^{0.5}[\varepsilon]^{0.5}[g]^{-0.5}$	$1/N^{0.5}$
	Vibrational frequency	$[L]^{-0.5}[\varepsilon]^{-0.5}[g]^{0.5}$	$N^{0.5}$
	Dynamic response acceleration	$[g]$	1
	Dynamic response stress	$[\rho][L][g]$	$1/N$
	Dynamic response strain	$[\varepsilon]$	1

3.2. Numerical Modelling

Using the CEL method, a numerical simulation analysis was conducted based on the aforementioned model test. The SFT tube body was set as a rigid body and was simulated using Lagrangian solid elements. The water body was simulated using Eulerian elements, with a density of 1000 kg/m^3, a viscosity of 0.001 Pa·s, and a speed of sound of 1480 m/s. The ground containing the shaking table was simulated using rigid shell elements, and the anchor cable system was simulated using simplified spring elements. The anchor cable and the shaking table surface were hinged, as was the submerged floating tunnel section. To simulate the relaxation of the anchor cable, bilinear stiffness was introduced.

$$K_i(\Delta L_i) = \begin{cases} k = \frac{EA}{L_n}, & \Delta L_i \geq 0 \\ 0, & \Delta L_i < 0 \end{cases} \tag{6}$$

The anchor cable tensile stiffness, denoted as L_n, is a constant which is determined using the anchor cable modulus of elasticity (E) and its cross-sectional area (A). When the instantaneous length (L_i) of the anchor cable was greater than the unstretched length (original length), L_n remained constant. However, when the transient length was less than the original length, the anchor cable relaxed, and L_n became zero.

The contact between the SFT tube and the fluid was defined as a smooth hard contact, which meant that no mutual intrusion was allowed in the normal direction. The tangential friction coefficient was set to 0.1. A fluid-free surface boundary was used on the surface of the water body, and the fluid velocity on all four sides of the water body was set to zero, ensuring that the fluid flowed only within the grid. An energy dissipation treatment was applied around the water body, and a fixed boundary was applied at the bottom of the water tank. The model dimensions and other physical parameters were the same as those reported by Sun and Chen [17,18]. The model comprised 158,006 elements and 169,020 nodes, and is shown in Figure 6.

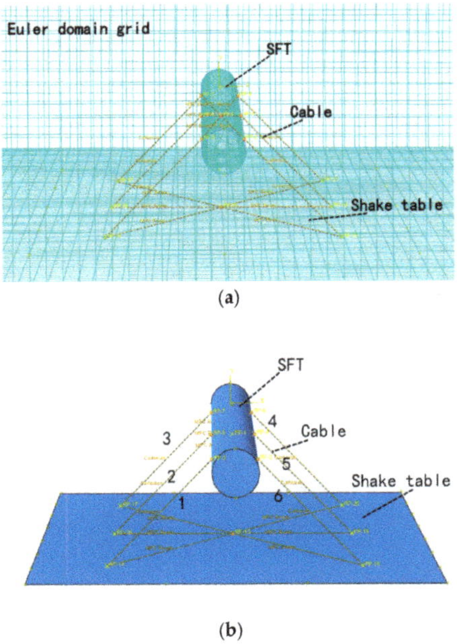

Figure 6. Numerical model diagram of SFTs. (**a**) Numerical model diagram of a submerged floating tunnel (SFT) simulated using the coupled Eulerian–Lagrangian method. (**b**) SFT model (numbers are anchor cable numbers).

3.3. Results Analysis

The numerical simulation was conducted in two steps, the first being a static equilibrium calculation, and the second being a dynamic seismic response calculation. In the latter, a sinusoidal seismic wave was applied horizontally to the bottom of the model, and a seismic time course analysis was conducted. Figure 7 shows comparisons of the time curve of the horizontal acceleration at the center of the SFT body, the horizontal hydrodynamic load–time curve on the outer surface of the tube body, and the time curve of the axial force of the anchor cable, with the experimental results reported by Sun and Chen [17,18]. The hydrodynamic load on the outer surface of the tube body mainly comprised frictional resistance and differential pressure resistance. The former represented the frictional force generated by the water body tangentially on the surface of the tube body and was a direct manifestation of the viscosity of the water body. The latter was an indirect manifestation of the viscosity of the water body, and it was largely caused by the reciprocal vibration of the tube body in the horizontal direction, which led to the water body alternately horizontally separating to the left and right of the surface of the tube body, thus producing pressure differences in the horizontal direction. The calculation results showed that the hydrodynamic load in the vertical direction at this time was much smaller than that in the horizontal direction, and was therefore not chosen as the hydrodynamic load in the vertical direction.

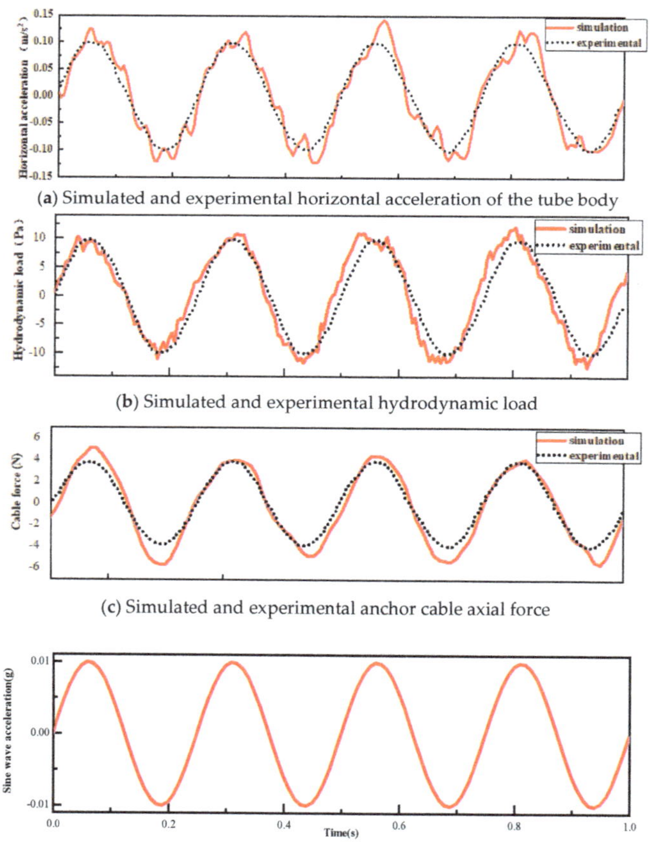

(**a**) Simulated and experimental horizontal acceleration of the tube body

(**b**) Simulated and experimental hydrodynamic load

(**c**) Simulated and experimental anchor cable axial force

(**d**) Sinusoidal acceleration amplitude curve

Figure 7. Comparison of numerical and experiment solutions.

4. SFT Dynamic Frequency Response Analysis

To investigate the frequency response characteristics of the submerged floating tunnel in water, a series of sinusoidal seismic waves, with frequencies ranging from 1 Hz to 15 Hz and an amplitude of 0.01× g, was applied to the bottom of the aforementioned numerical model. The dynamic response of the SFT model under each seismic wave frequency was analyzed, and the peak values of each response indicator were recorded and statistically analyzed. The results are presented in Figure 8.

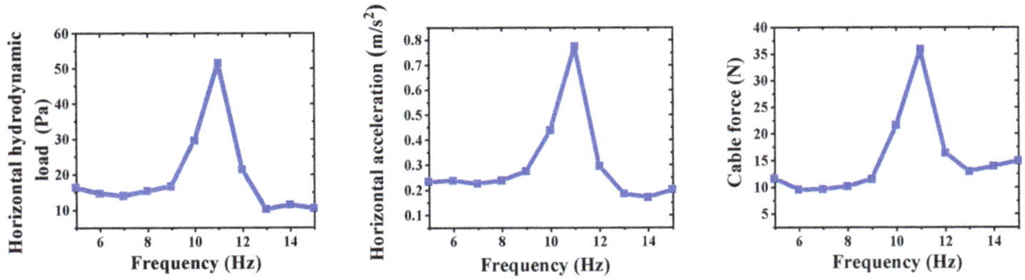

Figure 8. Peak value of each dynamic response index at different frequencies.

It is evident from the figure that, when the seismic wave frequency was 11 Hz, the horizontal hydrodynamic load, horizontal acceleration, and cable force of the tube body reached their maximum values (51.7 Pa, 0.777 m/s^2, and 36 N, respectively). This indicates that the horizontal self-oscillation frequency of the SFT model was approximately 11 Hz. According to the corresponding similarity constant, this implies that the horizontal self-oscillation frequency of the prototype of the tunnel model was approximately 1.5 Hz.

Next, the seismic response of the SFT under the same horizontal sinusoidal seismic wave, with a frequency of 4 Hz and a vibration time of 1 s, was investigated. The acceleration amplitude was adjusted to 0.03× g, 0.05× g, 0.07× g, and 0.09× g, and the peak horizontal hydrodynamic load, horizontal acceleration, and cable force were recorded and analyzed. The results are presented in Figure 9. We provide the tension time history curves of the cable under various conditions in Supplementary Materials Figures S8 and S9, further illustrating the dynamic changes in the tension.

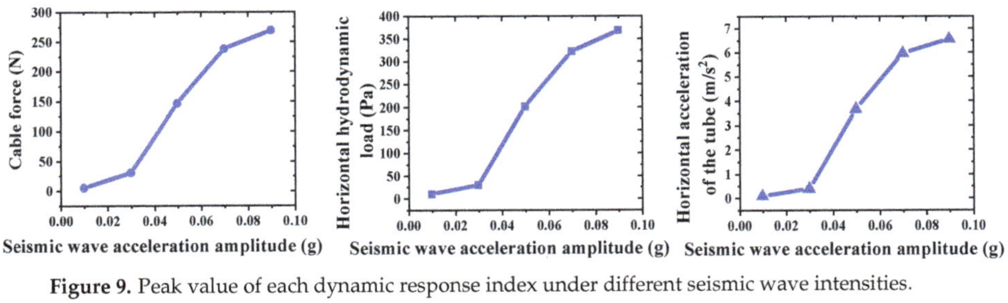

Figure 9. Peak value of each dynamic response index under different seismic wave intensities.

It can be observed that the dynamic response indices of the SFT exhibited a gradual increase when the seismic acceleration amplitude ranged from 0.01× g to 0.03× g. However, a noticeable inflection point and a sharp increase in the dynamic response indices occurred when the seismic acceleration amplitude was between 0.05× g and 0.08× g. When the seismic acceleration amplitude reached between 0.08× g and 0.09× g, the dynamic response indices of the SFT continued to increase, but did so more slowly than before. Based on the similarity constants between the simulation and the model test provided in Table 3, the following dynamic response indices of the prototype SFT were obtained.

$$\sigma_{\mathrm{m}} = F_{\mathrm{m}}/A_{\mathrm{m}} \tag{7}$$

$$\sigma_{\mathrm{p}} = N\sigma_{\mathrm{m}} \tag{8}$$

$$P_{\mathrm{p}} = NP_{\mathrm{m}} \tag{9}$$

$$a_p = a_m \tag{10}$$

where σ_m is the tensile stress of the model anchor cable, σ_p is the tensile stress of the prototype anchor cable, F_m is the model cable force, A_m is the cross-sectional area of the model cable, P_p is the seismic horizontal water pressure load of the prototype, P_m is the seismic horizontal water pressure load of the model, a_p is the acceleration of the prototype tunnel tube body, a_m is the acceleration of the model tunnel tube body, and N is the prototype/model similarity ratio.

Table 4 summarizes the tensile stress of the anchor cable in the prototype SFT under the influence of seismic waves of varying intensities, the horizontal ground-vibration water pressure load on the tube body, and the horizontal acceleration of the tube body, as calculated using Equations (7)–(10).

Table 4. Peak values of the dynamic response indices of the prototype and model under various seismic wave intensities.

Seismic Wave Acceleration ($\times g$)	Cable Force (MPa)		Horizontal Hydrodynamic Load (Pa)		Horizontal Acceleration of the Tube (m/s^2)	
	σ_m	σ_p	P_m	P_p	a_p	a_m
0.01	3.40	169.85	11	550	0.11	0.11
0.03	10.70	535.03	30.5	1525	0.41	0.41
0.05	83.51	4175.51	202	10,100	3.68	3.68
0.07	135.20	6760.08	322.8	16,140	5.98	5.98
0.09	152.92	7646.14	369.2	18,460	6.57	6.57

As shown in Table 4, the maximum tensile stress of the anchor cable in the prototype SFT was 535.03 MPa when the seismic wave acceleration strength was 0.03× g, which is within the ultimate tensile strength of the steel strand (1960 MPa; Li et al., 2022 [3]). However, the tensile stress of the anchor cable in the prototype SFT was 4175 MPa when the seismic wave acceleration strength was 0.05× g, which far exceeds the ultimate tensile strength of the steel strand. Therefore, the calculation shows that the cable tension component of this SFT was the most affected parameter under seismic action. Thus, in real-life applications of this SFT, the tensile strength of the cable would need to be appropriately increased to ensure that the cable had a sufficient level of safety redundancy. Additionally, based on the tensile strength performance of the steel strands during seismic activity, it was found that the seismic warning threshold for the SFT was 0.03× g. Therefore, in real-life applications of this SFT, an earthquake early warning alarm would be needed when the motion acceleration reached this level, allowing the operating company to have ample time to implement earthquake response measures.

The horizontal acceleration of the tube body significantly increased as the seismic wave strength increased, and the peak acceleration of the prototype SFT tube body was much larger than the peak acceleration of the corresponding seismic wave, resulting in an amplification effect. This might have been due to the interaction between the tube body and the anchor cable vibration under seismic action, causing an increase in the acceleration response of the tube body. Additionally, the interaction between a seabed-anchored cable and the seabed itself under the action of seismic waves would further increase the acceleration response of the tube body of an SFT. Therefore, the seismic design of an SFT should include measures to suppress this interaction, such as dampers applied to the anchor cables.

5. Conclusions

Few scholars have employed the CEL method to numerically simulate the dynamic response of SFTs. In this study, two simulated examples were explored to demonstrate the reliability of the CEL method for calculating both one-way and two-way fluid–structure coupling problems. Based on this, a CEL numerical model of an SFT was established, and its results agree well with those of previously published shaking table experiments, indicating that the CEL method can effectively simulate the dynamic response of SFTs under seismic effects. Finally, the dynamic response of an SFT under an earthquake was analyzed, and the following conclusions were drawn.

(1) The verification of the CEL method in two hypothetical cylindrical flow calculations and a comparison of the results with those of experimental model SFT tests demonstrated that the CEL method is capable of effectively simulating the seismic dynamic response of SFTs.

(2) The natural frequency of an SFT prototype in a water environment was found to be 1.5 Hz. Thus, when conducting the seismic design of the SFT, the response of the structure under earthquake action should be evaluated. In the analysis, it is necessary to consider the matching relationship between the natural frequency of the structure and the frequency of the earthquake wave motion in order to determine the resonance condition of the structure and its corresponding seismic response characteristics.

(3) The seismic warning threshold for the prototype SFT was $0.03\times g$. Thus, if an SFT based on this prototype is applied in a real-life setting, an earthquake early warning alarm would need to be issued once the ground motion acceleration reached this level. This would ensure that the anchor cable had a sufficient level of safety redundancy, and it would allow the operating company ample time to implement earthquake response measures.

(4) Under seismic conditions, the peak acceleration of the prototype SFT tube body significantly increased, leading to amplification effects. Therefore, in the seismic design of SFTs, the interaction between the tube body, anchor cable, and seabed should be the foremost consideration.

Overall, our findings reveal that the CEL method's potential applications in fluid–structure interaction analysis transcend traditional boundaries, demonstrating its remarkable adaptability in simulating the responses of intricate underwater structures under seismic conditions. This discovery not only expands the realm of CEL method applications, but also introduces a novel and promising numerical simulation approach for seismic research on submerged floating tunnels. However, it is worth emphasizing that our research results are preliminary and require further investigation and improvements. While the successful application of the CEL method in simulating seismic responses of submerged floating tunnels provides robust initial support for its reliability and effectiveness in engineering applications, further refinement and validation are crucial to ensure its widespread adoption and optimal performance.

The seismic analysis of SFTs is complex due to the vibration of their structures being affected not only by hydrostatic pressure, but also by dynamic water pressure. A water body can cause changes in the frequency, vibration pattern, and damping of SFT structures, resulting in coupling between structural vibration and the water body's seismic response. Additionally, the seismic effect of submarine soil deformation would directly affect the anchorage foundation of an SFT and its tunnel section. This research provides a strong technological foundation and data support for further studies of the dynamic interaction mechanisms between a water body, an SFT, its anchored foundations, and soil bodies under seismic conditions.

Supplementary Materials: The following supporting information can be downloaded at: https://www.mdpi.com/article/10.3390/app14093589/s1, Figure S1: Schematic Diagram of the Experimental [16,18]; Figure S2: Schematic Diagram of Numerical Models; Figure S3: Schematic Diagram of the Experimental and Numerical Models; Figure S4: Velocity field cloud images at different stages of bidirectional fluid-structure coupling of the model; Figure S5: Comparison of numerical and empirical drag force solutions; Figure S6: The schematic diagram of operator splitting for the Euler equations; Figure S7: The Volume of Fluid method involves the representation of material volume fraction and the schematic diagram of material boundaries; Figure S8: Variation curve of cable tension under the action of sinusoidal waves with different peak accelerations; Figure S9: Variation curve of cable tension under the action of sinusoidal waves with different frequencies; Table S1: Comparison of the empirical and simulated drag force calculation results (where V represents the horizontal velocity of the column).

Author Contributions: Conceptualization, X.C.; methodology, H.L.; software, H.L.; validation, X.C., H.L. and H.P.; formal analysis, H.L.; investigation, H.L.; resources, H.L.; data curation, H.L.; writing—original draft preparation, H.L.; writing—review and editing, X.C.; visualization, H.L.; supervision, X.C.; project administration, H.L.; funding acquisition, X.C., H.L. and H.P. All authors have read and agreed to the published version of the manuscript.

Funding: This research was funded by the Basic Research Business Special Fund of Central Public Welfare Research Institutes, with the project number DQJB23K41.

Institutional Review Board Statement: Not applicable.

Informed Consent Statement: Not applicable.

Data Availability Statement: The data that support the findings of this study are available from the corresponding author, upon reasonable request.

Conflicts of Interest: The authors declare no conflict of interest.

References

1. China Association for Science and Technology publishes 60 major scientific problems and engineering and technical problems in 12 fields. *J. China Railw. Soc.* **2018**, *40*, 24, 53, 87, 135.
2. Wu, Z.; Wang, D.; Ke, W. Experimental investigation for the dynamic behavior of submerged floating tunnel subjected to the combined action of earthquake, wave, and current. *Ocean Eng.* **2021**, *239*, 109911. [CrossRef]
3. Li, H.; Cheng, X.; Yu, H.; Fu, D. Seismic response simulation of submerged floating tunnel with tension legs: A case study on the Qiongzhou Strait Crossing Tunnel Project. *Mod. Tunnel. Technol.* **2022**, *59*, 146–154.
4. Xiang, Y.; Yang, Y.; Bai, B. Methods for Analyzing Seismic Response of Submerged Floating Tunnel and Research Development. In Proceedings of the 14th International Conference on Vibration Problems, Crete, Greece, 1–4 September 2019; Springer Nature: Singapore, 2019.
5. Wu, Z.; Ni, P.; Mei, G. Vibration response of cable for submerged floating tunnel under simultaneous hydrodynamic force and earthquake excitations. *Adv. Struct. Eng.* **2018**, *21*, 1761–1773. [CrossRef]
6. Fogazzi, P.; Perotti, F. The dynamic response of seabed anchored floating tunnels under seismic excitation. *Earthq. Eng. Struct. Dyn.* **2000**, *29*, 273–295. [CrossRef]
7. Martinelli, L.; Barbella, G.; Feriani, A. Modeling of Qiandao Lake submerged floating tunnel subject to multi-support seismic input. *Procedia Eng.* **2010**, *4*, 311–318. [CrossRef]
8. Martinelli, L.; Barbella, G.; Feriani, A. A numerical procedure for simulating the muti-support seismic response of SFT anchored by cables. *Eng. Struct.* **2011**, *33*, 2850–2860. [CrossRef]
9. Brancaleoni, F.; Castellani, A.; D'Asdia, P. The response of submerged tunnels to their environment. *Eng. Struct.* **1989**, *11*, 47–56. [CrossRef]
10. Martire, G.; Faggianoa, B.; Mazzolani, F.M.; Zollo, A.; Stabile, T. Seismic analysis of an SFT solution for the Messina Strait crossing. *Procedia Eng.* **2010**, *4*, 303–310. [CrossRef]
11. Xiao, J.; Huang, G. Transverse earthquake response and design analysis of SFT with various shore connections. *Procedia Eng.* **2010**, *4*, 233–242. [CrossRef]
12. Chen, W.; Huang, G. Seismic wave passage effect on the dynamic response of SFT. *Procedia Eng.* **2010**, *4*, 217–224. [CrossRef]
13. Chao, C.F.; Xiang, Y.Q.; Lin, J.P. Patterns of cable VIV suppression based on the fluid-structure interaction. *Mech. Eng.* **2015**, *37*, 725–730. (In Chinese)
14. Chao, C.F.; Xiang, Y.Q.; Yang, C. Experiments on dynamic fluid-structure coupled responses of anchor cables of submerged floating tunnel. *J. Vib. Shock* **2016**, *3*, 158–163. (In Chinese)

15. Sun, S.; Chen, J. Study on hydrodynamic pressure of SFT under earthquake-SV wave. *J. Disast. Prev. Mitig. Eng.* **2006**, *4*, 425–430. (In Chinese)
16. Sun, S.; Chen, J. Study on dynamic water pressure of undersea anchorage SFT-P wave. *J. Harbin Inst. Technol.* **2008**, *8*, 1292–1296. (In Chinese)
17. Chen, J.; Li, J.; Sun, S. Experimental and numerical analysis of submerged floating tunnel. *J. Cent. South Univ.* **2012**, *19*, 2949–2957. [CrossRef]
18. Sun, S.; Chen, J. Dynamic Response Analysis of SFT. Ph.D. Thesis, Dalian University of Technology, Dalian, China, 2008. (In Chinese)
19. Zhang, H.; Yang, Z. A global review for the hydrodynamic response investigation method of submerged foating tunnels. *Ocean Eng.* **2021**, *225*, 108825. [CrossRef]
20. Zhu, Y. *Wave Mechanics in Ocean Engineering*; Tianjin University Press: Tianjin, China, 1991.
21. Matsui, T.; San, K.C. Finite element slope stability analysis by shear strength reduction technique. *Soils Found.* **1992**, *32*, 59–70. [CrossRef]
22. Shiau, J.S.; Lyamin, A.V.; Sloan, S.W. Bearing capacity of a sand layer on clay by finite element limit analysis. *Can. Geotech. J.* **2003**, *40*, 900–915. [CrossRef]
23. Hirt, C.W.; Nichols, B.D. Volume of fluid (VOF) method for the dynamics of free boundaries. *J. Comput. Phys.* **1981**, *39*, 201–225. [CrossRef]
24. Noh, W.F. *CEL: A Time-Dependent Two Space-Dimensional, Coupled Eulerian-Lagrangian Code*; Lawrence Radiation Laboratory: Livermore, CA, USA, 1964.
25. Hajivalie, F.; Yeganeh-Bakhtiary, A.; Houshanghi, H.; Gotoh, H. Euler–Lagrange model for scour in front of vertical breakwater. *Appl. Ocean Res.* **2012**, *34*, 96–106. [CrossRef]
26. Mittal, V.; Chakraborty, T.; Matsagar, V. Dynamic analysis of liquid storage tank under blast using coupled Euler–Lagrange formulation. *Thin-Walled Struct.* **2014**, *84*, 91–111. [CrossRef]
27. Cheng, Z.; Chauchat, J.; Hsu, T.J.; Calantoni, J. Eddy interaction model for turbulent suspension in Reynolds-averaged Euler–Lagrange simulations of steady sheet flow. *Adv. Water Resour.* **2017**, *111*, 435–451. [CrossRef]
28. *ABAQUS*, Version 2016. EF Documentation. Dassault Systèmes: Vélizy-Villacoublay, France, 2016.
29. Roshko, A. *On the Development of Turbulent Wakes from Vortex Streets*; NASA: Washington, DC, USA, 1953.
30. Taneda, S. Downstream development of the wakes behind cylinders. *J. Phys. Soc. Jpn.* **1959**, *14*, 843–848. [CrossRef]
31. Sarpkaya, T. Vortex-induced oscillations: A selective review. *J. Appl. Mech.* **1979**, *46*, 241–258. [CrossRef]
32. Xiao, S. *Research on Constitutive Model and Numerical Simulation of Sand Rheology under Low Effective Stress*; Tsinghua University: Beijing, China, 2020. (In Chinese)

applied sciences

MDPI

Article

A Dynamic Analysis Method of Liquid-Filled Containers Considering the Fluid–Structure Interaction

Xibing Fang [1,2], Xin Bao [3,*], Fengjiang Yue [4] and Qiyuan Zhao [1,2]

1 Sanya Science and Education Innovation Park, Wuhan University of Technology, Sanya 572024, China; 106953@whut.edu.cn (X.F.); 274714@whut.edu.cn (Q.Z.)
2 School of Civil Engineering and Architecture, Wuhan University of Technology, Wuhan 430070, China
3 State Key Laboratory of Coastal and Offshore Engineering, Dalian University of Technology, Dalian 116000, China
4 Department of Civil Engineering, Xinjiang University, Urumqi 830047, China; yuefjzxj@126.com
* Correspondence: baoxin@dlut.edu.cn

Abstract: Based on acoustic fluid elements, a dynamic analysis of liquid sloshing modes and liquid-filled containers was undertaken, considering the effect of fluid–structure interactions (FSIs). The liquid sloshing modes in two-dimensional (2D) and three-dimensional (3D) containers were analyzed, and the results were compared with liquid sloshing modes measured in tests and theoretically calculated modes. This finding thus verifies the correctness of the simulation method based on acoustic fluid elements. Cylindrical liquid-filled containers with different water levels were subjected to a modal analysis and dynamic and time-history analysis. The results show that the finite element analysis (FEA) based on acoustic fluid elements can accurately simulate liquid sloshing modes in liquid-filled containers, as well as the vibration characteristics of these containers with different liquid levels. The vibration frequency of liquid-filled containers decreases with rising liquid levels. The liquid level significantly affects the distribution of the maximum displacement, maximum acceleration, and maximum von Mises stress on the sidewall of liquid-filled containers. Numerical simulations based on acoustic fluid elements provide an effective and reliable method for dynamic analysis of liquid-filled containers considering the effect of FSIs.

Keywords: fluid–structure interaction; acoustic fluid elements; liquid sloshing modes; liquid-filled containers; dynamic analysis

1. Introduction

The effect of fluid–structure interactions (FSIs) between solid structures and liquids is prevalent in engineering practice. For example, liquid-filled containers under seismic action or other vibration loads are a structure with the typical effect of FSIs [1]. Analysis of the FSI effect of liquid-filled containers, on the one hand, focuses on liquid sloshing therein, and on the other, considers the influence of liquid sloshing on these containers.

In order to tackle liquid sloshing problems, Dodge et al. [2–5] systematically expounded the theoretical and engineering applications of liquid sloshing modes. However, these methods pay more attention to theoretical and analytical methods and are only applicable to solving cases with regular shapes and simple external excitations, yet fail to solve vibration problems of liquid-filled structures of complex shapes [6,7]. Bao et al. [8,9] studied liquid modes based on potential-based fluid elements. When using this method, the liquid sloshing frequency is much lower than the structural vibration frequency, and the resulting first hundreds of modes are all liquid sloshing modes rather than modes of solid structures [10]. Therefore, it is difficult to apply the mode-superposition response spectrum method and time history analysis to any dynamic analysis of liquid-filled structures, and these methods are inapplicable to the analysis of dynamic responses of liquid-filled containers.

Structural engineers generally pay more attention to how liquid motion affects structural responses under seismic action, while they are not interested in the motion of liquids themselves. To consider the influence of liquid sloshing on the dynamic responses of liquid-filled containers and avoid complex liquid sloshing computation, the simplified equivalent mechanical model of fluid sloshing is generally used. On the basis of the potential flow theory, Graham [11] took the lead to establish an equivalent mechanical model of fluid sloshing in a rectangular container, which has been widely applied in the engineering field. Housner [12] deduced a simplified calculation formula based on an equivalent model of fluids based on physical intuition. The Housner model is a good approximation for the exact solution to the model proposed by Graham and has found extensive applications in civil and hydraulic engineering; however, the Housner model is not based on physical intuition and therefore is not an exact physical model, so its results are unreliable in some cases. Li [13] improved the Housner model on the basis of the theory of linear potential flow and provided a fitting solution to the equivalent model using a semi-analytical and semi-numerical method, whereas the improved Housner model has such complex parameters that it is only suitable for regular two-dimensional (2D) models and not complex 2D and three-dimensional (3D) models. Apart from proposing the equivalent methods based on the potential flow theory, Wstergaard [14] and Chopra [15] also developed the added-mass method. Rajasankar et al. [16] applied the added-mass method to the finite element method (FEM) and performed FSI analysis. Bao et al. [17] proposed an improved added-mass model based on the added-mass method and conducted dynamic analysis on an annular tank, providing reference for the design and application thereof. However, the distributed mass coefficient is difficult to determine in the above added-mass method and improved added-mass methods, so when using these methods to analyze structures of liquid-filled containers, the results are generally less reliable.

The conventional analytical methods are only applicable to cases with regular geometric shapes and simple external excitations, and FEM based on potential-based fluid elements is not applicable to the analysis of the dynamic responses of liquid-filled structures. Moreover, equivalent mechanical models based on fluid sloshing and the added-mass methods also have drawbacks. Considering this, it is necessary to use a method that is not only suitable for exploring liquid sloshing modes but is also applicable to assessing influences of liquid sloshing on the dynamic responses of liquid-filled containers, thus providing a reference for the engineering design and application of such containers. Therefore, the current research conducted a finite element analysis (FEA) on liquid sloshing modes at first and compared the analysis results with theoretical solutions and liquid sloshing frequencies and modes measured in previous tests, thus verifying the correctness of the FEM. However, the theoretical solutions and test models are only applicable to 2D models while the practical models are all 3D ones, so 3D models of liquid sloshing were also analyzed. Then, a dynamic analysis was conducted on liquid-filled containers with different liquid levels, considering the influences of the effect of FSIs and the intrinsic frequency and dynamic response of liquid-filled containers. The FEM based on acoustic fluid elements used in this research provides an effective and reliable method for the dynamic analysis of liquid-filled containers considering the effect of FSIs.

2. The FEA Theory Based on Acoustic Fluid Elements

A liquid-filled container and a liquid constitute an FSI system, in which the liquid and structure (liquid-filled container) are simulated using FEM. The FEM simulation methods of structures have been introduced elsewhere [18]. For FEA considering the effect of FSIs, one should use the fluid displacement as an unknown quantity [19] and harness the similarity between the fluid motion equation and the equation of motion of structural elastomers, which results in a finite element calculation model of fluid consistent with the finite element scheme; the other is to take the fluid pressure as an unknown quantity [20] and coordinate the displacement and pressure on the structure–fluid interface, from which the obtained mass and stiffness matrices are asymmetric matrices. When analyzing liquid sloshing problems using the pressure scheme based on acoustic fluid elements [21], the theory is as described below.

For the structure,

$$[M_S]\{\ddot{u}\} + [C_S]\{\dot{u}\} + [K_S]\{u\} - [R]\{p\} = \{f_S\} \tag{1}$$

where $[M_S], [C_S], [K_S]$ separately represent the mass, damping, and stiffness matrices of the structure; $\{\ddot{u}\}, \{\dot{u}\}, \{u\}$ separately denote the acceleration, speed, and displacement vectors at nodes of structural elements; $[R]$ denotes the coupling matrix at the structure–fluid interface; $\{p\}$ is the nodal pressure vector of fluid elements; and $\{f_S\}$ is the load vector of the structure.

For the fluid,

$$[M_F]\{\ddot{p}\} + [C_F]\{\dot{p}\} + [K_F]\{p\} + \overline{\rho}_0[R]^T\{\ddot{u}\} = \{f_F\} \tag{2}$$

where $[M_F], [C_F], \text{and} [K_F]$ separately represent the mass, damping, and stiffness matrices of the fluid; $\{\dot{p}\}, \{\ddot{p}\}$ are the first-order and second-order derivatives of nodal pressure of fluid elements; $\overline{\rho}_0$ denotes the fluid density; and $\{f_F\}$ is the load vector of the fluid.

The following can be obtained by combining Equations (1) and (2):

$$\begin{bmatrix} [M_S] & 0 \\ \overline{\rho}_0[R]^T & [M_F] \end{bmatrix} \begin{Bmatrix} \{\ddot{u}\} \\ \{\ddot{p}\} \end{Bmatrix} + \begin{bmatrix} [C_S] & 0 \\ 0 & [C_F] \end{bmatrix} \begin{Bmatrix} \{\dot{u}\} \\ \{\dot{p}\} \end{Bmatrix} + \begin{bmatrix} [K_S] & -[R] \\ 0 & [K_F] \end{bmatrix} \begin{Bmatrix} \{u\} \\ \{p\} \end{Bmatrix} = \begin{Bmatrix} \{f_S\} \\ \{f_F\} \end{Bmatrix} \tag{3}$$

By solving Equation (3), the liquid sloshing modes in the FSI system and the FSI dynamic responses can be obtained. The bounding surface of the water domain are modeled using various types of interface defined as follows:

(i) *Fluid–structure interaction:* Set a fluid–structure interaction on the boundary between fluid and structure.

(ii) *Free surface:* Set a free surface on the top surface of water domain, which can provide an approximate representation of the water surface wave.

(iii) *Rigid wall:* Set a rigid wall on the bottom and the lateral surface where the water cannot flow through the boundary.

As explained above, the surface of the water domain is a defined, specified boundary condition. In general, the software ANSYS 2023 can automatically generate fluid–structure interactions along the boundary between the fluid and structure.

3. Analysis of Liquid Sloshing Modes

3.1. Theoretical Solutions to 2D Liquid Sloshing Modes

Following the potential flow theory of ideal fluids, the fluid is assumed to be a non-viscous, irrotational, and incompressible ideal fluid, for which the influence of the free surface tension of the liquid is not considered. For general practical engineering problems, the above assumption is feasible. The corresponding eigenvalue equation is established by establishing the free sloshing equation for the liquid using the theoretical method. The liquid sloshing frequency and mode are attained by finding the eigenvalues.

3.1.1. Two-Dimensional Rectangular Container

For a regular rectangular container (Figure 1a), the jth sloshing frequency of the liquid is [2,3]:

$$\omega_j = \sqrt{g\left(\frac{j\pi}{2a}\right)\tanh\left(\frac{j\pi h}{2a}\right)} \quad j = 1, 2, 3, \ldots \tag{4}$$

where g is the gravitational acceleration; a represents the half width of the rectangular container; and h is the liquid level. According to Equation (4), the sloshing of the free liquid surface is attributed to gravity. The jth sloshing mode of the free liquid surface is

$$\phi_j(x, z) = \begin{cases} C_j \sin \frac{j\pi}{2a} x & j = 1, 3, 5, \ldots \text{ (antisymmetric mode)} \\ D_j \cos \frac{j\pi}{2a} x & j = 2, 4, 6, \ldots \text{ (symmetric mode)} \end{cases} \quad (5)$$

where C_j and D_j are constants. Figure 2 illustrates the first six sloshing modes of the free liquid surface. Therein, odd-order frequencies ω_1, ω_3, ω_5 correspond to the first three antisymmetric modes, while even-order frequencies ω_2, ω_4, ω_6 correspond to the first three symmetric modes, respectively.

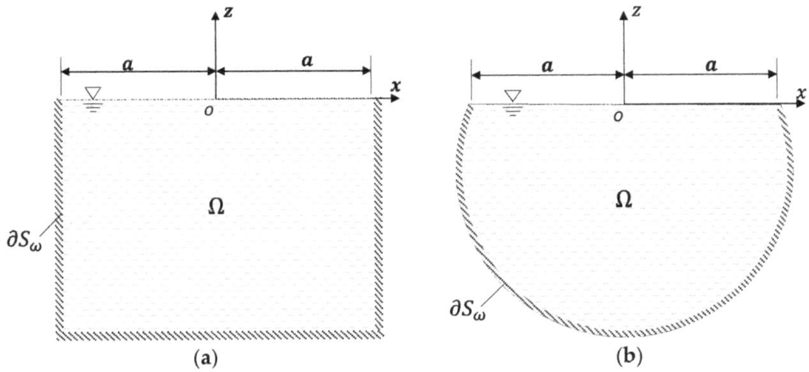

Figure 1. Two-dimensional containers. (**a**) Rectangular container. (**b**) Arbitrary section container.

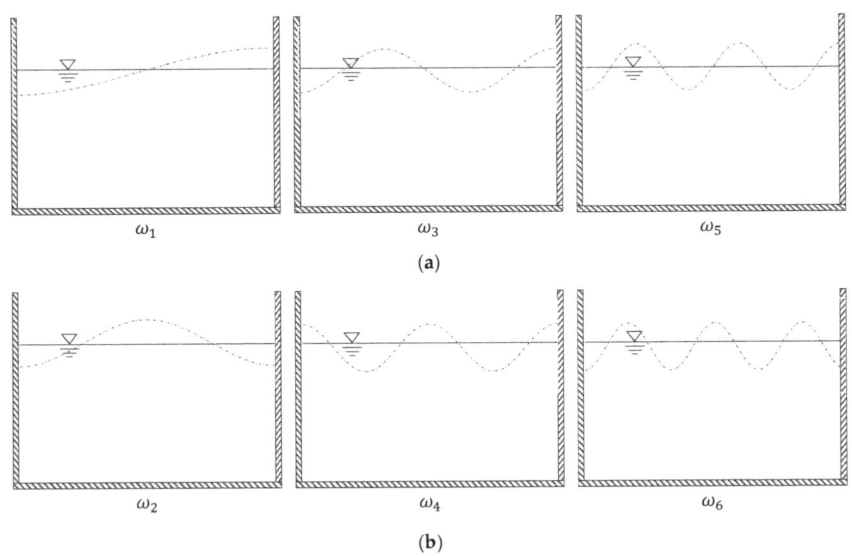

Figure 2. The first six sloshing modes of the free liquid surface in the rectangular container. (**a**) Antisymmetric (odd-order) modes ($j = 1, 3, 5$). (**b**) Symmetric (even-order) modes ($j = 2, 4, 6$).

3.1.2. Container with Arbitrary Cross Sections

A container of arbitrary shape is displayed in Figure 1b. It is challenging to solve the theoretical solution due to the irregular boundary. Considering this, the Ritz method is adopted to calculate the liquid sloshing frequency of containers of arbitrary shape [22]:

$$\omega_j^2 \cong \begin{cases} B_j\alpha_j g & j = 1,\ 3,\ 5,\ \ldots\ \text{(antisymmetric)} \\ D_j\alpha_j g & j = 2,\ 4,\ 6,\ \ldots\ \quad \text{(symmetric)} \end{cases} \tag{6}$$

$$B_j = \frac{-b_{Aj} - \sqrt{b_{Aj}^2 - 4a_{Aj}c_{Aj}}}{2a_{Aj}} \quad j = 1,3,5,\ldots \tag{7}$$

$$D_j = \frac{-b_{Sj} - \sqrt{b_{Sj}^2 - 4a_{Sj}c_{Sj}}}{2a_{Sj}} \quad j = 2,4,6,\ldots \tag{8}$$

$$\begin{cases} a_{Aj} = \iint_\Omega \left(\sin^2\alpha_j x + \sinh^2\alpha_j z\right)dxdz \\ b_{Aj} = 2\iint_\Omega (\sinh\alpha_j z \cdot \cosh\alpha_j z)dxdz - a/\alpha_j \\ c_{Aj} = \iint_\Omega \left(\cos^2\alpha_j x + \sinh^2\alpha_j z\right)dxdz \end{cases} \quad j = 1,3,5,\ldots \tag{9}$$

$$\begin{cases} a_{Sj} = \iint_\Omega \left(\cos^2\alpha_j x + \sinh^2\alpha_j z\right)dxdz \\ b_{Sj} = 2\iint_\Omega (\sinh\alpha_j z \cdot \cosh\alpha_j z)dxdz - a/\alpha_j \\ c_{Sj} = \iint_\Omega \left(\sin^2\alpha_j x + \sinh^2\alpha_j z\right)dxdz \end{cases} \quad j = 2,4,6,\ldots \tag{10}$$

where g denotes the gravitational acceleration; $\alpha_j = j\pi/2a$; a is the half width of the resting free liquid surface; and Ω is the liquid area.

3.2. FEA of 2D Liquid Sloshing Modes

The model adopted is a flat container, through which the 2D liquid sloshing is simulated. The test models and the finite element models are illustrated in Figures 3 and 4, respectively. The four-node acoustic fluid elements are used to model the water domain. The mesh densities of all finite element models were refined until the convergence of the results. Among them, the mesh size of the rectangular container is 0.005 m, and the mesh size of the circular container and U-shaped container along the diameter direction is 0.005 m. The water level and half width of the rectangular container are h = 0.12 m and a = 0.10 m; the water level and radius of the circular container are h = 0.16 m and R = 0.125 m; and the radius, water level, and cavity thickness of the U-shaped container are R = 0.10 m, h = 0.115 m, and 20 mm, respectively [23].

In the FEA, the containers are assumed to be rigid bodies. Generally, containers are all elastomers, while the elastic vibration of containers only slightly influences the overall liquid sloshing, so the influence of the stiffness of containers on liquid sloshing is ignored when analyzing liquid sloshing modes.

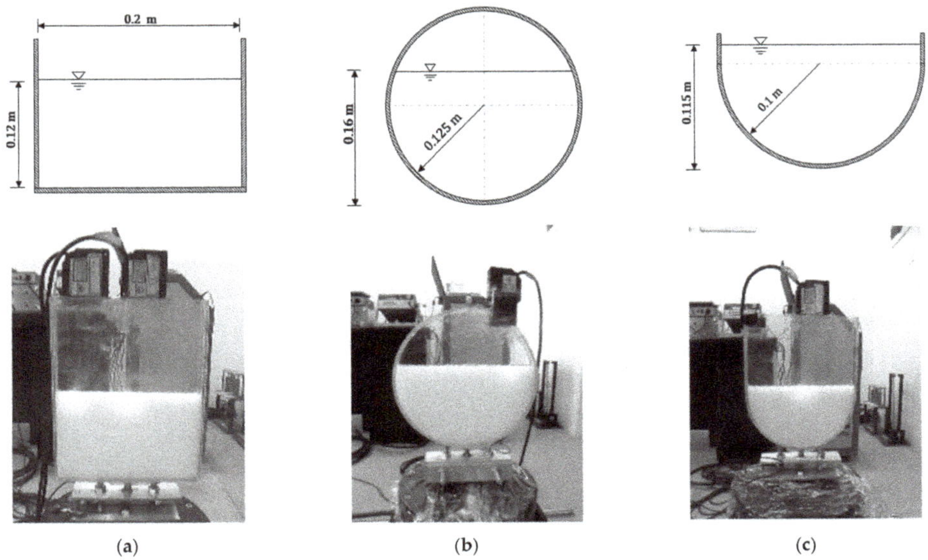

Figure 3. Test models [23,24]. (**a**) Rectangular container. (**b**) Circular container. (**c**) U-shaped container.

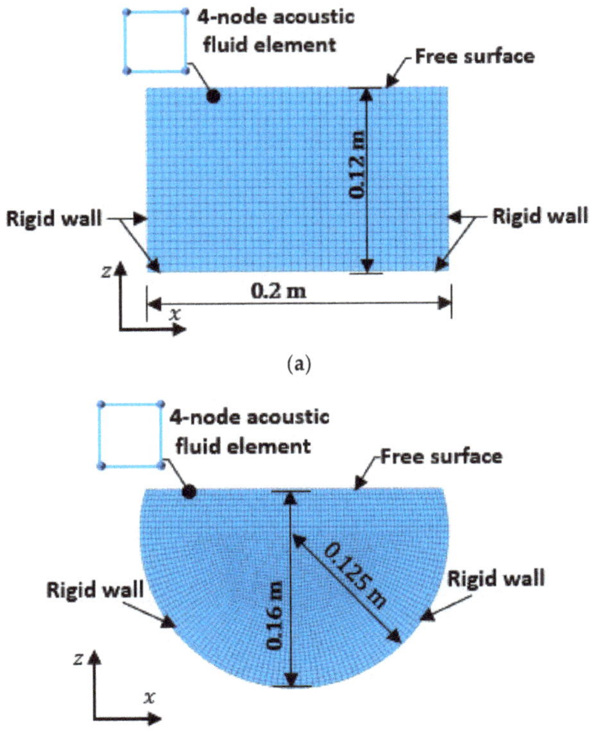

Figure 4. *Cont.*

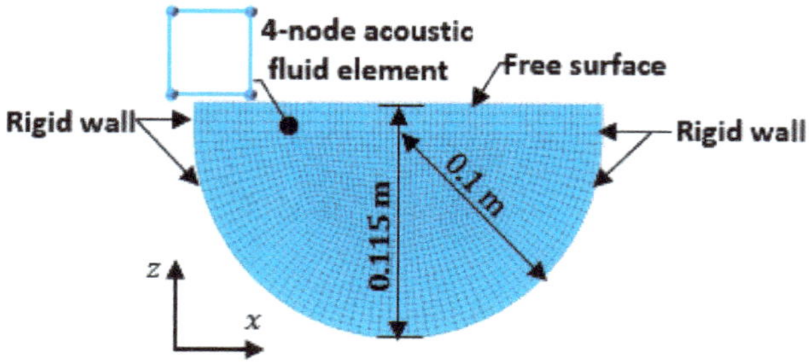

(c)

Figure 4. Finite element models. (**a**) Rectangular container. (**b**) Circular container. (**c**) U-shaped container.

3.2.1. Rectangular Container

The first four sloshing modes in the rectangular container obtained by FEA are shown in Figure 5; those measured in tests are illustrated in Figure 6. Table 1 presents that the liquid sloshing frequency attained by FEA is consistent with the results of theoretical calculation, with the maximum error within 0.5%. The maximum error in the sloshing frequency attained by FEA and that measured in tests is 3.1%. One of the causes for the error in the liquid sloshing frequency obtained using FEA and tests is that the liquid is assumed to be an ideal fluid in FEA, that is, incompressible fluid without viscosity, while the liquid in the tests is viscous. Furthermore, there is also an error (albeit small) arising in the test process. A comparison of Figure 2 with Figures 5 and 6 shows that the sloshing modes obtained by FEA agree well with the theoretically calculated modes, with a small difference in only the fourth order. The fourth sloshing modes obtained by the FEA and experiment are both even-order symmetric modes (Figure 2), so they are not, in the theoretical sense, different.

Table 1. Comparison of the first four orders in FEA, theoretical analysis, and tests (unit: Hz).

Order	Rectangular Container Water Level h = 0.12 m			Circular Container Water Level h = 0.16 m			U-Shaped Container Water Level h = 0.115 m		
	Theoretical Calculation	Test	FEA	Theoretical Calculation	Test	FEA	Theoretical Calculation	Test	FEA
1	1.93	1.89	1.93	1.77	1.68	1.77	1.90	1.85	1.89
2	2.79	2.73	2.79	2.55	2.52	2.57	2.78	2.73	2.78
3	3.42	3.40	3.43	3.12	3.09	3.15	3.42	3.35	3.42
4	3.95	3.94	3.97	3.61	3.51	3.64	3.95	3.87	3.95

Note: Test data in the table are taken from [23,24].

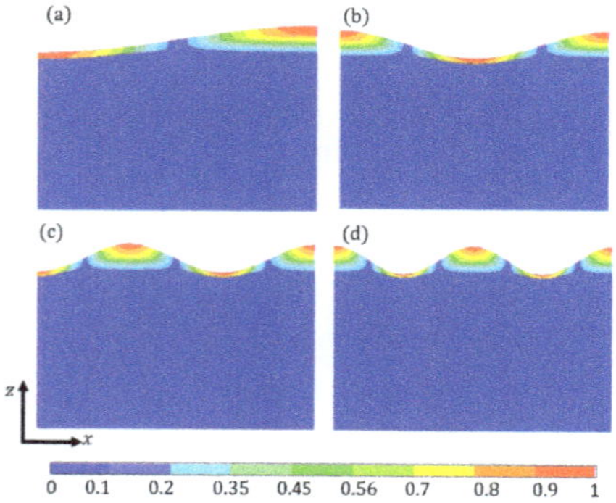

Figure 5. The first four liquid sloshing modes by FEA: (**a**) 1st; (**b**) 3rd; (**c**) 6th; (**d**) 9th.

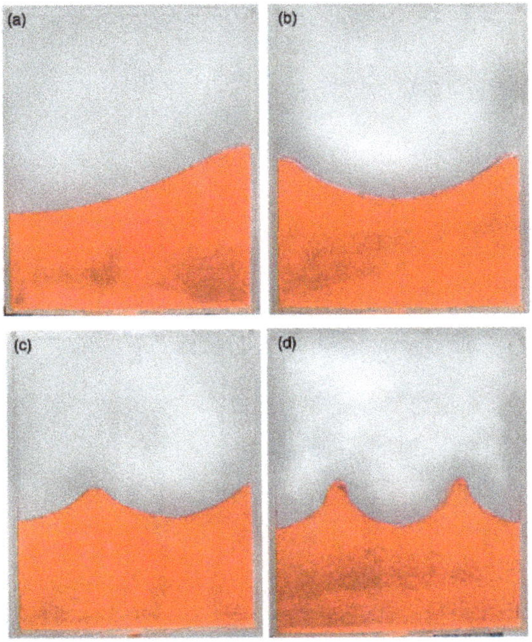

Figure 6. The first four liquid sloshing modes in the tests [23,24]: (**a**) 1st; (**b**) 2nd; (**c**) 3rd; (**d**) 4th.

3.2.2. Circular Container

The first four liquid sloshing modes in the circular container obtained by FEA are illustrated in Figure 7; those obtained in the tests are shown in Figure 8. Their comparison shows that the sloshing modes obtained by FEA match those in the tests, with a small difference in only the second order, whereas the second modes are both symmetric modes, so they can be regarded, in the theoretical sense, as consistent. Table 1 shows that the sloshing frequencies obtained by FEA are basically consistent with the first four sloshing frequencies attained by theoretical calculation, with the maximum error of 1.4%. The sloshing frequencies acquired by FEA have the maximum discrepancy of 5% with those measured experimentally, the source of which can be found in the error analysis as applied to the rectangular container.

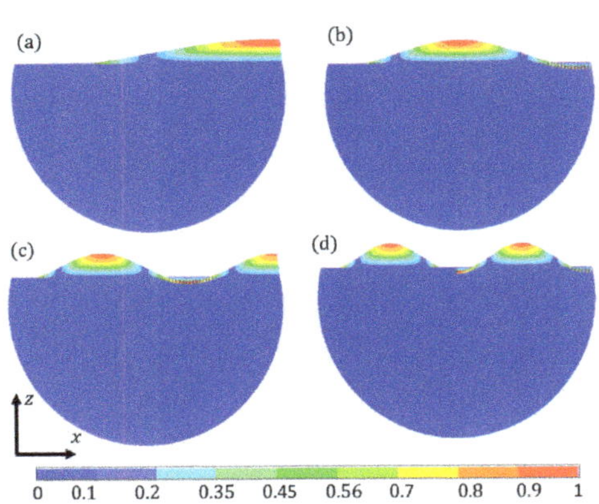

Figure 7. The first four liquid sloshing modes by FEA: (**a**) 1st; (**b**) 2nd; (**c**) 3rd; (**d**) 4th.

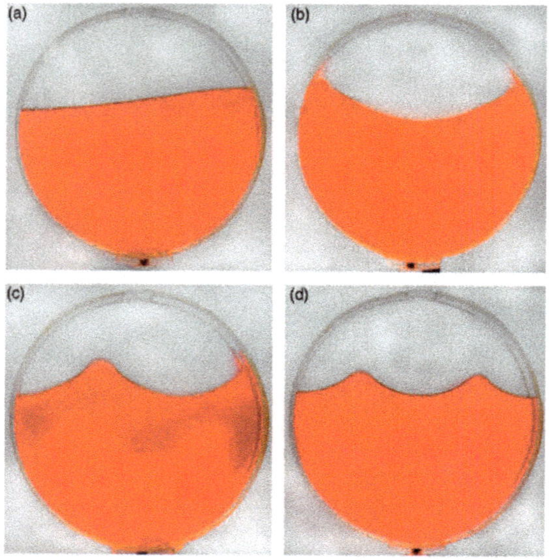

Figure 8. The first four liquid sloshing modes in the tests [23,24]: (**a**) 1st; (**b**) 2nd; (**c**) 3rd; (**d**) 4th.

3.2.3. U-Shaped 2D Container

The first four sloshing modes in the U-shaped 2D container attained by FEA are illustrated in Figure 9, while those measured in the tests are shown in Figure 10. Their comparison shows that the sloshing modes attained by FEA are consistent with those measured in the tests, with a subtle difference in only the fourth order, whereas the fourth modes are both symmetric modes, which are regarded as similar (in the theoretical sense). As displayed in Table 1, the sloshing frequencies predicted by FEA match those theoretically calculated for the first four frequencies, with differences of less than 0.5%. The sloshing frequencies obtained by FEA and those measured experimentally differ by no more than 2%, which indicates that the FEA results are reliable (again, the error analysis follows that applied to the rectangular container).

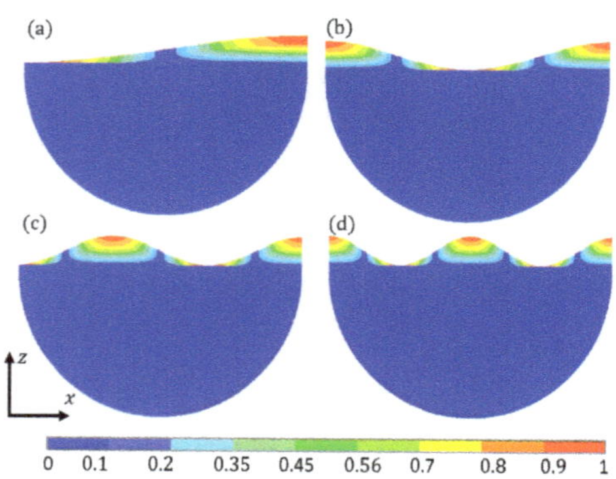

Figure 9. The first four liquid sloshing modes by FEA: (**a**) 1st; (**b**) 2nd; (**c**) 3rd; (**d**) 4th.

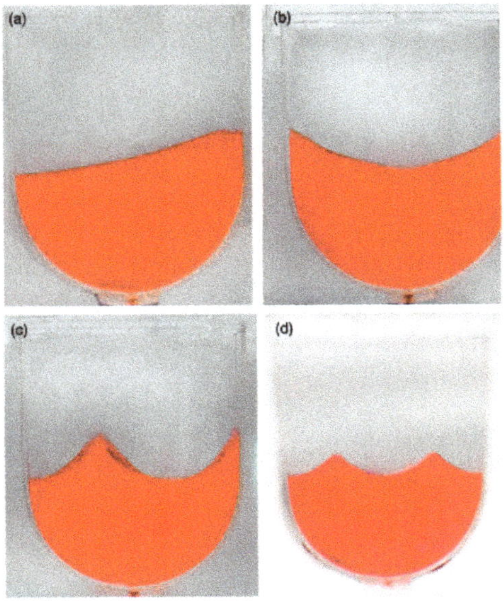

Figure 10. The first four liquid sloshing modes in the tests [23,24]: (**a**) 1st; (**b**) 2nd; (**c**) 3rd; (**d**) 4th.

Based on the above analysis results, the liquid sloshing frequencies conform well to the theoretical calculation and test results, despite small differences in several modes. This outcome indicates that the FEM based on acoustic fluid elements can be used to simulate the liquid sloshing modes in containers of different shapes.

3.3. FEA of 3D Liquid Sloshing Modes

To compare with the 2D models, the models of 3D containers were assigned identical geometrical dimensions to their 2D counterparts. The 3D diagram models and the 3D finite element models are shown in Figures 11 and 12, respectively. Compared to the 2D finite element model, the eight-node acoustic fluid elements are used to model the water domain, and the element size is reduced for enhancing computational efficiency. Among them, the mesh size of the rectangular container is 0.01 m, and the mesh size of the circular container and U-shaped-container along the diameter direction is 0.01 m. For the cuboid container, the water level is h = 0.12 m, and its length and width are $a = b$ = 0.20 m; the water level and radius of the spherical container are h = 0.16 m and R = 0.125 m; the water level and radius at the bottom of the U-shaped 3D container are h = 0.115 m and R = 0.10 m, respectively.

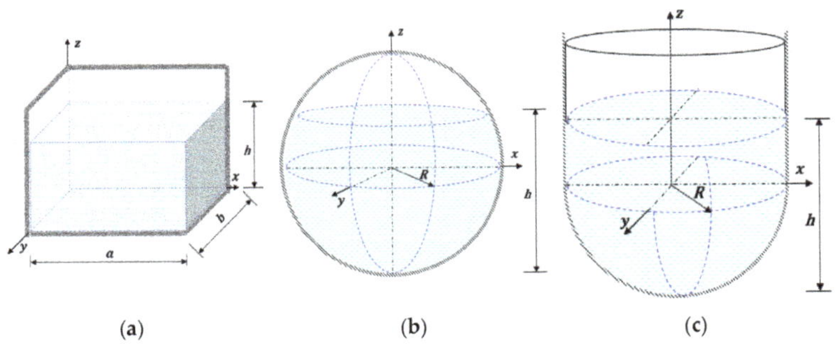

Figure 11. Three-dimensional diagram models. (**a**) Cuboid container. (**b**) Spherical container. (**c**) U-shaped 3D container.

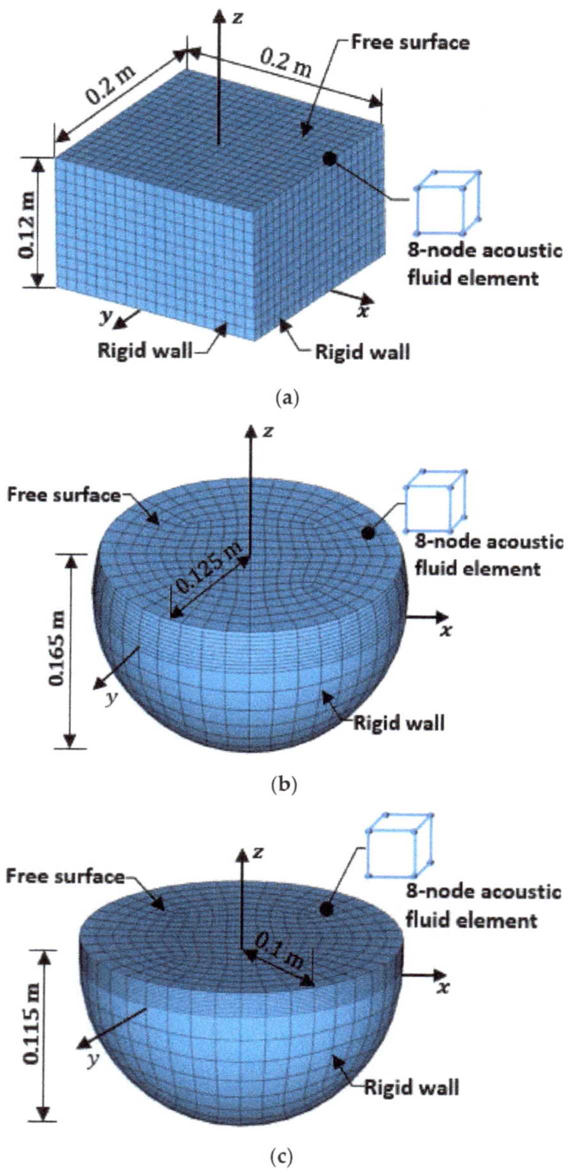

Figure 12. Three-dimensional finite element models. (**a**) Cuboid container. (**b**) Spherical container. (**c**) U-shaped 3D container.

3.3.1. Cuboid Container

The liquid sloshing modes in the cuboid container obtained by FEA are shown in Figures 13 and 14. Comparisons of Figures 5 and 13, combined with Tables 1 and 2, show that the first, third, sixth, and ninth liquid sloshing modes in the cuboid container separately correspond to the first, second, second, and fourth liquid sloshing modes in the rectangular container. The difference is that the liquid sloshing modes in the cuboid container are more complex than those in the rectangular container, and the frequencies are higher. It is more difficult to resolve the liquid sloshing modes in the cuboid container than those in the rectangular container in the theoretical sense. Therefore, the FEM based on acoustic fluid elements can better simulate liquid sloshing modes in the cuboidal container.

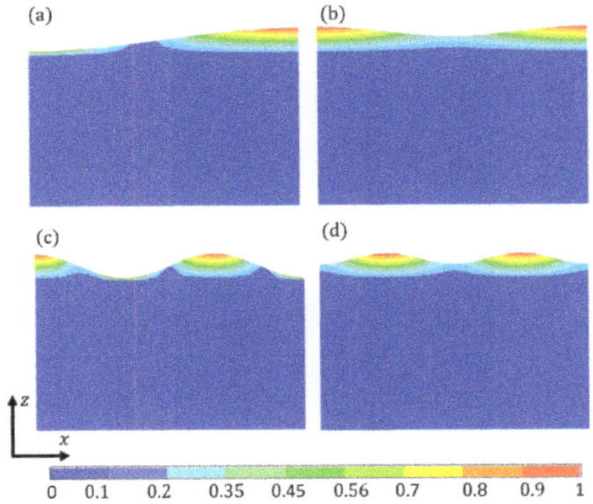

Figure 13. Two-dimensional view of liquid sloshing modes: (**a**) 1st; (**b**) 3rd; (**c**) 6th; (**d**) 9th.

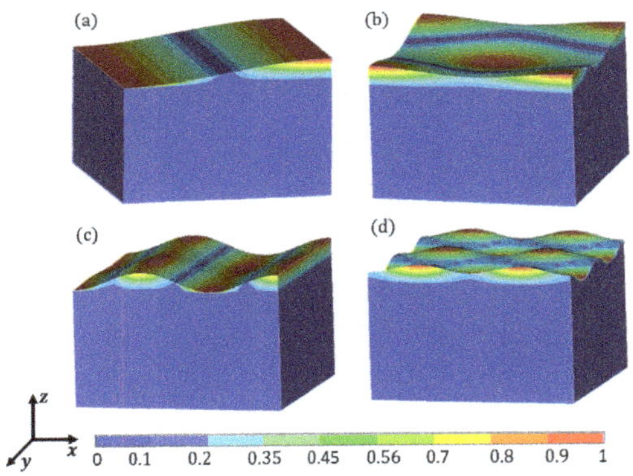

Figure 14. Three-dimensional view of liquid sloshing modes: (**a**) 1st; (**b**) 3rd; (**c**) 6th; (**d**) 9th.

Table 2. The first ten liquid sloshing frequencies (unit: Hz).

Container	Order									
	1	2	3	4	5	6	7	8	9	10
Cuboid container (Water level h = 0.12 m)	1.93	2.34	2.80	2.97	3.34	3.45	3.55	3.79	4.01	4.08
Spherical container (Water level h = 0.16 m)	1.95	2.57	2.85	3.04	3.37	3.44	3.79	3.80	3.88	4.12
U-shaped 3D container (Water level h = 0.115 m)	2.04	2.72	3.09	3.27	3.65	3.66	4.04	4.11	4.22	4.40

3.3.2. Spherical Container

The liquid sloshing modes in the spherical container acquired by FEA are shown in Figures 15 and 16. Comparisons of Figures 7 and 15, combined with Tables 1 and 2, show that, different from the cuboid and rectangular containers, the liquid sloshing modes in the spherical container do not correspond to those in the spherical container. This is because the length of the free liquid surface in the cuboid container ($a = b = 0.20$ m) remains identical to that in the rectangular container ($a = 0.20$ m), while the free liquid surface in the spherical container is not geometrically comparable to that in the spherical container.

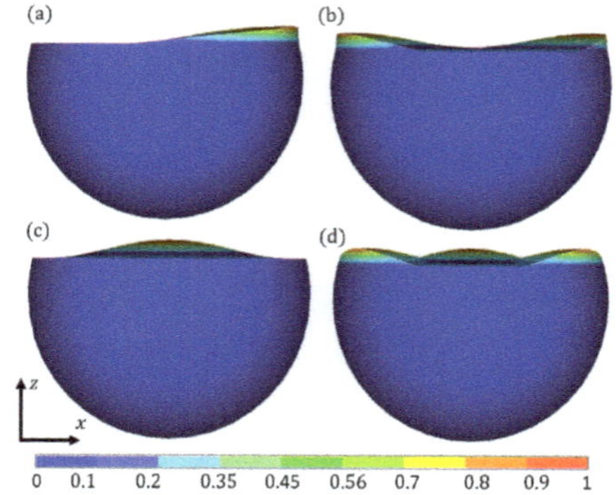

Figure 15. Two-dimensional view of liquid sloshing modes: (**a**) 1st; (**b**) 2rd; (**c**) 3rd; (**d**) 4th.

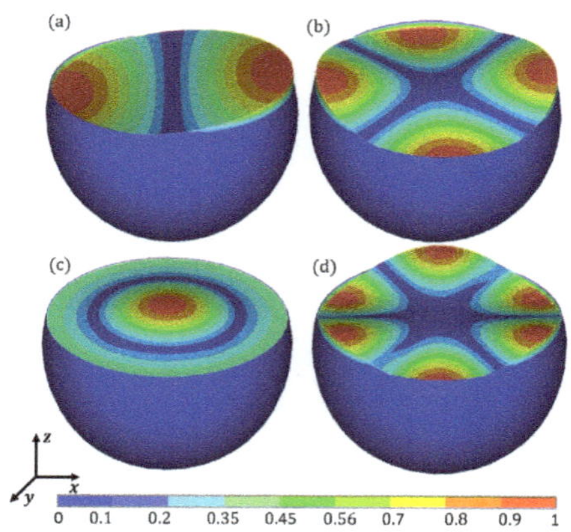

Figure 16. Three-dimensional view of liquid sloshing modes: (**a**) 1st; (**b**) 2rd; (**c**) 3rd; (**d**) 4th.

3.3.3. U-Shaped 3D Container

The liquid sloshing modes in the U-shaped 3D container obtained by FEA are shown in Figures 17 and 18. By comparing Figures 9 and 17 and combining with Tables 1 and 2, the liquid sloshing modes in the U-shaped 3D container do not correspond to those in the U-shaped 2D container, while they are similar to those in the spherical container. This is because the free liquid surface in the U-shaped 3D container demonstrates geometric similarity to that in the spherical container.

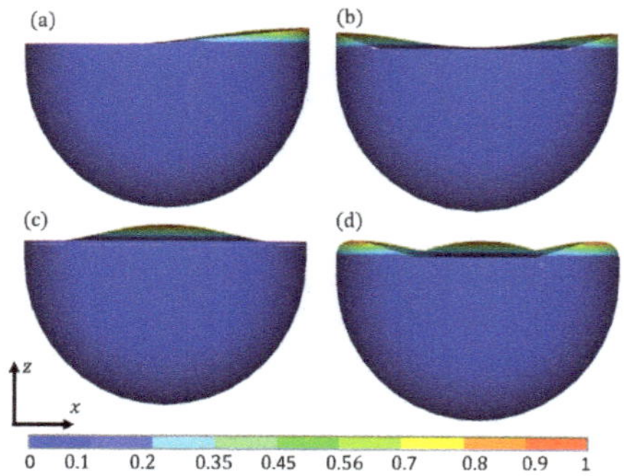

Figure 17. Two-dimensional view of liquid sloshing modes: (**a**) 1st; (**b**) 2rd; (**c**) 3rd; (**d**) 4th.

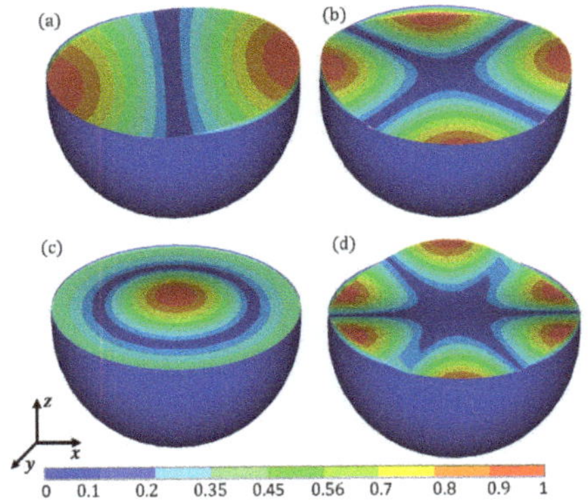

Figure 18. Three-dimensional view of liquid sloshing modes: (**a**) 1st; (**b**) 2rd; (**c**) 3rd; (**d**) 4th.

A comparison of the liquid sloshing modes in the 2D containers with those in the 3D containers reveals that the liquid sloshing modes in the 3D containers are more complex than those in the 2D containers, and the modes do not correspond to each other. The FEM based on acoustic fluid elements can be used to simulate liquid sloshing modes in 2D and 3D containers of different shapes.

4. Modal Analysis and Time-Historical Analysis of Cylindrical Liquid-Filled Containers

4.1. Liquid Modal Analysis

For the cylindrical liquid-filled container in Figure 19, the liquid sloshing modes in elastic and rigid containers were analyzed separately. As illustrated in Figure 20, the eight-node acoustic fluid elements are used to model the water domain, and the eight-node solid elements are used to model the cylindrical container shell. The model sizes and material parameters are listed in Table 3 [8].

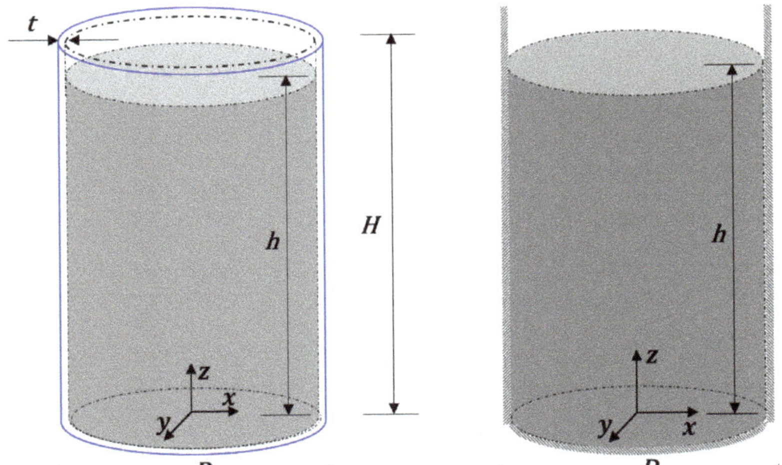

Figure 19. Cylindrical liquid-filled containers. (**a**) Elastic container. (**b**) Rigid container.

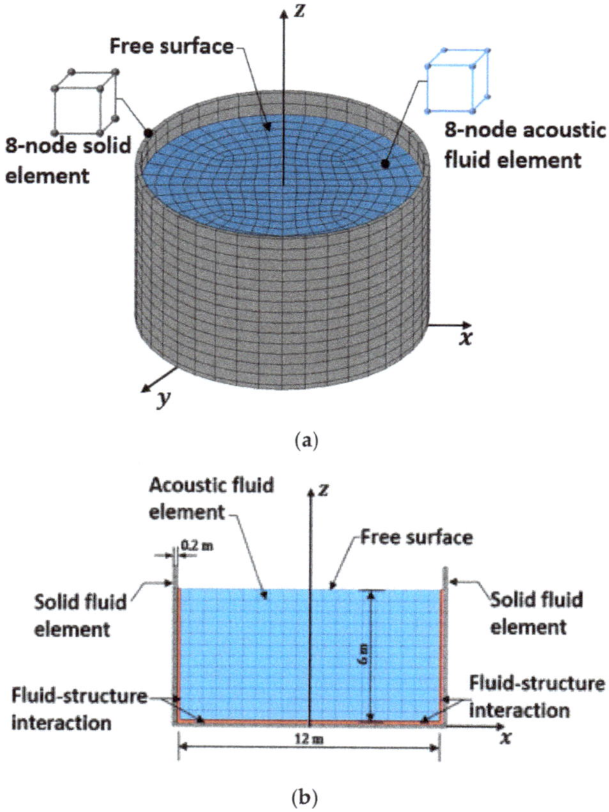

Figure 20. Finite element model of elastic container: (**a**) 3D view. (**b**) Two-dimensional elevation through a central cutting plane.

Table 3. Model sizes and material parameters.

Physical Qualities	Symbol	Value
Container diameter	D	12 m
Container height	H	7 m
Container wall thickness	t	0.2 m
Liquid level	h	6 m
Container elastic modulus	E	3.10×10^{10} Pa
Container density	ρ_1	2643 kg m^{-3}
Container Poisson's ratio	μ	0.15
Liquid density	ρ_2	1000 kg m^{-3}
Liquid acoustic velocity	c	1435 m s^{-1}

The liquid sloshing frequency in the cylindrical liquid-filled container is [2,3]

$$\omega_i = \sqrt{\lambda_i \frac{g}{R} \tanh\left(\lambda_i \frac{h}{R}\right)} \tag{11}$$

where λ_i is the ith root of the derivative of the family of Bessel functions; g is the gravitational acceleration; R is the radius of the liquid-filled container; and h denotes the liquid level in the liquid-filled container.

The first five liquid sloshing frequencies in the cylindrical liquid-filled container are listed in Table 4. The liquid sloshing modes in the cylindrical liquid-filled container and the vibration modes of this container are shown in Figures 21 and 22, respectively. The FEA numerical solution conforms to the theoretical solution, which indicates that it is reliable and can be used to simulate liquid sloshing modes in containers. Moreover, the FEA predictions of liquid sloshing frequencies in rigid and elastic containers are same for this example, suggesting that the influence of structural stiffness on the liquid sloshing modes can be ignored.

Table 4. First five liquid sloshing frequencies of cylindrical containers (unit: Hz).

	Order				
	1	2	3	4	5
FEA numerical solution (rigid container)	0.27	0.35	0.40	0.42	0.47
FEA numerical solution (elastic container)	0.27	0.35	0.40	0.42	0.47
Theoretical solution	0.27	0.35	0.40	0.42	0.47

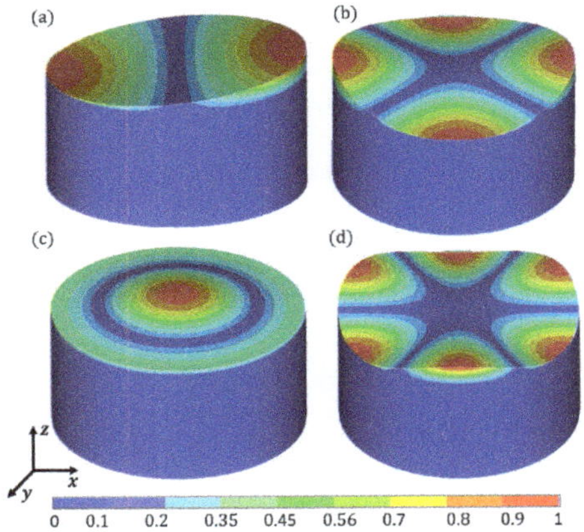

Figure 21. Liquid sloshing modes in the cylindrical liquid-filled container: (**a**) 1st; (**b**) 2rd; (**c**) 3rd; (**d**) 4th.

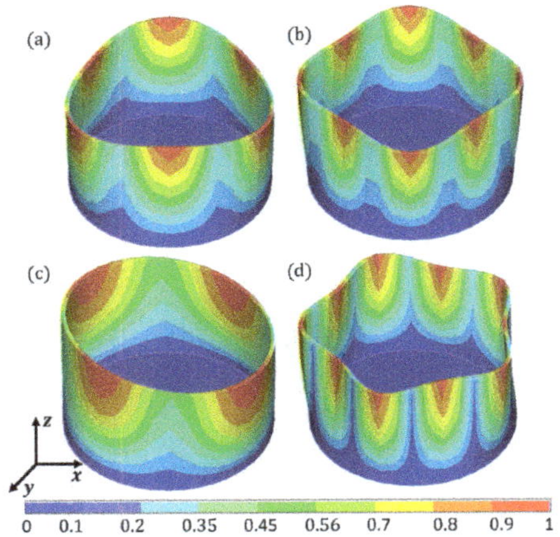

Figure 22. Vibration modes of the cylindrical liquid-filled container: (**a**) 1st; (**b**) 2rd; (**c**) 3rd; (**d**) 4th.

4.2. Modal Analysis of Cylinder Containers

In practical engineering application, the vibration frequencies of liquid-filled containers themselves attract more attention, in addition to liquid sloshing modes in liquid-filled containers. Therefore, it is necessary to study containers filled to different levels.

In order to measure the influence of different water depths on the frequency of the cylindrical container, the dimensionless coefficient is used to represent the influence coefficient of the frequency of the cylindrical container caused by water depth:

$$R_f = \frac{f_i - f_0}{f_0} \tag{12}$$

where f_i is the frequency of the cylindrical container at different water depths, and f_0 is the frequency of the cylindrical container without water.

The first five frequencies of liquid-filled containers with different liquid levels are displayed in Table 5 and Figure 23. The results show that the liquid exerts significant influences on the intrinsic frequency of cylindrical containers. As the liquid level rises in the liquid-filled containers, the vibration frequency of cylindrical containers decreases. If the liquid level in the liquid-filled containers is 6 m, the first frequency reduces by 31.24%. Therefore, the influences of the effect of FSIs on the intrinsic frequency and dynamic response of liquid-filled containers should be considered in the engineering design of such containers.

Table 5. The first five frequencies of cylindrical containers (unit: Hz).

Order	Liquid Level						
	0 m	1 m	2.0 m	3.0 m	4.0 m	5.0 m	6.0 m
1	19.94	19.94	19.88	19.45	18.05	15.93	13.71
2	20.11	20.11	20.02	19.48	18.23	16.30	14.20
3	26.81	26.81	26.72	26.13	24.39	21.46	18.37
4	29.29	29.28	29.09	27.85	24.93	21.88	19.41
5	38.30	38.30	38.16	37.10	34.10	30.44	26.34

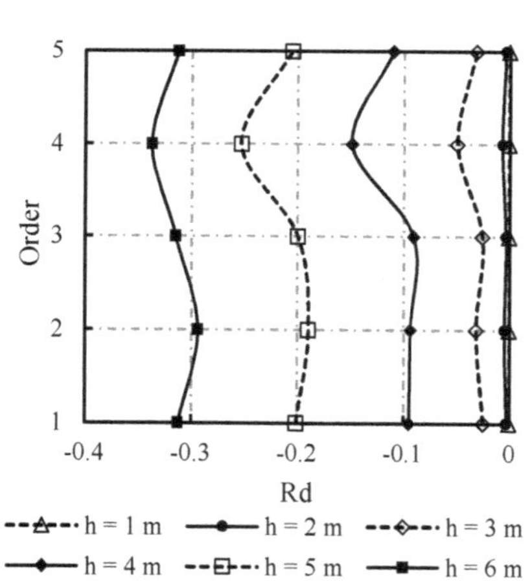

Figure 23. The first five R_f of cylindrical containers.

4.3. Time-Historical Analysis of Cylindrical Liquid-Filled Containers

Time-historical analysis was conducted by taking cylindrical liquid-filled containers with different liquid levels. The El-Centro (1940), Kobe (1995), and Loma Prieta (1989) waves were selected and applied in the x direction, with the peak acceleration of 0.1 g. The acceleration time-history curves and Fourier spectra are shown in Figure 24. Structural analysis generally focuses on the displacement, acceleration, and stress of structures. Hence, the maximum displacement, maximum acceleration, and maximum von Mises stress on the sidewall at different heights of liquid-filled containers at $x = D/2$ and $y = 0$ were monitored.

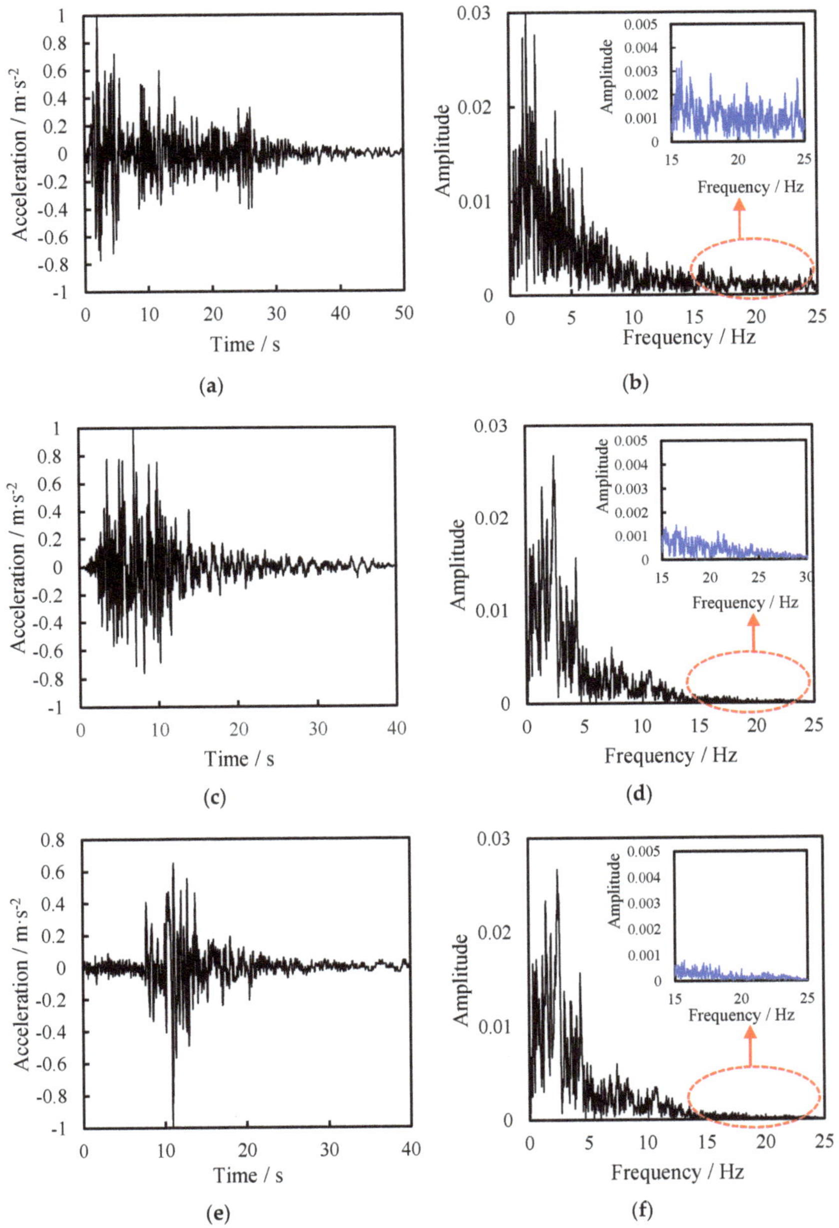

Figure 24. Acceleration time-history curves and Fourier spectra. (**a**) El-Centro wave. (**b**) Fourier spectrum under the El-Centro wave. (**c**) Kobe wave. (**d**) Fourier spectrum under the Kobe wave. (**e**) Loma Prieta wave. (**f**) Fourier spectrum under the Loma Prieta wave.

In order to measure the influence of different water depths on the seismic response of the container, the dimensionless coefficient is used to represent the influence coefficient of the dynamic response of the container caused by water depth, and the coefficient of displacement along the sidewall height is as follows:

$$R_d = \frac{d_i - d_0}{d_0} \tag{13}$$

where d_i is the maximum displacement of the sidewall at different water depths, and d_0 is the maximum displacement of the sidewall without water. The coefficient of acceleration along the sidewall height is

$$R_a = \frac{a_i - a_0}{a_0} \tag{14}$$

where a_i is the maximum acceleration of the sidewall at different water depths, and a_0 is the maximum acceleration of the sidewall without water. The coefficient of von Mises stress along the sidewall height is

$$R_\sigma = \frac{\sigma_i - \sigma_0}{\sigma_0} \tag{15}$$

where σ_i is the maximum von Mises stress of the sidewall at different water depths, and σ_0 is the maximum von Mises stress of the sidewall without water.

The distribution of the maximum displacement and Rd on the sidewall of containers with different liquid levels along the height are displayed in Figures 25 and 26. As the liquid level rises, the maximum displacement and Rd on the sidewall of cylindrical liquid-filled containers increase. When the liquid levels are 0 (no liquid contained), 1, and 2 m, the maximum displacement on the sidewall appears on the top of the container; under conditions with liquid levels of 3 and 4 m, the maximum displacement on the sidewall appears at the height of 2.5 m; if the liquid levels are 5 and 6 m, the maximum displacement on the sidewall occurs at the height of 3 m. Unlike the distribution of the maximum displacement on the sidewall, the position of the maximum Rd remains largely unchanged.

The maximum displacement and Rd on the sidewall have a similar distribution along the height at different liquid levels under the three input seismic waves, and they always increase on the sidewall of the cylindrical liquid-filled container with rising liquid levels, whereas the values of the maximum displacement and Rd are different. Taking the liquid level of 6 m as an example, the maximum displacements on the sidewall are separately 0.052, 0.049, and 0.046 mm under the El-Centro, Kobe, and Loma Prieta waves, and they all appear at a height of 3 m on the sidewall. The Rd values on the sidewall are, separately, 4.90, 4.95, and 4.62 under the El-Centro, Kobe, and Loma Prieta waves, and they all appear at a height of 1 m on the sidewall.

The distribution of the maximum acceleration and Ra on the sidewall of containers with different liquid levels along the height are similar to that of the maximum displacement on the sidewall. Figures 27 and 28 show that as the liquid level rises, the maximum acceleration and Ra on the sidewall of cylindrical liquid-filled containers constantly grow. In the case that the liquid levels are 0, 1, and 2 m, the maximum acceleration on the sidewall appears on the top of the liquid-filled container; if the liquid levels are 3 and 4 m, the maximum acceleration on the sidewall is found at the height of 2.5 m; when the liquid levels are 5 and 6 m, the maximum acceleration on the sidewall occurs at a height of 3 m. The position of the maximum Rd remains almost unchanged.

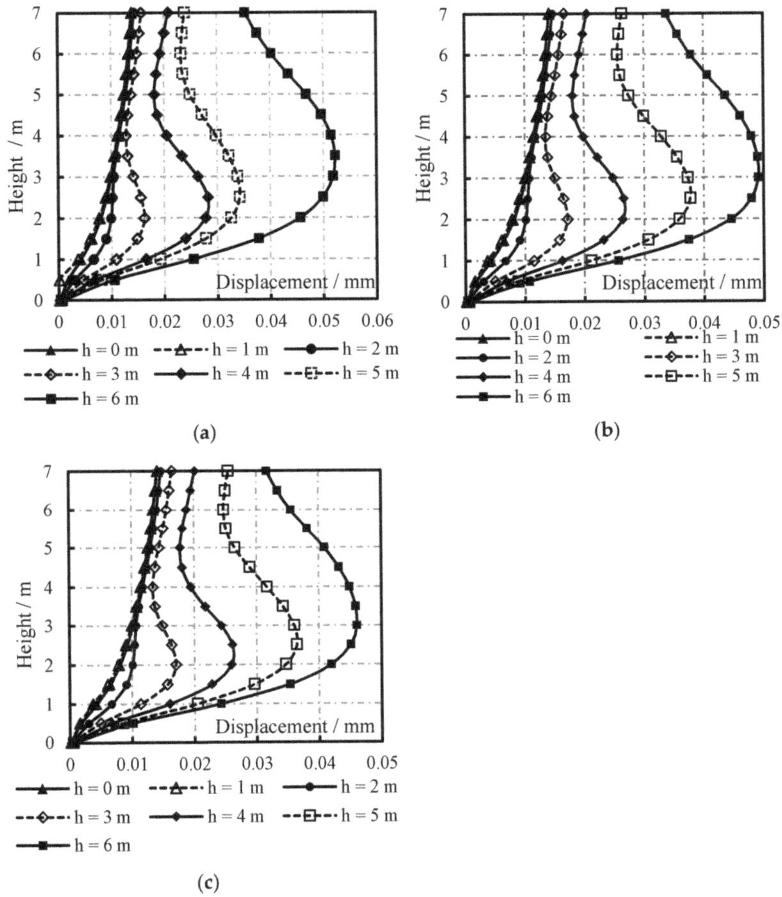

Figure 25. The distribution of the maximum displacement on the sidewall of containers. (**a**) El-Centro wave. (**b**) Kobe wave. (**c**) Loma Prieta wave.

Under action of the three seismic waves, the maximum acceleration and Ra on the sidewall of containers with different liquid levels have a similar distribution along the height. The maximum acceleration and Ra on the sidewall of cylindrical liquid-filled containers always increase with rising liquid levels. Taking the liquid level of 6 m as an example, the maximum accelerations on the sidewall are 0.46, 0.20, and 0.058 m/s^2 under the El-Centro, Kobe, and Loma Prieta waves, respectively, and they all appear at a height of 3 m. The Ra values on the sidewall are, separately, 4.72, 8.58, and 6.35 under the El-Centro, Kobe, and Loma Prieta waves, and they all appear at a height of 1 m on the sidewall.

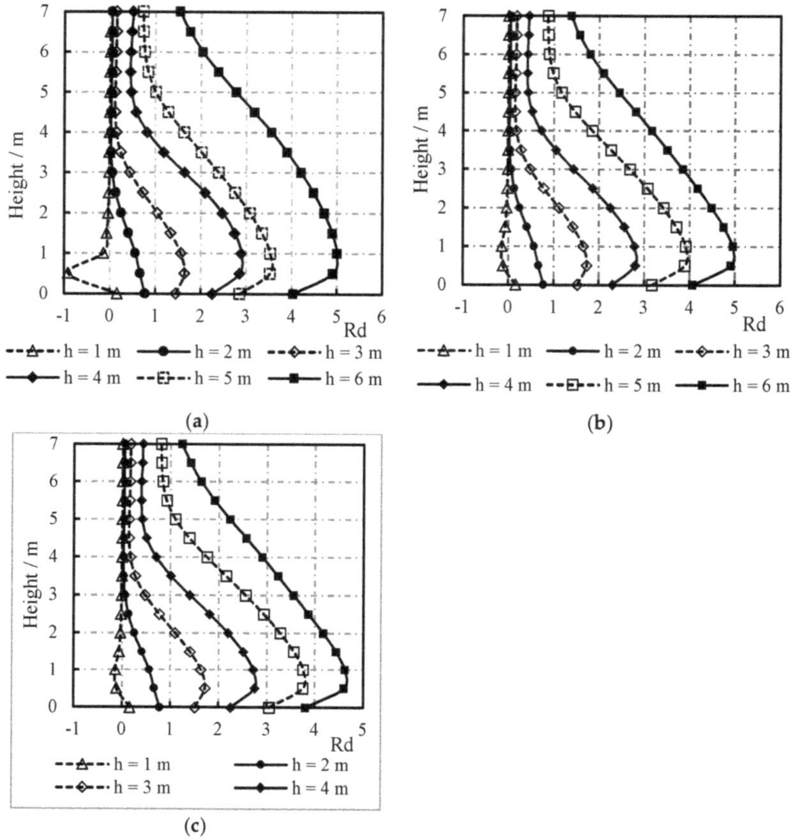

Figure 26. The distribution of Rd on the sidewall of containers. (**a**) El-Centro wave. (**b**) Kobe wave. (**c**) Loma Prieta wave.

Under the action of different seismic waves, the maximum acceleration on the sidewall of containers with different liquid levels differs greatly in the distribution height. On the one hand, this is because the frequencies of the input seismic waves differ greatly from the structural frequency of vibration. The main frequencies of the El-Centro, Kobe, and Loma Prieta waves are all concentrated within 5 Hz (Figure 24). In the case of different liquid levels, the minimum and maximum first frequencies of the cylindrical container are 13.17 Hz (liquid level of 6 m) and 19.94 Hz (liquid level of 0 m), respectively. This suggests an unobvious acceleration amplification effect. On the other hand, the El-Centro wave has more high-frequency components (above 15 Hz) compared with the Kobe and Loma Prieta waves, so the maximum acceleration on the sidewall under the El-Centro wave is greater than those under the Kobe and Loma Prieta waves.

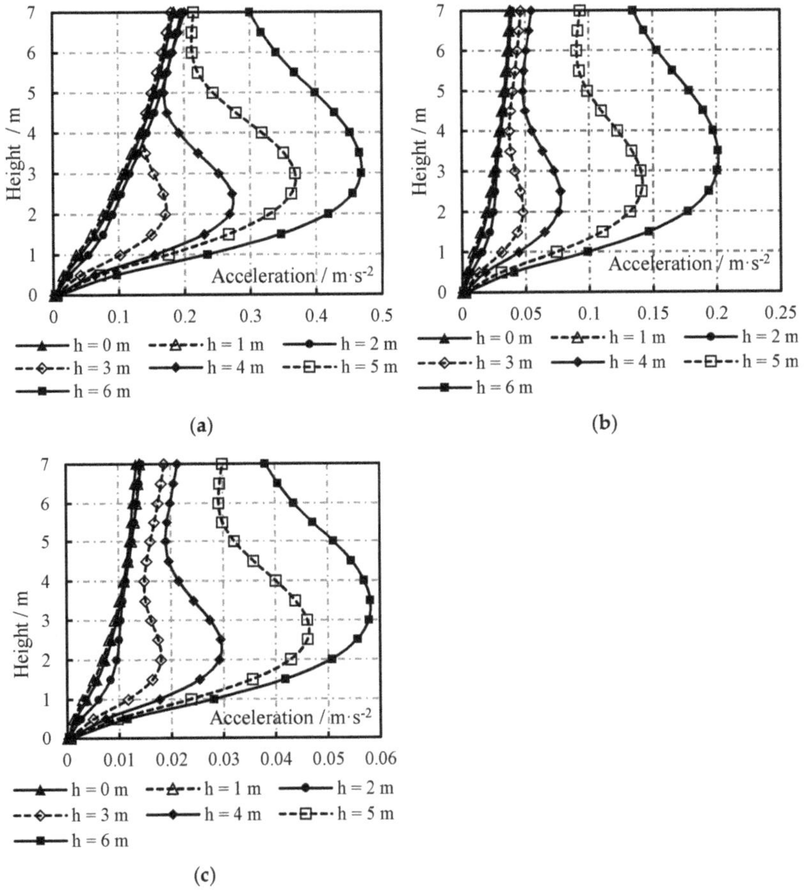

Figure 27. Distribution of the maximum acceleration on the sidewall of containers. (**a**) El-Centro wave. (**b**) Kobe wave. (**c**) Loma Prieta wave.

The distribution of the maximum von Mises stress and R_σ on the sidewall of containers with different liquid levels along the height are shown in Figures 29 and 30. The maximum von Mises stress is always found at the bottom of the liquid-filled containers. The maximum R_σ is always found at approximately 2 m of the liquid-filled containers. Apart from that at the bottom, the maximum von Mises stress on the sidewall of cylindrical liquid-filled containers constantly increases as the liquid rises. When the liquid levels are 0, 1, and 2 m, the maximum von Mises stress on the sidewall appears at a height of 1.0 m; if the liquid levels are 3 and 4 m, the maximum von Mises stress on the sidewall occurs at a height of 2.0 m; in the case that liquid levels are 5 and 6 m, the maximum von Mises stress on the sidewall occurs at a height of 2.5 m.

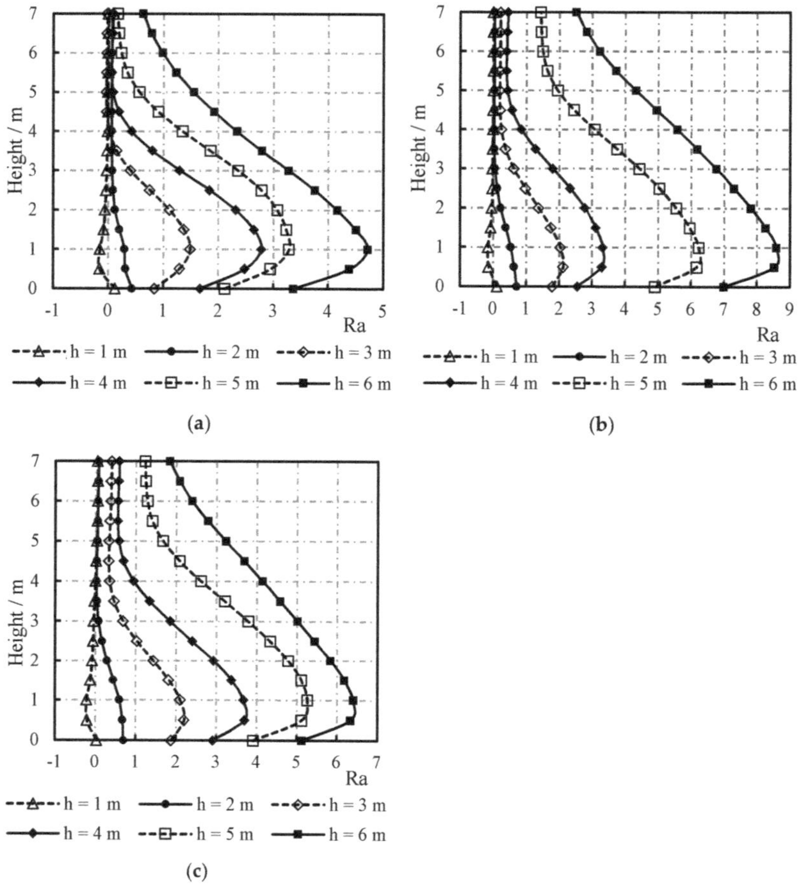

Figure 28. The distribution of Ra on the sidewall of containers. (**a**) El-Centro wave. (**b**) Kobe wave. (**c**) Loma Prieta wave.

Under action of the three seismic waves, the maximum von Mises stress and R_σ on the sidewall show a consistent distribution along the height of containers filled to different levels. As the liquid level rises, the maximum von Mises stress on the sidewall of cylindrical liquid-filled containers increases, while the values of the maximum von Mises stress and R_σ differ. Taking the liquid level of 6 m as an example, under the El-Centro, Kobe, and Loma Prieta waves, the maximum von Mises stresses on the sidewall are, separately, 0.150, 0.144, and 0.135 MPa, and they are all found at a height of 2.5 m. The R_σ values on the sidewall are, separately, 6.67, 6.35, and 5.92 under the El-Centro, Kobe, and Loma Prieta waves, and they all appear at a height of 2 m on the sidewall.

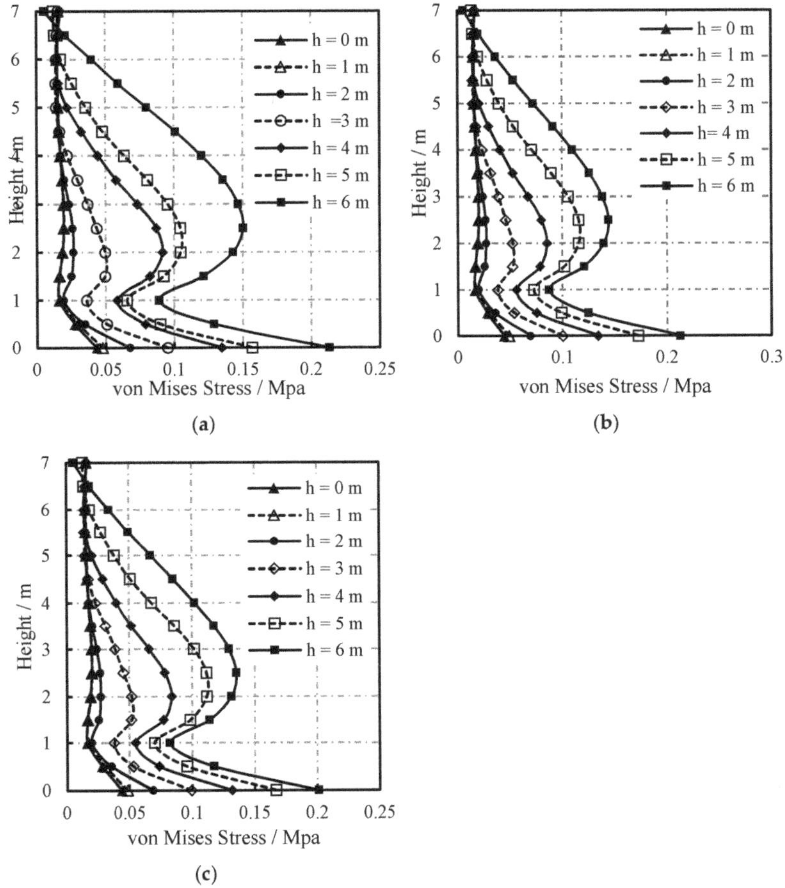

Figure 29. Distribution of the maximum von Mises stress on the sidewall of containers. (**a**) El-Centro wave. (**b**) Kobe wave. (**c**) Loma Prieta wave.

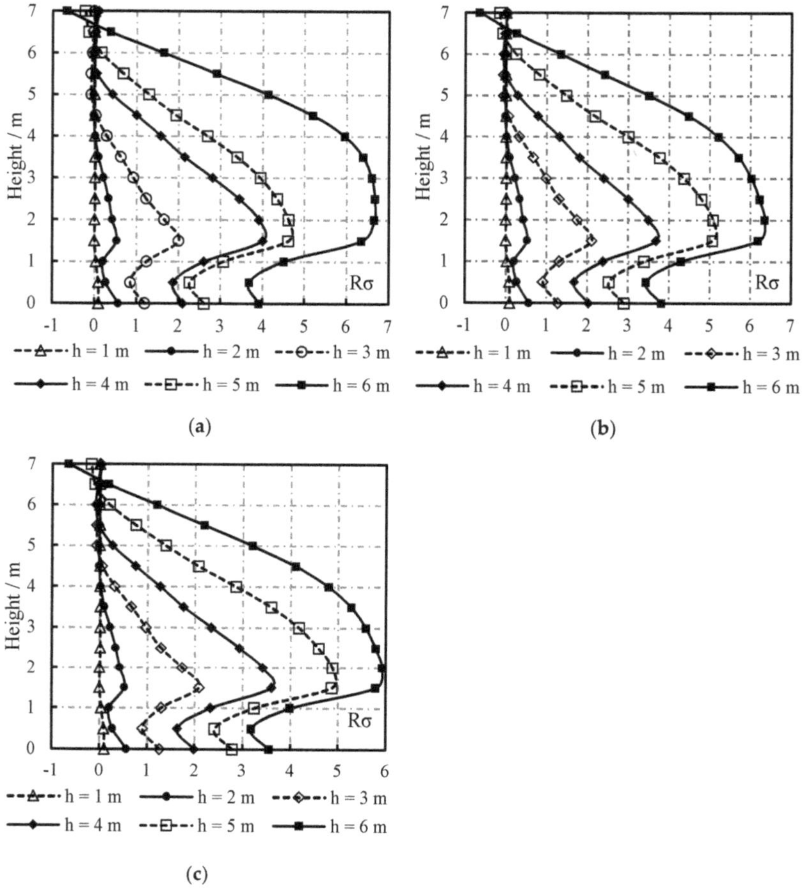

Figure 30. The distribution of R_σ on the sidewall of containers. (**a**) El-Centro wave. (**b**) Kobe wave. (**c**) Loma Prieta wave.

5. Conclusions

A dynamic analysis was performed on liquid sloshing modes in liquid-filled containers and the liquid-filled containers themselves using the FEM based on acoustic fluid elements. The following conclusions can be drawn:

(1) The liquid sloshing modes in 2D and 3D liquid-filled containers of regular shapes and arbitrary cross sections were analyzed and compared with theoretical solutions and test results. The results reveal that the FEM based on acoustic fluid elements is accurate;

(2) The liquid level exerts significant influences on the intrinsic frequency of liquid-filled containers. As the liquid level in liquid-filled containers rises, the vibration frequency of cylindrical containers decreases. When the liquid level in liquid-filled containers is 6 m, the first frequency decreases by 31.24%. During the engineering design of such liquid-filled containers, the influence of the effect of FSIs on the intrinsic frequency of these containers should not be ignored. FEM based on acoustic fluid elements can be used to model such an effect;

(3) For the cylindrical liquid-filled containers in this research, the liquid level essentially did not influence the displacement, acceleration, and stress of the liquid-filled containers under horizonal seismic action if the liquid level was low. As the liquid level rises, the displacement and acceleration of, and stress on, such liquid-filled containers increase

significantly. The acceleration responses of liquid-filled containers are particularly significantly affected by the spectral characteristics of the input seismic wave.

Author Contributions: Conceptualization, X.F. and X.B.; methodology, X.F.; software, X.F.; validation, F.Y., Q.Z. and X.B.; formal analysis, X.F.; investigation, Q.Z.; resources, X.F.; data curation, X.B.; writing—original draft preparation, X.F.; writing—review and editing, X.B.; All authors have read and agreed to the published version of the manuscript.

Funding: This research was funded by [Hainan Special PhD Scientific Research Foundation of Sanya Yazhou Bay Science and Technology City] grant number [HSPHDSRF-2022-03-008].

Data Availability Statement: The raw data supporting the conclusions of this article will be made available by the authors on request.

Acknowledgments: The author is very grateful to Wenhui Wei for his careful guidance of this paper.

Conflicts of Interest: The authors declare that they have no known competing financial interests or personal relationships that could have appeared to influence the work reported in this paper.

References

1. Ju, R.; Zeng, X. *Coupled Vibration Theory of Elastic Structures and Liquids*; Seismological Press: Beijing, China, 1983. (In Chinese)
2. Dodge, F.T. *The New Dynamic Behavior of Liquids in Moving Containers*; Southwest Research Institute: San Antonio, TX, USA, 2000.
3. Abramson, H.N. *The Dynamic Behavior of Liquids in Moving Containers*; NASA SP-106; NASA Special Publication: Washington, DC, USA, 1966.
4. Ibrahim, R.A. *Liquid Sloshing Dynamics: Theory and Applications*; Cambridge University Press: Cambridge, UK, 2005.
5. Faltinsen, O.M.; Timokha, A.N. *Sloshing*; Cambridge University Press: Cambridge, UK, 2009.
6. Wang, J.; Li, Q.; Zhu, Z.; Lu, M.; Zhang, X. A review of numerical methods for fluid-structure interaction with large free surface sloshing. *Shanghai Mech.* **2001**, *4*, 447–454. (In Chinese)
7. Wen, D.; Zheng, Z.; Sun, H. Development of aseismic research on liquid storage tanks. *China Adv. Mech.* **1995**, *1*, 60–76. (In Chinese)
8. Bao, X.; Liu, J. Dynamic finite element analysis methods for liquid container considering fluid-structure interaction. *Nucl. Power Eng.* **2017**, *38*, 111–114. (In Chinese)
9. Liu, J.; Bao, X.; Tan, H.; Wang, D.; Li, X. Dynamic analysis of annular liquid containers. *Chin. Eng. Mech.* **2018**, *35*, 66–73. (In Chinese)
10. Xu, G.; Ren, W.; Zhang, W.; Reimerdes, H.G.; Dafnis, A.; Korsch, H. Dynamic characteristic analysis of liquid-filled tanks as a 3-d fluid-structure coupling system. *China J. Mech.* **2004**, *3*, 328–335. (In Chinese)
11. Graham, E.W.; Rodriguez, A.M. The characteristics of fuel motion which affect airplane dynamics. *J. Appl. Mech.* **1952**, *19*, 381–388. [CrossRef]
12. Li, Y.; Wang, J. A supplementary, exact solution of an equivalent mechanical model for a sloshing fluid in a rectangular tank. *J. Fluid. Struct.* **2012**, *31*, 147–151. [CrossRef]
13. Housner, G.W. *Earthquake Pressures on Fluid Containers*; California Institute of Technology: Pasadena, CA, USA, 1954; pp. 1–39.
14. Westergaard, H.M. Water Pressures on Dams during Earthquakes. *Trans. Am. Soc. Civ. Eng.* **1933**, *98*, 418–432. [CrossRef]
15. Hall, J.F.; Chopra, A.K. Hydrodynamic effects in the dynamic response of concrete gravity dams. *Earthq. Eng. Struct. Dyn.* **1982**, *10*, 333–345. [CrossRef]
16. Rajasankar, J.; Iyer, N.R.; Rao, T.V.S.R.A. A new 3-D finite element model to evaluate added mass for analysis of fluid-structure interaction problems. *Int. J. Numer. Methods. Eng.* **1993**, *36*, 997–1012. [CrossRef]
17. Bao, X.; Liu, J.; Yang, Y.; Tan, H.; Wang, D. Comparative study on added mass methods for dynamic analysis of annular water tank in nuclear engineering. *J. Build. Struct.* **2018**, *39*, 130–139. (In Chinese)
18. Wang, X. *Finite Element Method*; Tsinghua University Press: Beijing, China, 2003. (In Chinese)
19. Chen, H.; Taylor, R.L. Vibration analysis of fluid-solid systems using a finite element displacement formulation. *Int. J. Numer. Methods. Eng.* **1990**, *29*, 683–698. [CrossRef]
20. Everstine, G.C. Finite element formulations of structural acoustics problems. *Comput. Struct.* **1997**, *65*, 307–321. [CrossRef]
21. *Acoustic Analysis Guide Release 2023 R2*; ANSYS, Inc.: Canonsburg, PA, USA, 2023.
22. Li, Y.; Wang, Z. An Approximate Analytical Solution of Sloshing Frequencies for a Liquid in Various Shape Aqueducts. *Shock Vib.* **2014**, *2*, 118–124. [CrossRef]

23. Li, Y.; Wang, Z. Unstable characteristics of two-dimensional parametric sloshing in various shape tanks: Theoretical and experimental analyses. *J. Vib. Control* **2016**, *22*, 4025–4046. [CrossRef]
24. Li, Y. *Fundamentals of Fluid Sloshing Dynamics*; Science Press: Beijing, China, 2017. (In Chinese)

applied sciences

MDPI

Article

Engineering-Driven Approach for the Structural Design of Geometrically Complex Modular Artificial Reefs

Fábio Cruz *, Isabel B. Valente, Dmytro Maslov, Tiago Miranda and Eduardo B. Pereira *

Department of Civil Engineering, Advanced Production and Intelligent Systems (ARISE), Institute of Science and Innovation for Bio-Sustainability (IB-S) and Institute for Sustainability and Innovation in Structural Engineering (ISISE), University of Minho, 4800-058 Guimarães, Portugal; isabelv@civil.uminho.pt (I.B.V.); dmytro.maslov@civil.uminho.pt (D.M.); tmiranda@civil.uminho.pt (T.M.)

* Correspondence: id8024@alunos.uminho.pt (F.C.); eduardo.pereira@civil.uminho.pt (E.B.P.)

Abstract: Over the years, artificial reefs with diverse forms and functions have become increasingly important in maritime engineering and ecosystem restoration plans, with socio-economic and ecological impacts on marine ecosystems. However, due to the increasing complexity of designs and durability requirements, the current structural design approaches lack adequate documentation and standardization. This work addresses this challenge by detailing an engineering-driven approach for the structural design of artificial reef structures made of reinforced concrete or composite steel and concrete. This study establishes the premises for the structural design, followed by the quantification of actions based on standards and recommendations for marine structures. Hydrodynamic and numerical models were utilized to assess the effects of these actions on the structure. A cross-section organic design was then implemented, followed by a parametric study exploring various structural and material combinations for optimization. This study validates the developed design methodology combining hydrodynamic actions and strength analysis for complex modular artificial reef structures of 3 to 15 m size, specifically tailored for open waters. The results confirm the applicability and reliability of the developed design methodology, demonstrating its potential for guiding future numerical and experimental studies on modular artificial reef structures in open waters.

Keywords: modular multifunctional artificial reef; structural design; cross-section analysis; numerical model; parametric study; optimization

1. Introduction

In recent years, there has been a growing interest in the development of marine infrastructures and technologies to harness maritime resources [1–6]. These efforts aim at supporting the expanding global population and promoting the sustainability of human activities [7–9]. Significant strides have been made, ranging from the creation of modular floating structures for energy generation and artificial energy islands, to the development of modular artificial reef structures dedicated to the protection and restoration of marine ecosystems [10–14]. In fact, artificial reefs (ARs) might contribute to multiple other purposes such as increasing fisheries yield [15], recreational dive tourism [16], improving surf conditions [17], coastal protection [2,18,19], and preventing bottom trawling [20]. Some authors have highlighted the importance of designing ARs specifically for the ecosystems projected to be enhanced [8,12]. However, according to the literature, the purpose for which artificial reefs have been created is seldom revealed [6].

Despite all the advantages, these innovative structures face numerous challenges characteristic of a new era of coastal and maritime engineering [21–23]. These challenges include structural complexity and durability requirements, complex hydrodynamic interactions with ocean currents, and suitability for sea operations. There is an urgent need for new approaches that not only address these problems effectively but also prioritize ecologically friendly solutions, bio-receptiveness, efficiency, and resilience. This is essential for fostering a sustainable and balanced relationship between human activities and the environment [4,21,24–28].

Over time, the design of ARs has predominantly relied on empirical methods, guided by subjective judgments and assessments. In some cases, this approach was limited to sinking structures without the adoption of a clear design process. However, growing interest in artificial reef technology led to the development of several guidelines and recommendations [29–32], as well as frameworks for the development of purpose-oriented ARs. In general, these frameworks address functional aspects and provide a global view of the entire process and, to some extent, also address the structural design, albeit superficially. Recently, Vogler proposed a conceptual framework to design artificial reefs, which includes a design component that focuses mainly on geometric design, materials, and manufacturing [33,34]. Carral et al. (2022) proposed a new general and integrated conceptual framework to guide the design of AR units by adopting an ecosystem ecology perspective [35]. The flowchart is composed of three main parts: (i) marine ecosystem ecology model, (ii) AR-ecosystem index transformation, and (iii) the stability analysis. Although there is no detailed framework for the structural design, the general methodology highlights this aspect. The authors presented also a methodology based on geographic information systems (GISs) to determine the best area for installing ARs in a specific estuary considering the geomorphological, hydrodynamic, and bathymetric characteristics of the area, as well as the distance to the harbour and economic factors [36]. ARs designed for surfing show a robust and heavy morphology, similar to breakwaters, following similar design principles [17]. A preliminary design achieved step by step, making use of the theory and state of the art of multifunctional artificial reefs for surfing, was proposed by Voorde [17,26,37,38]. On the other hand, the geometry complexity of ARs is also an important aspect in other cases, and some works have found that increasing reef complexity has a significant and positive impact on fish species flourishing, density, and increased biomass [39,40]. Therefore, in recent years there has been a demand for modular, slender, and complex geometries for ARs. As a result, more sophisticated structures need to be designed based on engineering methods using analytical, numerical, and experimental analysis to produce structures capable of withstanding the harsh conditions typically encountered in marine environments.

The development of artificial reefs from an engineering, reliability, and feasibility standpoint remains relatively unexplored in the literature. Given the growing importance of these structures in the rehabilitation of marine ecosystems, it is essential to integrate clear and technically sound engineering methodologies. While aspects related to geometric design are well established—covering conceptual requirements [30,41–43], as well as hydrodynamic and morphodynamic performance using numerical simulation [18,44–52] and physical modelling [26,53,54]—the structural design based on primary analytical and numerical methodologies is still insufficiently addressed. Existing studies that explore structural behaviour typically focus on advanced fluid–structure interaction (FSI) simulations [49,55,56]. This may hinder the replication of such structures in different geographies and scenarios, due to the lack of understanding of the overall relationships between general requirements including materials and action-effect relationships, and the design approach itself.

This work addresses this challenge by implementing an engineering-driven approach for the design of a geometrically complex and modular multifunctional artificial reef (MMFAR), allowing the safety verification and possibly the generalization of the methodology to other similar projects. The developed approach may also support subsequent stages of analysis based on CFD and FSI for detailed functional analysis.

2. Materials and Methods

2.1. Engineering-Driven Methodology

In this study, a systematic and engineering-driven approach was implemented having the structural design as the central process, based on analytical assumptions and simplified numerical models. The deployment site and the MMFAR requirements and functions were also considered as problem variables. Subsequent numerical analysis (CFD and FSI) and experimental characterization for detailed characterization of innovative materials and structural systems were also acknowledged. This approach is represented in Figure 1, and in this research the focus is mostly on processes 1 to 5, directly related to the structural design, summarized as follows:

1. **Concept and geometric design:** in this stage, with the MMFAR objectives established, the process selects the site location and collects environmental data on physical and biological aspects. The design of shapes and dimensions is undertaken with the objective of meeting functional requirements. Stakeholders, including experts and the local community, participate actively.
2. **Constructive approach:** the development of cross-sections and 3D models is initiated. It considers materials, structural systems, constructive processes, and durability specifications. These aspects must prevent marine pollution and establish a support that meets technical and functional requirements. Logistics and construction techniques are also considered.
3. **Site location and sea conditions:** the analysis of the site location results in the identification of specific actions affecting the structure, such as dead, live, environmental, and accidental. Environmental actions are among the most important. The actions and load combinations follow the recommendations from ISO, API, DNV, and NORSOK [2–5]. Historical site data and prediction models are used to account for time and space variability.
4. **Structural model:** the effects of the loads are obtained from a simplified numerical simulation. For that, the numerical model is discretized in bar elements, and a linear-elastic quasi-static analysis is performed. The materials, section properties, and supports are defined, along with load cases and combinations. The results are the axial (N), shear (V), flexural (M), and torsional (T) stresses for safety verifications.
5. **Cross-section design:** determines the load capacity based on material strength and cross-section dimensions, checking limit states to assess performance and feasibility. A parametric study explores various structural combinations in order to identify the promising solutions. This approach optimizes the cross-section, eliminates less promising options, and encourages new structural systems.

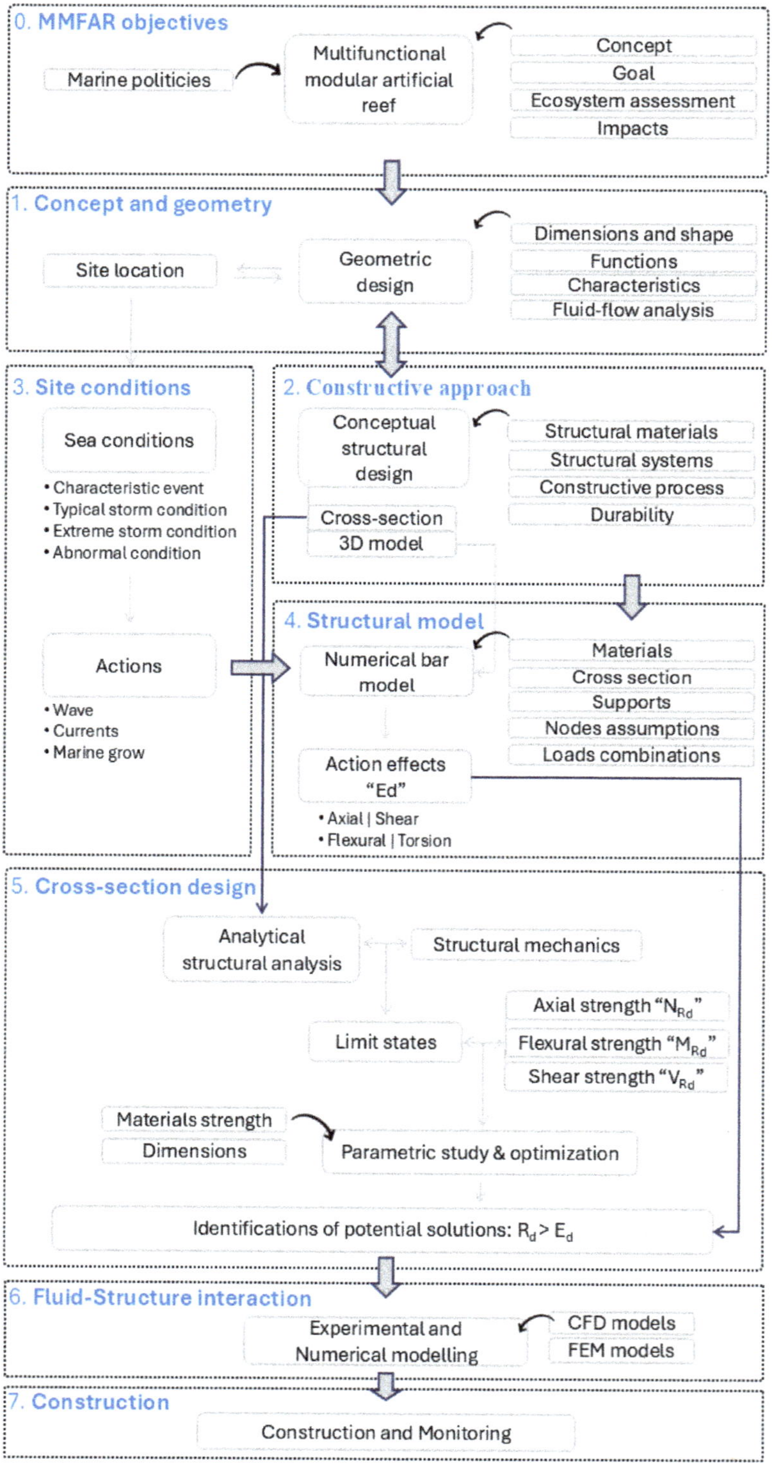

Figure 1. Design flowchart.

2.2. Concept and Geometric Design

The concept of the MMFAR and the geometrical design were developed within the NEXTSEA project, as detailed by Maslov et al. (2024) [57]. The MMFAR was aimed at supporting marine restoration, enhancing fisheries, and promoting recreational diving [45,57]. In particular, the recovery of key species such as sea bass and seabream was targeted [57]. The geometric design considered also the marine protected area requirements, including site characteristics, bottom properties and habitats, and characteristics of the MMFAR resulting from its interaction with the flow, such as upwelling zones, velocities, and wake zones [45,57]. Figure 2 illustrates the evolution of the reef module geometry from module 1 to module 3. To ensure ease of transportation, the weight of each individual unit must not exceed 150 kg. The limb presents a variable cross-section, with a diameter of approximately 150 mm at the smallest section (at the connection end), increasing to around 196.5 mm at the junction with the central bulb. The modular units, united by connectors, allow various MMFAR configurations depending on the assembly scheme (niche, barrier, tunnel). The final assembled structures are reticulated, slender, and expandable, enabling the execution of various configurations from small niches to full reef barriers.

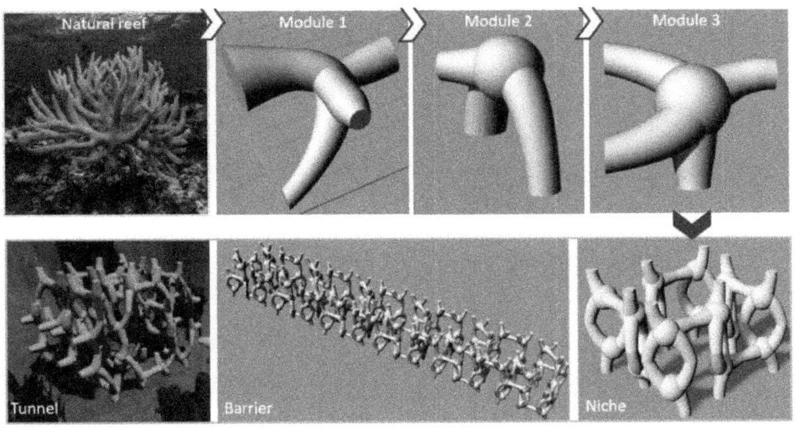

Figure 2. Geometric design using novel approaches and design processes for the 3D network of individual modules of the MMFAR (adapted from [57]).

2.3. Constructive Approach

Considering the requirements and complex geometry of the MMFAR, it was necessary to design an efficient structural system to (i) allow adequate adjustment of the reinforcing elements within the module and ensure uniform cross-sectional stress distribution as well as a minimum nominal reinforcement cover; (ii) ensure continuity of the internal metallic reinforcement elements and between elements, to facilitate electrical conduction for added functionalities (e.g., self-healing techniques); and (iii) the need to adopt a modular and adaptable construction system to allow the practical and simple assemblage, compatible to underwater operations.

Algorithmic thinking using parametric modelling tools as well as structural design approaches was used for the iterative design and analysis, as shown subsequently. Concrete was chosen for its versatility and cost-effectiveness, as well as its proneness to serve as a colonizing substrate [41,43]. Its brittle nature was countered by the incorporation of ductile reinforcements, such as hot-dip galvanised (HDG) steel which is technically and financially attractive while offering high corrosion resistance. Other options may include self-healing cementitious composites and fibre-reinforced polymers, although the design of these materials was not addressed in this study.

In this sense, different structural systems for the reef module were proposed, as shown in Figure 3: (i) a reinforced concrete structure approach (RC) based on the use of galvanized steel rebars along each limb of the unit and (ii) a composite steel and concrete structure (CS) where a HDG steel tube was adopted, aligned with the geometric centroid of each limb. The tube was perforated to allow the concrete to flow and fill the inner part of the tube, increasing the bond between the materials and providing a dead weight to counteract buoyancy.

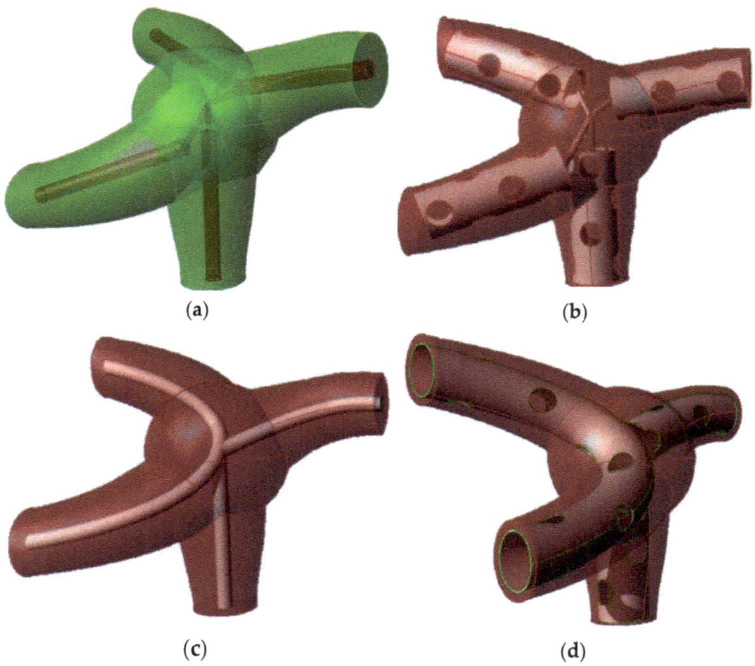

Figure 3. Structural systems: (**a**) RC unit with four galvanized steel rebars interconnected using external rods by welding, (**b**) CS unit based on the use of four galvanized steel tubes with holes along the length, interconnected using external rods by welding, (**c**) RC with three central galvanized steel rebars interconnected at the centre of the unit by welding or tying, and (**d**) CS based on the use of three central galvanized steel tube profiles with holes along the length, interconnected by welding at the intersection.

The adoption of straight reinforcement elements was tried; however, the accommodation of the reinforcement elements in the geometry was found to be hard, in particular trying to guarantee the minimum concrete cover. Curved reinforcement elements, although requiring additional forming procedures, were easier to adapt to the complex geometry of the module, ensuring better distribution of the reinforcement area along the cross-section and providing adequate reinforcement cover.

The extension of the tube or rebar beyond the edges of each module was considered to support the connection system, particularly favourable in the case of the solution adopting the steel tubes as shown in Figure 4.

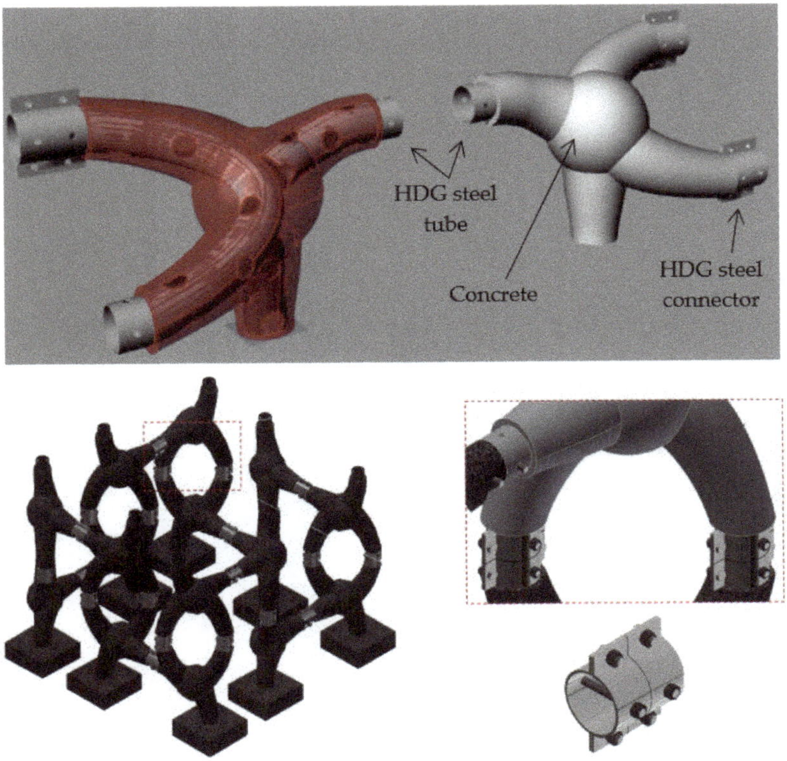

Figure 4. 3D model of MFMAR with the connectors system.

2.4. Site Location and Sea Conditions

As a case study, the developed MMFAR was assumed to be deployed in a Marine Protected Area (MAP), in Esposende—Portugal. This site represents a promising location for the implementation of MMFAR for ecosystems restoration and protection, and for recreational or scientific diving. Figure 5 shows the geographical location of the Littoral North Marine Park, as well as the bathymetry in the area. The ideal installation depth of the MMFAR was defined as 20 m, allowing recreational diving and avoiding the direct effect of high waves. The present study is related mainly to the structural design of an underwater structure; therefore in Table 1 the typical wave parameters in the area for different sea conditions of the study area are presented. Further details may be found in [58–61].

Table 1. Wave height and period for different sea conditions [58–61].

Characteristic Event E_w	Typical Storm Condition $E_{w,storm}$	Extreme Storm Condition $E_{w,máxstorm}$	Abnormal Extreme Condition $E_{w,e}$
$H_S = 2.20$ m $T = 7.89$ s	$H_{S,storm} = 4$ m $T = 10$ s	$H_{S,máxstorm} = 9$ m $T = 14.6$ s	$H_{S,40} = 15.61$ m $T_{p,40} = 20$ s

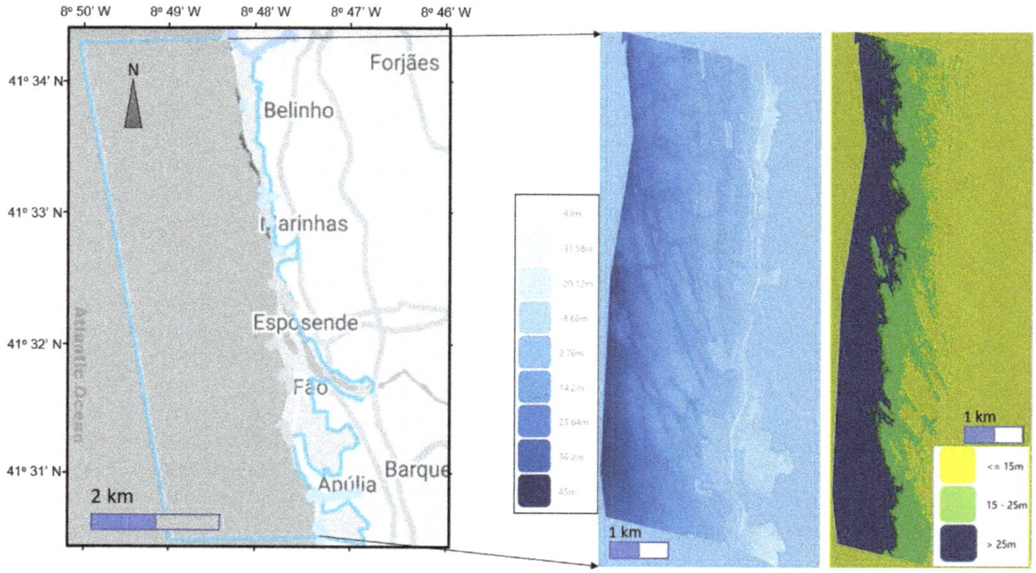

(a) (b) (c)

Figure 5. (**a**) Geographical location of the Littoral North Marine Park in Esposende, Portugal; (**b**) detailed bathymetry distribution, and (**c**) zone with acceptable MMFAR depth [45].

2.5. Structural and Hydrodynamic Model

The research conducted considered the development of a MMFAR niche, whose structure was composed of 16 units, each of which occupied a cube with an edge of 65 cm (Figure 6). With a width of approximately 2 by 2 m, the structure represents a bottom projection area of 4.807 m^2 and a total volume of 1.76 m^3 [57]. The full assemblage was numerically studied based on a reticulated model of bars to evaluate the stresses for the structural mechanics analysis. For that, a finite element tool, with a 3D graphical user interface object-oriented, was used for the linear static analysis of the structure. In this process, the 3D elements were transformed into bar elements connected by nodes. The connection zone was not considered by simplification, and the structure was treated as monolithic with a perfect continuity between modules. The base was restricted with double supports, resembling real support conditions. Concrete and steel were considered for the sectional properties.

The structure was subjected to the self-weight and the hydrodynamic load (dF), which in turn was the result of the environmental actions (E) including waves (Ew), tidal currents (Ec), and marine growth (Em). In this process, the fluid velocities and accelerations were resolved into drag (dF_D) and inertial forces (dF_I) and applied as distributed loads on the members of the numerical model. The velocities (u_{wave}; $u_{current}$) and accelerations ($\partial u/\partial t$) of the fluid were calculated using the Stokes Wave Theory (Table 2) [62,63]. The local current profile was combined vectorially with the wave kinematics to determine local fluid velocities and accelerations adopting the Morison Equation (1).

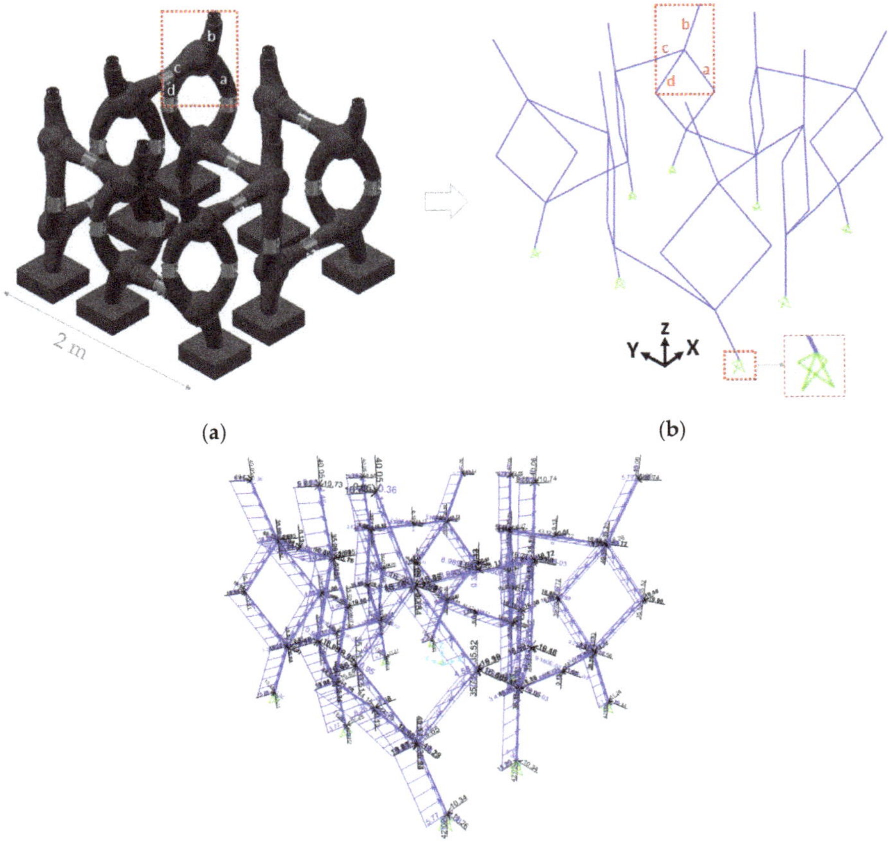

Figure 6. MMFAR structure: (**a**) 3D model, (**b**) numerical bar model, (**c**) forces applied to the numerical bar model. The marks "a, b, c and d" show the visual correspondence between the elements of the 3D model and the bar model.

$$
\begin{gathered}
dF = dF_D + dF_I \\
dF_D = C_d \times ½ \times \rho \times D \times (u_{wave} + u_{current})^2\, dz \\
dF_I = C_m \times \rho \times (\pi \times D^2)/4 \times (\partial u/\partial t)\, dt
\end{gathered}
\tag{1}
$$

In Equation (1), ρ represents water density, dz the depth interval, and D the diameter of the bar element. The hydrodynamic drag coefficient (C_d) and hydrodynamic inertia coefficient (C_m) were adopted following the recommendation of API [63,64]. The water current velocity profile ($u_{current}$) assumed a constant velocity of 1.2 m/s between the water surface level and half of the depth (z/2) and a constant velocity of 0.3 m/s between z/2 and the seabed. According to DNV-RP-C205, a value of 0.1 m was assumed for the marine growth [65]. Buoyant loads were also taken into account. The resulting forces were applied in x direction (d_{XX}). The load combinations followed the recommendations of ISO 19903:2006 [66]. The crest wave phase was considered since it corresponds to the maximum positive horizontal velocity and to zero horizontal acceleration. The vertical velocity is zero, and the vertical acceleration is maximum (in negative value).

For the analysis the characteristic event, the typical storm condition and the extreme storm condition were considered. In addition, abnormal environmental extreme events were also considered as accidental loads.

Table 2. 1st and 2nd order components of stokes wave theory [62,63].

Quantity	1st Order Component	2nd Order Component
Dispersion Relationship, c^2	$\frac{g}{k}\tanh kd$	$\frac{g}{k}\tanh kd$
Wave profile, η	$\frac{H}{2}\cos(kx-\omega t)$	$\frac{\pi H^2}{8L}\frac{\cosh kd}{\sinh^3 kd}[2+cosh\,2kd]\cos 2(kx-\omega t)$
Horizontal velocity, u	$\frac{\pi H}{T}\frac{\cosh ks}{\sinh kd}\cos(kx-\omega t)$	$\frac{3}{4c}\left(\frac{\pi H}{T}\right)^2\frac{\cosh 2ks}{\sinh^4 kd}\cos 2(kx-\omega t)$
Vertical velocity, v	$\frac{\pi H}{T}\frac{\sinh ks}{\sinh kd}sin(kx-\omega t)$	$\frac{3}{4c}\left(\frac{\pi H}{T}\right)^2\frac{\sinh 2ks}{\sinh^4 kd}\sin 2(kx-\omega t)$
Horizontal acceleration, $\dot{u}$	$\frac{2\pi^2 H}{T^2}\frac{\cosh ks}{\sinh kd}sin(kx-\omega t)$	$\frac{3\pi}{2L}\left(\frac{\pi H}{T}\right)^2\frac{\cosh 2ks}{\sinh^4 kd}\sin 2(kx-\omega t)$
Vertical acceleration, $\dot{v}$	$\frac{2\pi^2 H}{T^2}\frac{\sinh ks}{\sinh kd}cos(kx-\omega t)$	$\frac{3\pi}{4L}\left(\frac{\pi H}{T}\right)^2\frac{\sinh 2ks}{\sinh^4 kd}\cos 2(kx-\omega t)$
Dynamic pressure, p	$pg\frac{H}{2}\frac{\cosh ky}{\cosh kd}\cos[k(x-ct)]$	$\frac{3}{4}pg\frac{\pi H^2}{L}\frac{1}{\sinh 2kd}\left[\frac{\cosh 2ks}{\sinh^2 kd}-\frac{1}{3}\right]\cos 2(kx-\omega t)$ $-\frac{1}{4}pg\frac{\pi H^2}{L}\frac{1}{\sinh kd}[\cosh 2ks-1]$

c: wave celerity, g: gravitational acceleration, k: wave number ($k = 2\pi/L$), L: wavelength, d: water depth, H_w: wave height, η: wave profile (surface elevation of the water), w: wave angular frequency ($\omega = 2\pi f$, where f is the wave frequency), T: wave period, x: horizontal coordinate, t: time, u: horizontal velocity of the fluid, v: vertical velocity of the fluid, ρ: fluid density, p: dynamic pressure.

2.6. Actions and Effects

Table 3 shows the results obtained for each of the load cases and load combinations, according to the limit states identified as the most severe, in terms of maximum axial (N), shear (V), flexural (M), and torsional (T) stresses. Figure 7 identifies the critical sections, i.e., the locations where the maximum stresses occur. As shown, the worst scenario is the abnormal extreme condition. In this context, the maximum deflection obtained was 1.789 mm, the maximum axial stress was −86.05 kN, the maximum flexural stresses were −4.70 kN·m and 12.81 kN·m, the maximum negative and positive shear stresses were 36.43 kN, and the maximum negative and positive torsional stresses were 1.94 kN·m.

Table 3. Results in terms of the maximum efforts obtained for different load combinations.

Load Combination	U_r	N^-	N^+	$V_{2\text{-}2}{}^-$	$V_{2\text{-}2}+$	$V_{3\text{-}3}{}^-$	$V_{3\text{-}3}+$	$M_{2\text{-}2}{}^-$	$M_{2\text{-}2}+$	$M_{3\text{-}3}{}^-$	$M_{3\text{-}3}+$	T^-	T^+
	[mm]	[kN]		[kN]		[kN]		[kNm]		[kNm]		[kNm]	
Type of support: double support													
LC1_ULS(A)_E_{wcm}_dXX	0.09	−26.89	-	−1.17	1.00	−0.29	0.44	−0.15	0.13	−0.24	0.37	−0.08	0.06
LC2_ULS(B)_E_{wcm}_dXX	0.14	−46.37	-	−1.62	1.38	−0.73	1.08	−0.37	0.25	−0.24	0.52	−0.11	0.08
LC3_ULS(A)_$E_{wcm,storm}$_dXX	0.12	−28.32	-	−1.96	1.97	−0.51	0.81	−0.27	0.21	−0.47	0.61	−0.13	0.08
LC4_ULS(B)_$E_{wcm,storm}$_dXX	0.22	−49.32	-	−3.22	3.33	−1.14	1.76	−0.59	0.38	−0.84	1.16	−0.20	0.15
LC5_ULS(A)_$E_{wcm,máxstorm}$_dXX	0.42	−38.23	-	−8.53	8.53	−2.32	3.03	−1.07	1.02	−2.75	2.93	−0.44	0.43
LC6_ULS(B)_$E_{wcm,máxstorm}$_dXX	0.79	−67.99	-	−15.43	15.51	−4.05	5.88	−2.00	1.89	−5.08	5.47	−0.81	0.81
LC7_ALS_E_{wcme}_dXX	1.79	−86.05	-	−36.43	36.43	−11.14	12.65	−4.70	4.61	−12.46	12.81	−1.94	1.94
Critical section	-	Support leg		Connection Leg zone		Support leg		Central node		Connection Leg zone		Connection leg zone	

U_r—maximum global displacement; N—maximum axial stress; $V_{2\text{-}2}$—maximum shear stress in 2-2; $V_{3\text{-}3}$—maximum shear stress in 3-3; $M_{2\text{-}2}$—maximum flexural stress in 2-2; $M_{3\text{-}3}$—maximum flexural stress in 3-3; T—maximum torsional stress.

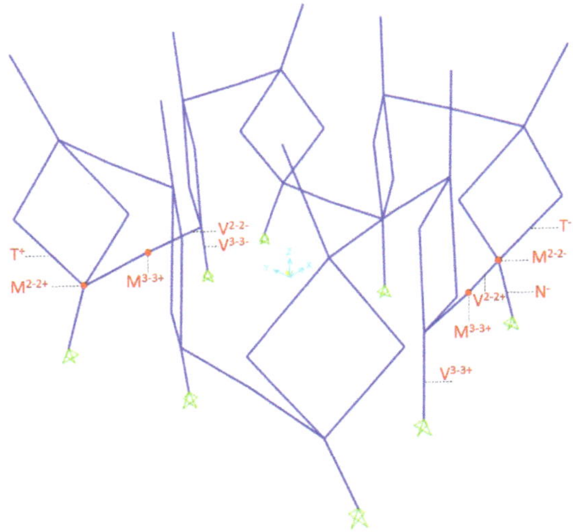

Figure 7. Location of the maximum stresses in the numerical model.

2.7. Design Considerations for Durability

According to NP-EN 1992-1-1 [67], the exposure class to be considered is XS2 (parts of permanently submerged marine structures), and according to NP-EN 1990 [68] the structural class 4 (S4) was considered, corresponding to a service life of at least 50 years. The durability of concrete structures depends on its composition and performance [67], and Annex E of NP-EN 1992-1-1 suggests that concrete for corrosion protection should comply with higher strength classes, C35/45 for exposure condition XS2. However, the National Annex for Portugal specifies C30/37* (or *C40/50 for CEM I or CEM II/A). According to NP-EN 206-1 Annex F [69], the recommended concrete composition includes a maximum w/c ratio of 0.45, minimum strength class C35/45, and a minimum cement dose of 320 kg/m^3. The concrete nominal cover (c_{nom}) of 50 mm was adopted, in accordance with the National Annex of NP-EN 1992-1-1 for Portugal.

2.8. Cross-Section Design

The cross-section design was conducted considering the two alternative structural systems studied, the RC and the CS, and adopting the organic sectional analysis based on the behaviour of the materials involved for ultimate limit states (ULSs).

Due to the geometric complexity and varying cross-sections of the members extending from the central core (sphere), the analysis focused on a typical cross-section that is geometrically representative (Figure 8). The critical cross-section, selected for its likely impact on design limitations, was analysed to determine its axial, flexural, and shear strengths, based on the area, geometry, and material properties of both structural systems. A parametric study was conducted varying the strength class of both materials and tube dimensions. NP-EN 1992-1-1 [67] was used for the analysis of RC cross-section, and NP-EN 1994-1-1 was applied for the CS cross-section [70]. Following, the analytical methodology is presented, and a parametric study developed.

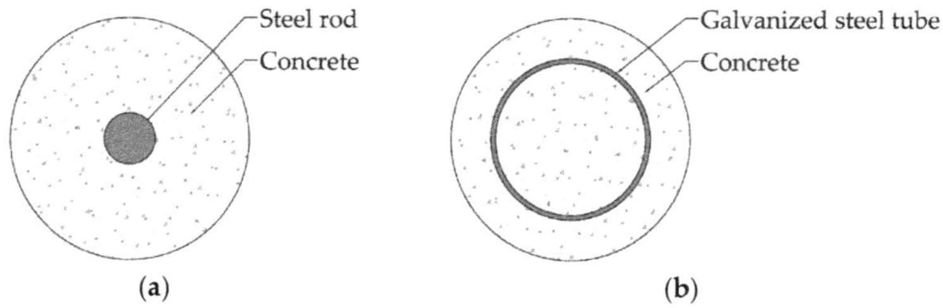

Figure 8. Representative cross section of the alternative structural systems: (**a**) RC structure and (**b**) CS structure.

2.8.1. Axial Strength

Considering that the most unfavourable scenario is the one where the entire section is under tension, the required amount of reinforcement in the cross section was obtained, ensuring that the stress in the reinforcement remains below the yield stress after concrete cracking, according to Equation (2).

Regarding the compressive strength ($N_{pl,Rd,comp}$), the contribution of both materials (concrete and steel) was considered. For the tensile strength ($N_{pl,Rd,trac}$), only the contribution of the steel was considered. $N_{pl,Rd,comp}$ and $N_{pl,Rd,trac}$ were obtained for smallest and largest cross-section areas, according to Equations (3) and (4) (Table 4), respectively.

Table 4. Axial strength of the cross-section according to the structural system.

	RC	CS	
	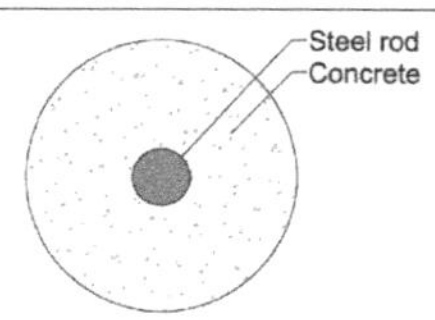	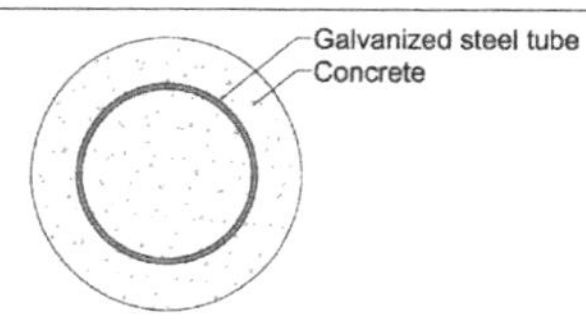	
$A_{s,min}$	$f_{ctm} \times \left[A_c + \left(\frac{E_s}{E_c} - 1\right) \times A_{s,min}\right] = A_{s,min} \times f_{yk}$		(2)
$N_{pl,Rd,trac}$	$f_{yd} \times A_s$	$A_a\, f_{yd} + A_s\, f_{sd}$	(3)
$N_{pl,Rd,comp}$	$(E_s \times \mathcal{E}_s) \times A_s + (A_c \times f_{cd})$	$A_a\, f_{yd} + \alpha_{cc} A_c f_{cd} + A_s\, f_{sd}$	(4)

A_a—steel profile area; **A_c**—concrete area; **A_s**—steel reinforcement area; **E**—modulus of elasticity (s—steel, c—concrete); ε—strain. **f_{ctm}**—concrete tensile strength; **f_{cd}**—compressive failure stress of the concrete; **f_{yk}**—characteristic yield strength; **f_{yd}**—design yield strength of the structural steel; **f_{sd}**—yield stress of the longitudinal reinforcement; α_{cc} was equal to 1.0 (tubular steel section covered and filled with concrete).

Given that the cross-sections were subjected to approximately centred stresses, the average compressive strain was limited to $\mathcal{E}_{c2}$. Note that, in the case of class C45/55 concrete, then $\mathcal{E}_{c2} = 2‰$. Thus, by strain compatibility, $\mathcal{E}_s = 2‰$. Considering that the reinforcement exhibits elastic behaviour until yielding, the maximum tension to which the reinforcement can be subjected is given by $\sigma_s = E_s \times \varepsilon_s$.

Additionally, the steel contribution ratio (δ), which must lie between 0.2 and 0.9 in the case of the CS structures, was also a criterion. This is one of the aspects that differentiates the design of a RC structure from a CS structure.

2.8.2. Shear Strength

In the absence of specific transverse reinforcement in a given section, the effectiveness of the concrete's contribution to shear strength diminishes under higher loads. Consequently, the longitudinal reinforcement assumes an important function in the transfer of shear forces. This scenario is particularly relevant in cases where the implementation of shear reinforcement is either impracticable due to spatial constraints or challenging to execute, such as in narrow elements or sections with intricate geometries, as is the case of the MMFAR structure under study.

Thus, the shear strength, V_{Rd}, of the cross-section was obtained by considering the contributions of concrete and longitudinal reinforcement, based on NP-EN 1992-1-1 [67], as detailed in Table 5 by Equations (5)–(7). In this context, the smallest cross-section, located near the connection zone, was considered the critical cross-section, as it is where the maximum shear strength occurs, as indicated in Figure 7.

Table 5. Shear strength of the cross-section for both structural systems.

V_{Rd}	$V_{Rd,c} + V_{Rd,sl}$	(5)
$V_{Rd,c}$	$\left[C_{Rd,c}\cdot k(100\cdot\rho_l\cdot f_{ck})^{\frac{1}{3}} + k_1\cdot\sigma_{cp}\right]\cdot b_w\cdot d$	(6)
$V_{Rd,sl}$	$A_{sl}\cdot f_{yd}$	(7)

$\mathbf{V_{Rd,c}}$—design shear resistance of the member without shear reinforcement; $\mathbf{V_{Rd,sl}}$—design "shear" resistance of the longitudinal reinforcement, $\mathbf{C_{Rd,c}}$—coefficient of shear strength of concrete ($0.18/\gamma_c$, with γ_c = 1.5), $\mathbf{r_l}$—geometric ratio of longitudinal reinforcement (A_{sl}/b_wd), $\mathbf{b_w}$—cross-section diameter, **d**—effective section height, $\mathbf{k_1}$—shear behaviour coefficient, $\mathbf{k} = 1 + \sqrt{(200/d)} \leq 2$, $\boldsymbol{\sigma_{cp}}$—compressive stress of concrete due to normal forces ($N_{Ed}/A_c < 0.2f_{cd}$), $\mathbf{N_{Ed}}$—axial force in the cross-section due to loading, **Ac**—area of concrete cross section; A_{sl}—area of longitudinal steel reinforcement $\mathbf{f_{ck}}$—characteristic compressive strength, $\mathbf{f_{cd}}$—compressive failure stress of the concrete; $\mathbf{f_{yd}}$—design yield strength of the structural steel.

2.8.3. Flexural Strength

Considering the need to guarantee ductile failure, the minimum reinforcement $A_{s,min}$ was obtained according to Equation (9). The corresponding cracking flexural stresses (M_{cr}) were computed by Equation (8). Table 6 shows the main equations and validation criteria related to the flexural behaviour of the cross-section for each structural system. Figure 9 shows the strains and stresses developed in both cross-section types assuming ULS and the simplified Bernoulli conditions for simple flexure.

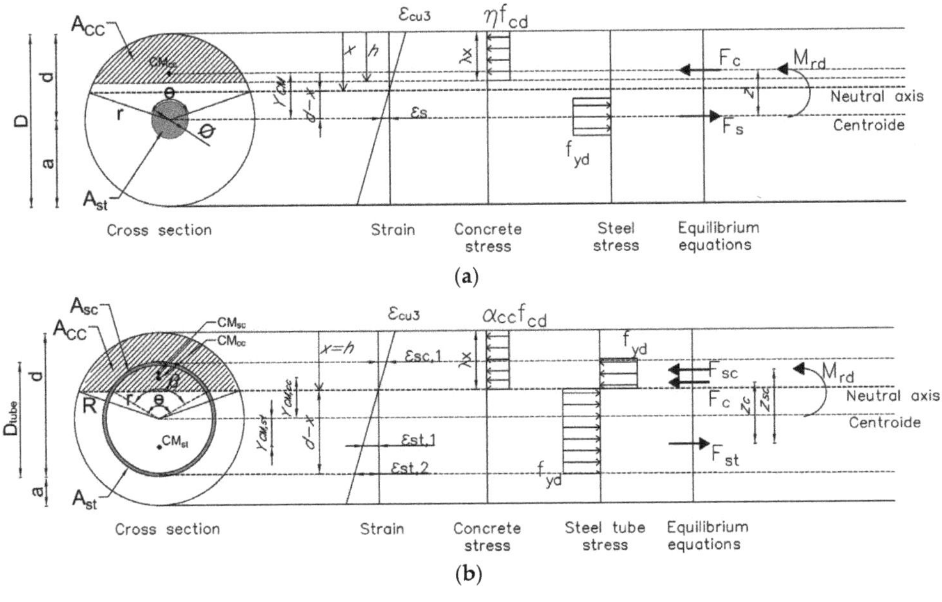

Figure 9. Schematics representing the simplified strains and stresses at ULS in flexure for (**a**) RC and (**b**) CS cross-sections.

Table 6. Flexural cracking strength (M_{cr}), flexural strength (M_{Rd}) and minimum reinforcement ($A_{s,min}$) of the cross-section for both RC and CS structural systems.

<table>
<tr><th></th><th>RC Structure</th><th>CS Structure</th><th></th></tr>
<tr><td>M_{cr}</td><td colspan="2">$M_{cr} = \frac{f_{ctm} \times \overline{I_x}}{\overline{y_g}} = \frac{f_{ctm} \times \frac{\pi D^4}{64}}{\frac{A_c \times \frac{D}{2} + (\alpha - 1) A_s \times a}{\pi \times r^2 + (\alpha - 1) A_s}}$</td><td>(8)</td></tr>
<tr><td>$A_{s,min}$</td><td colspan="2">$A_{s,min} = \frac{A_{cc} \times \left\lfloor \frac{4r}{3} \left(\frac{\sin\left(\frac{\theta}{2}\right)^3}{\theta - \sin\theta} \right) - (d - x) \right\rfloor}{\alpha \times (d - x)}$</td><td>(9)</td></tr>
<tr><td>M_{Rd}</td><td>$F_c \times z$
❖ $F_c = \left[r^2 \cos^{-1}\left(1 - \frac{h}{r}\right) - (r - h)\sqrt{2rh - h^2}\right]\eta f_{cd}$
❖ $z = \frac{4r}{3}\left(\frac{\sin\left(\frac{\theta}{2}\right)^3}{\theta - \sin\theta}\right)$
❖ $\theta = 2\cos^{-1}\left(\frac{r - x}{r}\right)$</td><td>$F_c \times z_c + F_{sc} \times z_{sc}$
❖ $F_c = \left(\left[R^2 \cos^{-1}\left(1 - \frac{h}{R}\right) - (R - h)\sqrt{2rh - h^2}\right] - \left[2\cos^{-1}\left(\frac{R - h}{r}\right) rt\right]\eta f_{cd}$
❖ $F_{sc} = 2\cos^{-1}\left(\frac{R - h}{r}\right) rt \times f_{yd}$
❖ $z_c = \frac{4R}{3}\left(\frac{\sin\left(\frac{\theta}{2}\right)^3}{\theta - \sin\theta}\right) + \left(\frac{r^2(\sin\beta - \sin(-\beta))}{r(2\pi - \beta)}\right)$
❖ $z_{sc} = \frac{r \times \sin\left(\frac{\beta}{2}\right)}{\frac{\beta}{2}} + \left(\frac{r^2(\sin\beta - \sin(-\beta))}{r(2\pi - \beta)}\right)$
❖ $\theta = 2\cos^{-1}\left(\frac{R - x}{R}\right)$
❖ $\beta = 2\cos^{-1}\left(\frac{R - x}{r}\right)$</td><td>(10)</td></tr>
<tr><td>A_s</td><td>$A_s = \frac{F_c}{\sigma_s}$</td><td>$A_{st} = \frac{F_c + F_{sc}}{\sigma_s}$</td><td>(11)</td></tr>
<tr><td>Condition</td><td>$\varepsilon_{cu3} = 3.5‰ \mid \varepsilon_s = "\infty" \mid \lambda = 0.8 \mid \eta = 0.9$</td><td>$\varepsilon_{cu3} = 3.5‰ \mid \varepsilon_s = "\infty" \mid \lambda = 1.0 \mid \eta = 0.85$</td><td></td></tr>
<tr><td>Validation</td><td colspan="2">$\varepsilon_c = \varepsilon_{cu3}$ and $\varepsilon_{s,i} > \varepsilon_{yd}$</td><td></td></tr>
</table>

Mcr—cracking flexural stress; $\mathbf{f_{ctm}}$—mean concrete axial tensile strength; $\overline{\mathbf{I_x}}$—equivalent moment of inertia of the section; $\overline{\mathbf{y_g}}$—distance from the centroid of the section to the outermost fibre; **Ac**—cross sectional area of concrete; **D**—diameter of the circular section, **r** or **R**—radius of the concrete circular section; **t**—wall tube thickness; **Ei**—materials Young's Modulus; **α**—Es/Ec, a—distance from the centroid of the reinforcement to the geometric centre of the section; **As**—cross sectional area of reinforcement, **Acc**—compressive area of the concrete section; **Ast**—tensile area of the steel section; **θ**—angle of the compressed concrete section; **β**—angle of the compressed steel section; **r**—tube radius; **d**—effective depth of a cross-section; x—neutral axis depth; **h** and **l**—effective height of the compression zone; **Fc**—concrete compressive force; **Fsc**—steel compressive force; **fcd**—design compressive strength of concrete; **η**—effective strength; **fyd**—design tensile strength of steel; **z**—lever arm of internal forces; $\boldsymbol{\sigma_s}$—steel stress; $\boldsymbol{\varepsilon_{cu3}}$—ultimate compressive strain in the concrete; $\boldsymbol{\varepsilon_s}$—strain in the steel; $\boldsymbol{\varepsilon_c}$—strain in the concrete.

The approach consisted of determining the equilibrium of the cross-section stresses that maximizes the flexural strength to verify the safety condition, $M_{Rd} > M_{Ed}$. The problem was solved through the implementation of a flexural inverse analysis, which identified the critical cross-section as the one located in the proximity of the connection zone, since it shows simultaneously the minimum area and the maximum stress, as illustrated in Figure 7.

The value of $\mathcal{E}_s$ was adopted to ensure the plastic regime and the optimal position of the neutral axis could be calculated. For that, the rectangular stress block for the distribution of stresses in the concrete and an elastic-perfectly plastic diagram for steel ($\mathcal{E}_{cu3}$, $\mathcal{E}_{s=\infty}$, l and h) were assumed. Subsequently, the required reinforcement to balance the tensile and compression forces in the cross-section was analysed, and the M_{Rd} was calculated according to Equations (10) and (11). The phenomenon of cross-section steel saturation was considered, whereby an increase in A_s does not necessarily result in a proportional increase in flexural strength due to cross-section limitations. The binary arms z, z_c, and z_{sc} and the parameter that translates the geometric relationship between the concrete and steel cross sections and the position of the neutral axis θ and β are shown in Table 6. Design assuming reinforcement in the elastic regime should be avoided, as it risks uneconomical solutions and brittle failure.

3. Results and Discussion

3.1. Parametric Study of RC and CS Design Without Cross-Sectional Diameters Modifications

The results of the parametric study conducted in terms of axial and flexural strength for both structural systems, where concrete and steel classes were varied, are shown in Table A1 (Appendix A) and Table A2 (Appendix B). Figures 10 and 11 illustrate the correlation between the concrete compressive strength (f_c) and steel class and the obtained flexural ($M_{pl,Rd}$) and normal ($N_{pl,Rd}$) strengths based on Tables A1 and A2. The plotting colour-code in the graphs is combined with the reinforcement configuration according to steel class.

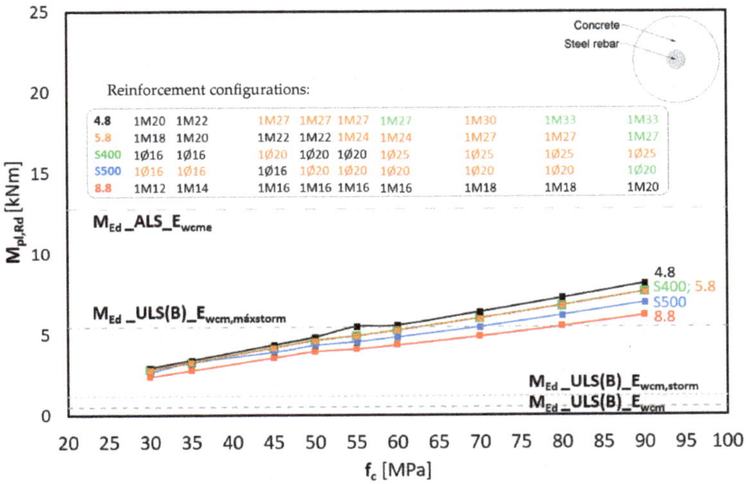

(a)

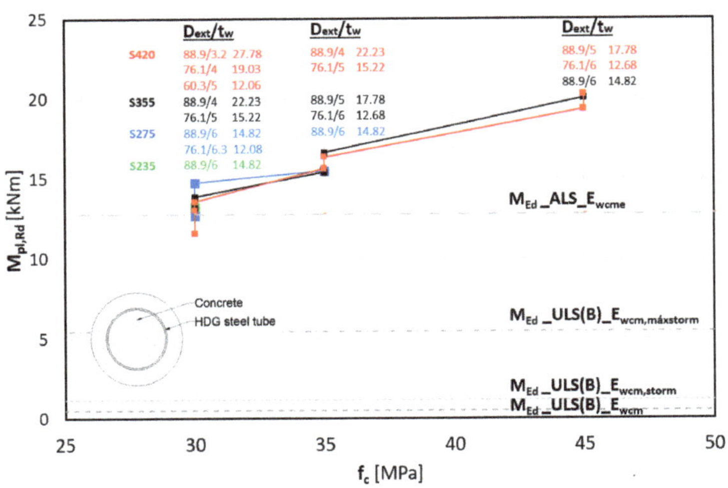

(b)

Figure 10. Flexural strength, $M_{pl,Rd}$, versus concrete strength for various structural combinations for cross-section types (**a**) RC and (**b**) CS.

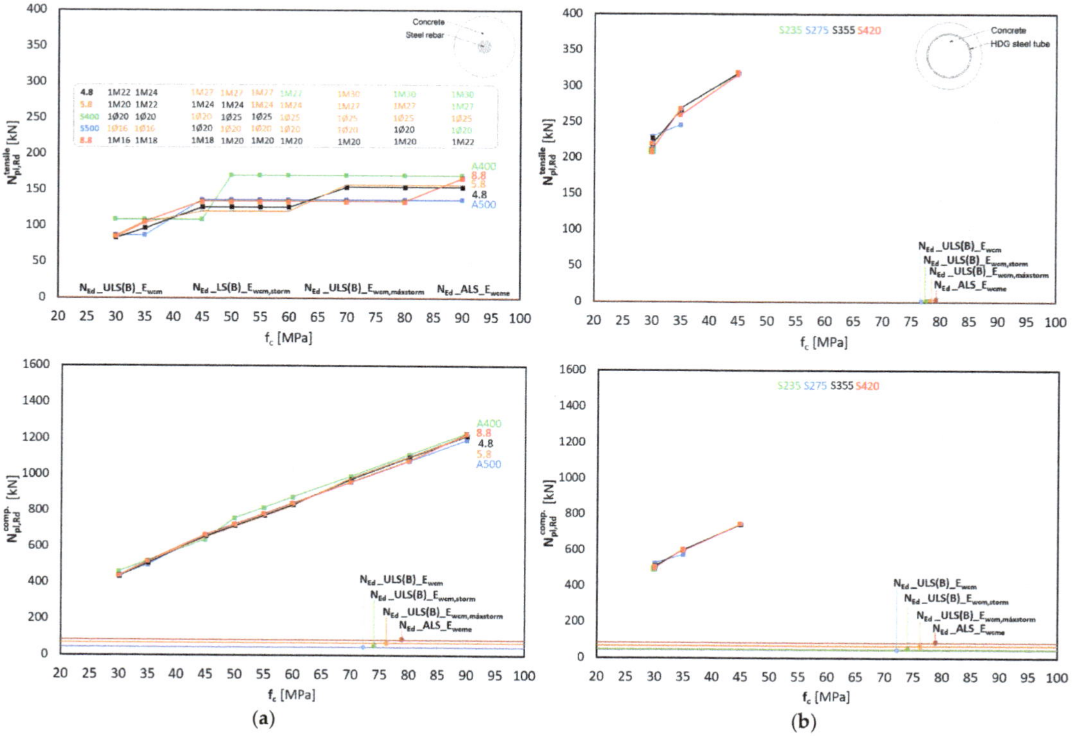

Figure 11. Tensile (**above**) and compressive (**below**) axial strength, $N_{pl,Rd}$, versus concrete strength for cross-section types (**a**) RC and (**b**) CS.

- **RC cross-section**

The reinforcement rebar (1ØD) was of conventional steel classes S400 and S500 used in civil construction, and threaded rods (1MD) of DIN 4.8, 5.8, and 8.8 steel class. The minimum concrete strength class was C30/37.

In terms of axial strength, it appears that the most unfavourable combination of materials for determining $A_{s,min}$ was concrete C90/105 and DIN 4.8 steel rebar. These require a minimum reinforcement of 5.02 cm^2, which can be fulfilled by a 30 mm threaded steel rod. As demonstrated, for all scenarios, the minimum reinforcement guaranteed that the compressive and tensile strengths ($N_{pl,Rd}$) (Figure 11) clearly exceed the maximum axial stress N_{Ed} (−86.05 kN, refer to Table 3) obtained from the numerical model.

In terms of flexural strength, the minimum reinforcement could be fulfilled using 1Ø8 mm (M_{cr} = 1.65 kN·m). The obtained $A_{s,min}$ was in all cases below the A_s required for ULS. Figure 10 presents the flexural strength versus concrete strength for various structural combinations of RC cross-section. In general, the reinforced concrete critical cross-section showed limitations in terms of flexural strength, since it did not achieve safety for abnormal extreme conditions and, in some cases, extreme storm action (Figure 10). The lowest flexural strength, $M_{pl,Rd}$ = 2.37 kN, was obtained for the combination of concrete class C30/37 and DIN 8.8 steel rebar, ensuring strength capacity only for the load combinations E_{wcm} and $E_{wcm,storm}$. The highest $M_{pl,Rd}$, 6.32 kN, was achieved for the concrete class C90/105 and DIN 4.8 steel rebar, meeting all requirements except the ALS_E_{wcme} load combination. As shown in Figure 10a, the $M_{pl,Rd}$ decreases as the steel strength increases for the same class of concrete, as higher-strength steel reduces the saturation reinforcement area ($A_{s,satur}$) of the cross-section, reducing the lever arm of internal forces and decreasing the section's flexural strength. This trend is attributed to the restriction of the cross-section diameter to a minimum of 150 mm, imposed by the geometric design. This highlights the importance of balancing steel reinforcement layout with steel strength in small sections to optimize structural performance. The most promising combinations of materials were those that resulted in $A_{s,satur}$ (flexural) $\geq A_{s,eff}$ (axial) and $Ø_{satur}$ (flexural) $\geq Ø_{eff}$ (axial), identified in green shading in Table A1. While the combinations identified in yellow coincide in diameter ($Ø_{eff} = Ø_{satur}$), these exhibit $A_{s,eff} >> A_{s,satur}$, disrupting the cross-section equilibrium and preventing reinforcement yielding. In this case, it would be necessary to reconfigure and optimize the cross-section to accommodate the reinforcement imposed by $A_{s,min}$, as discussed in Sections 3.2 and 3.3.

Regardless of the concrete class, it seems that the structural combinations showing the best compromise between the strength of the materials, the position of the neutral axis, and the normal and flexural strength, include the use of DIN 4.8, as evidenced by the highest $M_{pl,Rd}$ values shown in Table A1 and Figure 10a. In fact, the threaded rebar DIN 4.8 steel resulted in the most significant overlap between steel reinforcement diameters that satisfy both axial and flexural reinforcement requirements. For S500 steel, there is a central range in the graph where the effective steel area (A_s) remained constant, irrespective of the concrete strength (f_c), specifically within the range of f_c = 50 MPa to 90 MPa. This occurred due to a significant gap between $A_{s,pl}$ and $A_{s,eff}$, making the choice of S500 inefficient for lower concrete strength classes where $A_{s,pl}$ is significantly higher than $A_{s,eff}$. Furthermore, a higher concrete strength class results in the raise of the position of the neutral axis, increasing the $A_{s,pl}$ and respective M_{Rd}.

Summarizing, the results showed that the tensile behaviour of the cross-section limits the amount of reinforcement that must be provided in the cross-section. It was observed that the minimum reinforcement required for a RC element in tension with a variable cross-section (largest section being the most critical) is higher than the one obtained for flexure ($A_{s,satur}$) (smallest section being the most critical). However, the flexural strength of the section, although optimised internally, appears to be the limiting strength, primarily due to the reduced cross-sectional dimension. The results indicate that it may be more advantageous to optimize the reinforcement area rather than merely increasing the steel strength, particularly in the case of sections with reduced diameter. This analysis shows that stronger materials do not always lead to proportional structural gains, especially when geometric constraints and internal force distributions are significant.

- **CS cross-section**

Tube diameters 60.3 mm, 76.1 mm, and 88.9 mm were selected to comply with the concrete cover and to offer a range of possible solutions for adapting the steel section to specific strength requirements. The minimum thickness of the tube wall was chosen to comply with the A_s, as well as to verify the requirement of a minimum contribution of steel (δ) for the non-perforated and perforated variants (two opposite holes with $D_{hole} = D_{ext}/2$). The minimum strength class for concrete was set to C30/37. In this case it was deemed unnecessary to expand the study to include high-strength concrete classes, as the combinations achieved with lower-strength concrete classes were found to be favourable. The minimum reinforcement was not a limitation when compared to the ULS analysis since $A_{s,min}$ (0.41 cm^2) was substantially lower when compared to $A_{s,pl}$.

In general, this structural solution easily achieved flexural strengths exceeding 12.81 kNm, ensuring compliance with the $M_{Rd} > M_{Ed}$ criterion, even for the lower-strength concretes and steel classes as shown in Table A2 and Figure 10. As shown, for a given steel strength class and tube diameter, an increase in the concrete class resulted in a design that required a thicker tube, yielding an increase in $M_{pl,Rd}$. However, this increase was less relevant, since the material combinations analysed for lower-class concretes (e.g., C30/37) already met the safety criteria.

The geometric configuration that ensured the greatest nominal cover were associated to the tube with a diameter of 60.3 mm, which is important to prevent the corrosion caused by continuous exposure to seawater. However, in most cases, the adoption of this diameter and the maximum available thickness of 5 mm did not meet the required steel contribution ratio (δ). This criterion was only fulfilled when C30/37 concrete and S420 steel were combined. The parametric study showed that tube configurations with larger diameters (especially 88.9 mm) provided a greater number of efficient combinations, which is crucial for ensuring structural stability when the structure is subjected to significant horizontal forces. However, concrete cover in the connection region was small, and steel protection was reduced. Implementing additional protective measures or adjusting the section in this area should be considered.

The variety of structural solutions available for CS configuration resulted in a wider multitude of alternatives, withstanding the most extreme actions and ensuring optimal structural efficiency. A significant factor in this type of system is the D/t ratio, which directly impacts the $M_{pl,Rd}$. Additionally, the increase of fc was linked to the need for increased t_w, thus ensuring harmonious integration of materials and sufficient sectional resilience. Nevertheless, to have the section classified as mixed concrete–steel according to NP-EN 1994-1-1 [67], for increased fc the use of lower strength steels, such as S235, is not feasible.

The lowest $M_{pl,Rd}$ values were observed in sections comprising lower strength steels, such as S235, while the highest values were observed in sections comprising high-strength steels, such as S420. The highest steel classes (S355 and S420) resulted in the most efficient use of the full cross-section, providing greater strength and better efficiency in materials usage. Moreover, higher compressive strengths of concrete (f_c) resulted in higher normal plastic strengths ($N_{pl,Rd}$), with a nearly linear relationship (Figure 11b).

Figure 12 presents the shear strength of both structural systems obtained for the most critical scenarios. As observed, the shear strength clearly exceeded the applied shear stress, therefore ensuring shear safety.

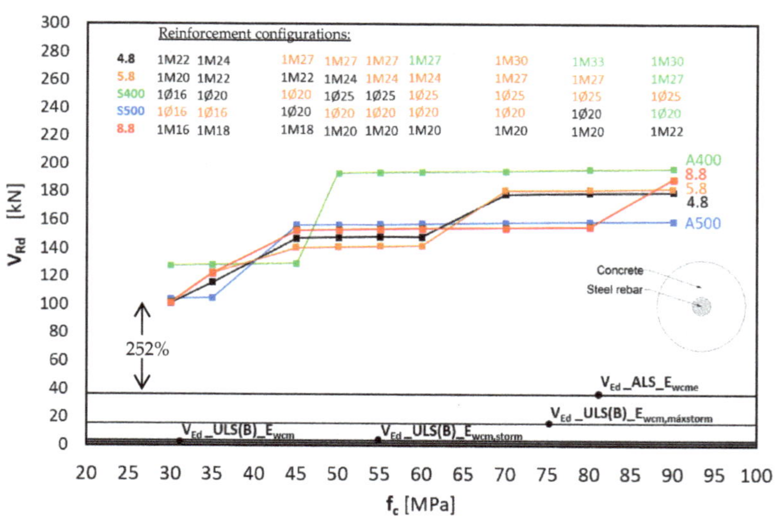

(a)

Figure 12. *Cont.*

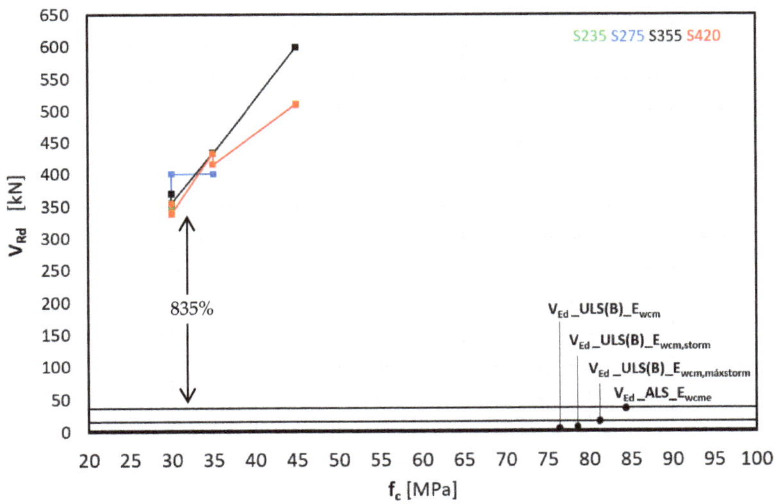

(**b**)

Figure 12. Shear strength (V_{Rd}) versus concrete strength for cross-section types (**a**) RC and (**b**) CS.

3.2. Optimization of the RC Cross-Section

In the previous sections it has been shown that the majority of combinations designed to optimise the structural efficiency of the original geometry invariably result in a RC cross-section imbalance. This, in turn, compromises the structure's ability to withstand applied loads and increases the risk of obtaining failure modes where the reinforcement is in the elastic regime. In order to address this challenge, two optimisation approaches were explored, where the variation of the initial cross-section size was allowed. The first was aimed at minimising modifications to the cross-section, albeit with certain trade-offs. The second approach allowed the free resizing of the cross-section to meet the most demanding scenario. Considering the variable geometry of the cross-section, a minimum and maximum diameter will be defined for the smallest and largest cross-sections of the reef branch, respectively.

In **approach 1**, a diameter of 196.5 mm was considered to the cross-section, corresponding to the largest cross-section found the initial design. The $A_{s,min}$ was then calculated, and an initial solution was determined (1ØD). This resulted in a real steel area ($A_{s,real}$), which served as a reference for establishing a maximum limit for the section diameter (D_{max}). For the flexural analysis, which was critical when the smaller cross-section was considered, the minimum viable diameter (D_{min}) was calculated to ensure that $A_{s,pl}$ (flexural) matched $A_{s,real}$ (axial). For each configuration, the flexural ($M_{Pl,Rd}$) and axial ($N_{pl,Rd}$) strengths were obtained.

Figure 13 presents the concrete and reinforcement combinations that meet the constraints of this approach for the RC cross-section. The upper graph illustrates the changes in minimum (D_{min}) and maximum (D_{max}) cross-sectional diameters correlated with different reinforcement configurations and concrete strengths. The lower graph displays the corresponding $M_{Pl,Rd}$ and M_{Ed}. The plotting colour-code in the graphs distinguished the steel class.

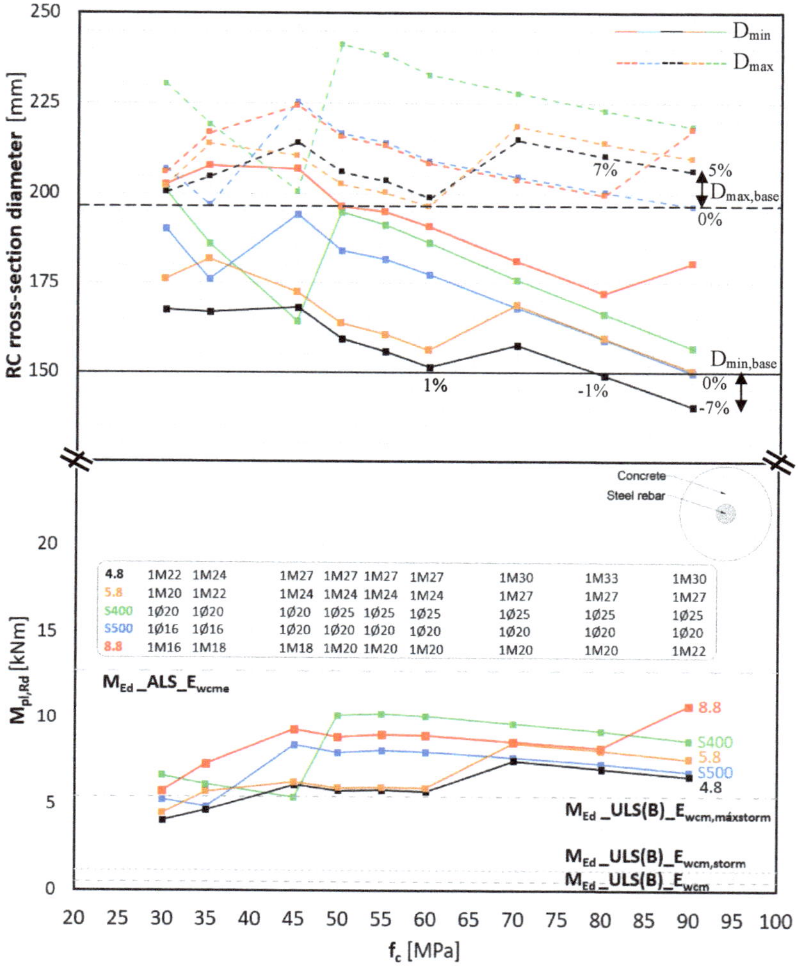

Figure 13. Correlation between the minimum (D_{min}) and maximum (D_{max}) cross-section diameters and the flexural strength, $M_{pl,Rd}$, as a function of concrete compressive strength, f_c, and reinforcement configuration (approach 1).

The results show that the alteration of the cross-section dimensions resulted in a significant improvement of $M_{pl,Rd}$. Most of the material combinations exceeded the requirements of the load combination ULS(B)_$E_{wcm,máxstorm}$, but none met the abnormal combination requirements, ALS_E_{wcme}.

A common trend across most combinations is the need to increase the diameter of the initial cross-section to optimize performance. The analysis identifies combinations that result in minimal changes (−7% to 7%) in the cross-sectional diameters, maintaining the original design's geometric effect. This reduction in material contributes to a more efficient and economical structural solution.

The maximum value found for D_{min} was 208 mm, occurring in the C35/45 + DIN 8.8 combination. This represents a 38.7% increase compared to the initial section (150 mm). On the other hand, the minimum value found for D_{min} was 140.19 mm, obtained for the C90/105 + 4.8 combination. D_{max} varied between 196.5 mm, for the C60/75 + DIN 5.8

combination, and 241.44 mm, for the C50/60 + S400 combination. The latter combination showed a 22.8% increase compared to the baseline design.

In **approach 2**, a free optimization was carried out to find the smallest cross-section and structural combination that guarantees the most challenging scenario.

For each combination of concrete and steel types, the reinforcement solution (1ØD) was imposed, and the minimum cross-section diameter (D_{min}) was found, ensuring that $A_{s,pl}$ matches $A_{s,real}$, while satisfying both axial and flexural strength criteria. Subsequently, the maximum diameter (D_{max}) was calculated considering that the minimum reinforcement area ($A_{s,min}$) determined by axial stresses did not exceed the real reinforcement area ($A_{s,real}$).

Figure 14 illustrates the RC cross-section diameter obtained for each concrete strength class adopting the aforementioned procedure, showing the minimum and maximum section diameters and flexural strengths for different material combinations.

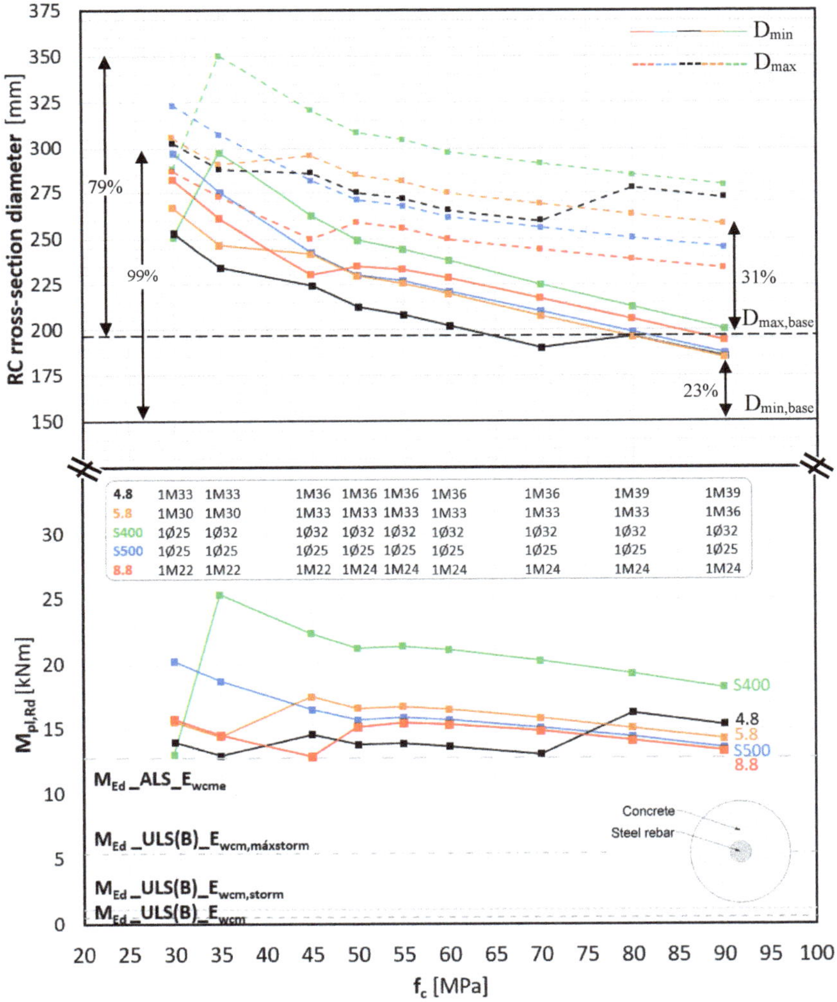

Figure 14. Correlation between the minimum (D_{min}) and maximum (D_{max}) cross-section diameters and the flexural strength, $M_{pl,Rd}$, as a function of concrete compressive strength, fc, and reinforcement configurations (approach 2).

The minimum diameter (D_{min}) required decreases with the increase in the concrete strength class. The influence of the concrete strength on the section size is significant, with increments ranging between 23% (for material combinations including high fc) and 99% (for material combinations with lower fc). The smallest admissible diameter was 185 mm, corresponding to the materials combination C90/105 + DIN 5.8. The largest diameter was 298 mm for combination C35/45 + S400 and C30/37 + S500. Moreover, the obtained steel reinforcement in the new cross-section leads to a new maximum diameter for the element in order to comply with $A_{s,min}$ requirement, when subjected to a normal tensile stress. In this context, a minimum diameter of 234.32 mm (+19%) was found for the C90/105 + DIN 8.8 combination and 350.76 mm (+79%) for the C35/45 + A400 combination. The combinations that best preserve the original dimensions, i.e., those requiring the smallest average diameter increase, correspond to higher concrete classes, namely C90/105 (S500, DIN 5.8, and 8.8), with average increases of 25%, 27%, and 24%, respectively.

3.3. RC Versus CS Cross-Section

The RC unit was in general found to be easier to build and adapt, especially regarding rebar shaping. However, creating an effective and simple connection system between modules to ensure structural continuity can be challenging, as documented by [71]. While using a single rebar offers benefits such as better reinforcement cover and consistent mechanical strength across all loading directions, there are disadvantages that cannot be neglected. The small cross-sectional dimensions limit optimization, especially under high flexural stresses, such as extreme storm or abnormal conditions. The close proximity of compression and tension forces reduces the torque arm and, consequently, flexural strength. Additionally, the original RC cross-section faces saturation of the steel, where increasing reinforcement does not significantly improve flexural strength. To avoid this, reinforcement ratios must be respected, steel distribution optimized, or section dimensions increased to ensure safety and ductility. In this context, using compression reinforcement in a polar array and a tubular profile with the appropriate diameter and thickness, as shown in Figure 15, is advantageous.

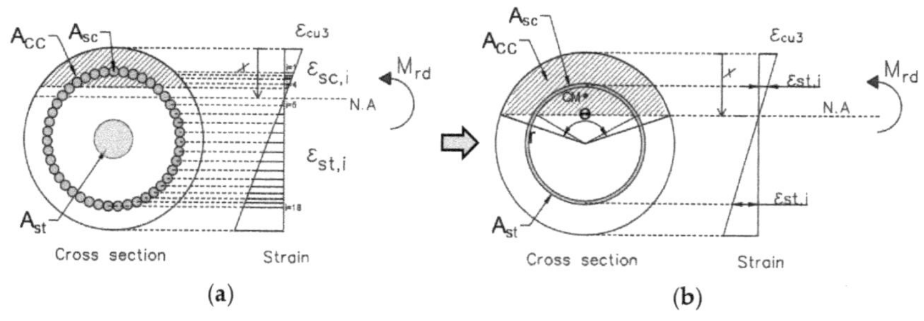

Figure 15. (**a**) RC structure with polar array of compression steel reinforcement and (**b**) CS structure.

Although the CS solution is more complex to construct, it offers significant advantages in terms of structural behaviour due to its geometric flexibility and material strength. Additionally, the incorporation of bidirectional holes along the reinforcing tube improves material adherence and facilitates concrete filling, as suggested in [28]. The combination of materials allows for better optimization of the cross-section, increasing structural efficiency, especially under bending conditions. This system also provides a practical solution for connecting modules as stated by [71]. However, the need for specific equipment to bend the steel tubes is a disadvantage in the construction process.

3.4. Cost-Effectiveness

Figure 16 illustrates the correlation between the weight and cost of the reef module, based on the CS cross-section geometry and the optimized RC cross-section that ensures structural safety under the most demanding loading scenario. The analysis was conducted to assess the cost implications of strengthening the RC modules in comparison to the CS module, addressing concerns related to cost and construction feasibility. In the case of RC solution, only configurations resulting in a sectional increase of less than 35% in the critical area (D_{min}) were considered, and these were identified as the green zone in Figure 16. In addition, the maximum dimension of the limb near the spherical zone was constrained to maintain the original geometric ratio between D_{max} and D_{min}, thereby ensuring consistency with the initial design morphology. This aimed to preserve the slenderness of the geometric form, as a bulkier structure could significantly alter the local hydrodynamic behaviour, potentially requiring further studies in that domain.

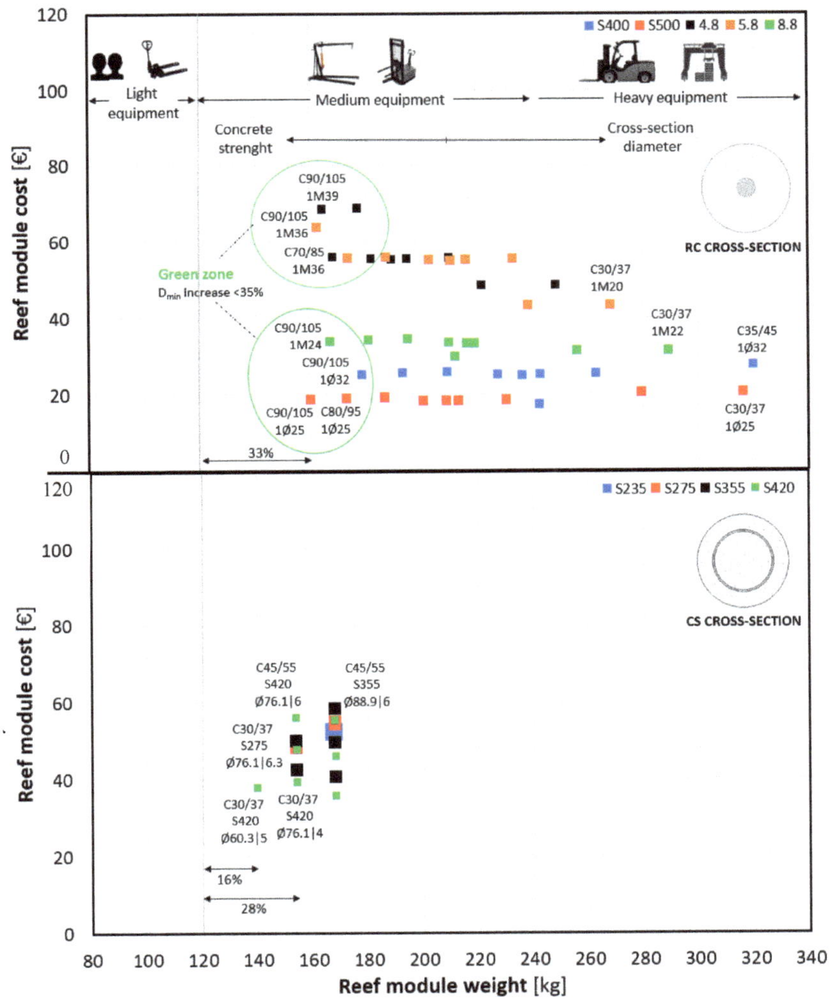

Figure 16. Correlation between reef module weight and cost for CS and optimized RC configurations.

The analysis of the reinforced concrete (RC) and composite steel (CS) cross-sections reveals distinct trade-offs between weight, cost, and structural efficiency (Figure 16). Most RC solutions have a weight of 160 kg, exceeding in 33% the initially desired 120 kg, which requires the use of medium equipment for handling the units. The cost of these RC solutions ranges from 18.7€ to 68.8€ per module. Conventional reinforcement results in lower costs, while the use of threaded bars significantly increases the unit cost. The RC combination with the best cost-construction feasibility ratio is the one using C90/105, S500, and 1Ø25, which has a weight of 159.8 kg and costs 18.7€ per unit. This solution offers an optimal balance between concrete usage and cost-efficiency, making it a more sustainable choice in terms of material optimization.

In comparison, CS solutions, which utilize a composite concrete and steel structure, present a lower increase in unit weight but come at a significantly higher cost (Figure 16). The lowest cost for CS solutions is approximately double that of the RC unit. Several CS solutions, with weights around 150 kg, have costs ranging from 39.23€ per unit to 56€ per unit. A notable example is the combination of C30/37, S420, Ø76.1, and t_w 4, which weighs 154 kg and costs 39.23€ per unit. Whilst these CS solutions may exhibit a reduced weight relative to RC solutions, they necessitate a higher financial investment due to the materials employed. However, they can utilise lower-strength, more readily available concretes. The solution that results in the lowest possible weight for the CS cross-section uses the combination of C30/37, S420, Ø60.3, with a 5 mm thick wall, which achieves a weight of 139.72 kg (+16%) and costs 37.81€. However, it is important to note that this solution has a slightly lower $M_{pl,Rd}$ than required, meaning additional care is needed during application.

The cost-effectiveness analysis reveals that the optimized RC solution offers the most economical and structurally adequate configuration for the reef module. However, this comes at the expense of increased volumetry, which may alter local hydrodynamic behaviour and potentially impact biological colonization. If maintaining the original geometry is a priority—whether for ecological performance, fluid-structure interaction, or design constraints—the composite (CS) solution becomes the preferable alternative. Despite its higher unit cost, it provides a structurally efficient design while using lower-strength, cement-reduced concretes, contributing to lower carbon emissions and greater sustainability. Thus, the choice between RC and CS depends on project priorities: cost and simplicity favour RC, while environmental performance and geometric fidelity favour CS.

4. Conclusions

This work introduces the concept of different structural systems and presents a design methodology for developing and analysing modular and geometrically complex artificial reef structures. This study is based in a cross-sectional analytical approach based in the principles of structural mechanics supported by numerical analyses to evaluate actions and their effects on the structure.

By detailing the geometric design, structural conception, and relevant environmental actions, an approach was proposed for the evaluation of possible structural configurations. The analytical methodology, along with the parametric study, provided crucial insights into the performance and feasibility of different design configurations. In particular, and for the specific case of the cross-sections developed, the main findings were as follows:

- Reduced fixed diameters establish an important geometric limit for optimizing the cross-section of the RC type.
- Tensile strength emerged as the primary limiting factor, particularly in the RC type with variable cross-sections, where the minimum reinforcement was constrained by the larger section.

- Flexural strength was not a limiting factor for the CS structural solution, as the variety of structural combinations indicates sufficient robustness under flexural stresses.
- In terms of axial and flexural strength, the CS system provided significantly higher design strengths when compared to RC sections. Additionally, the CS structure did not require the use of high-strength concrete, simplifying construction and potentially reducing costs and environmental impact.
- Shear strength was not a limiting factor for either structural solution.
- Full optimization of the cross-section to withstand the most demanding loading scenario would require a minimum size increase of 23%, making it bulky, heavier, and less attractive.
- If cost is not a limiting factor and preserving the original geometry is a priority, the CS solution stands out for its high structural efficiency, reduced carbon footprint, and minimal impact on hydrodynamic and ecological performance.
- The methodology needs to include factors such as fatigue or long-term plastic deformation, which are crucial for structures exposed to high load cycles, like artificial reefs.

These findings emphasize the importance of a systematic engineering-driven analysis to the design and construction of artificial reefs, as it allows a comprehensive assessment of structural viability before advancing to prototyping and more complex numerical analyses. This stage enhances efficiency and ensures the most promising designs are pursued, optimizing resources and efforts. The design methodology has limitations due to the incomplete consideration of axial and flexural stresses interactions, which may affect performance and lead to an underestimation of the forces acting on the structure. This highlights the need for further analysis in future work to ensure robustness under real exposure conditions. Additionally, in situ monitoring strategies, such as wave height measurements, pressure sensors, strain gauges, accelerometers, or acoustic emission, need to be explored to assess real-time structural performance and detect signs of damage or degradation [72].

Given that marine structures are exposed to dynamic ocean loads over extended periods, future research will address the cumulative fatigue effects. This will require dedicated studies to characterize fatigue resistance as a function of the number of cycles and stress levels in critical sections, based either on the Model Code 90 framework or on experimental data. These investigations will build upon the structural insights developed in this initial study and aim to enhance the durability and reliability of the proposed modular reef system.

Author Contributions: Conceptualization, F.C., E.B.P., I.B.V., T.M. and D.M.; methodology, F.C., E.B.P. and I.B.V.; software, F.C.; validation, F.C. and E.B.P.; formal analysis, F.C. and E.B.P.; investigation, F.C.; resources, F.C., E.B.P., I.B.V. and T.M.; data curation, F.C. and E.B.P.; writing—original draft preparation, F.C.; writing—review and editing, F.C., E.B.P., I.B.V., T.M. and D.M.; supervision, E.B.P. and I.B.V.; project administration, E.B.P., I.B.V. and T.M.; funding acquisition, E.B.P., I.B.V. and T.M. All authors have read and agreed to the published version of the manuscript.

Funding: This work is financed by FCT—Foundation for Science and Technology, under grant agreement SFRH/BD/137220/2018 and SFRH/BD/129583/2017 attributed to the 1st and 3rd author, respectively. The work was partly by projects with references "NORTE-01-0145-FEDER-000032"; POCI-01-0247-FEDER-046092 and PRR C644876810-00000019.

Institutional Review Board Statement: Not applicable.

Informed Consent Statement: Not applicable.

Data Availability Statement: The data presented in this study are available on request from the corresponding authors.

Acknowledgments: This work is financed by national funds through FCT—Foundation for Science and Technology, under grant agreement SFRH/BD/137220/2018 and SFRH/BD/129583/2017 attributed to the 1st and 3th author, respectively. This work was partly financed by FCT/MCTES through national funds (PIDDAC) under the R&D Unit Institute for Sustainability and Innovation in Structural Engineering (ISISE), under reference UIDB/04029/2020 (doi.org/10.54499/UIDB/04029/2020), and under the Associate Laboratory Advanced Production and Intelligent Systems ARISE under reference LA/P/0112/2020. This work was partially funded by the research projects "NEXT-SEA: Next Generation Monitoring of Coastal Systems in a Scenario of Global Change", NORTE-01-0145-FEDER-000032; "ASTRIIS—Atlantic Sustainability Through Remote and In-situ Integrated Solutions", POCI-01-0247-FEDER-046092, financed by FEDER funds; and "R2UTechnologies—modular systems", PRR C644876810-00000019, funded by Plano de Recuperação e Resiliência.

Conflicts of Interest: The authors declare no conflicts of interest.

Appendix A. RC Cross-Section

Table A1. Results of the parametric analysis of RC alternatives.

Materials (Concrete + Steel)	$A_{s,min}$			Tensile Strength	Compressive Strength	Flexural Strength			M_{Ed}			
	$A_{s,req}$	$Ø_{eff}$	$A_{s,eff}$	$N_{pl,Rd}$	$N_{pl,Rd}$	$A_{s,satur}$	$M_{pl,Rd}$	$Ø_{satur}$	ULS(B) E_{wcm}	ULS(B) $E_{wcm,storm}$	ULS(B) $E_{wcm,maxstorm}$	ALS E_{wcme}
	cm^2	mm	cm^2	kN	kN	cm^2	kNm	mm	kNm	kNm	kNm	kNm
C30/37 + S400	2.28	1Ø20	3.14	109.22	462.65	1.75	2.77	1Ø16	0.52	1.16	5.47	12.81
C30/37 + S500	1.81	1Ø16	2.01	87.39	440.82	1.25	2.60	1Ø16	0.52	1.16	5.47	12.81
C30/37 + DIN 4.8	2.88	1M22	2.99	83.20	436.63	2.40	2.92	1M20	0.52	1.16	5.47	12.81
C30/37 + DIN 5.8	2.28	1M20	2.41	83.83	437.26	1.75	2.77	1M18	0.52	1.16	5.47	12.81
C30/37 + DIN 8.8	1.41	1M16	1.54	85.70	439.13	0.83	2.37	1M12	0.52	1.16	5.47	12.81
C35/45 + S400	2.52	1Ø20	3.14	109.22	521.55	2.01	3.23	1Ø16	0.52	1.16	5.47	12.81
C35/45 + S500	2.00	1Ø16	2.01	87.39	499.73	1.46	3.23	1Ø16	0.52	1.16	5.47	12.81
C35/45 + DIN 4.8	3.19	1M24	3.46	96.28	508.61	2.80	3.40	1M22	0.52	1.16	5.47	12.81
C35/45 + DIN 5.8	2.52	1M22	2.99	104.00	516.33	2.01	3.23	1M20	0.52	1.16	5.47	12.81
C35/45 + DIN 8.8	1.55	1M18	1.89	105.18	517.52	0.98	2.76	1M14	0.52	1.16	5.47	12.81
C45/55 + S400	3.01	1Ø20	3.14	109.22	639.36	2.62	4.16	1Ø20	0.52	1.16	5.47	12.81
C45/55 + S500	2.39	1Ø20	3.14	136.52	666.67	1.87	3.89	1Ø16	0.52	1.16	5.47	12.81
C45/55 + DIN 4.8	3.81	1M27	4.52	125.77	655.92	3.60	4.37	1M27	0.52	1.16	5.47	12.81
C45/55 + DIN 5.8	3.01	1M24	3.46	120.35	650.49	2.62	4.16	1M22	0.52	1.16	5.47	12.81
C45/55 + DIN 8.8	1.85	1M18	2.41	134.12	664.27	1.27	3.55	1M16	0.52	1.16	5.47	12.81
C50/60 + S400	3.26	1Ø25	4.91	170.78	759.83	2.91	4.62	1Ø20	0.52	1.16	5.47	12.81
C50/60 + S500	2.58	1Ø20	3.14	136.52	725.57	2.08	4.33	1Ø20	0.52	1.16	5.47	12.81
C50/60 + DIN 4.8	4.12	1M27	4.52	125.77	714.82	4.00	4.86	1M27	0.52	1.16	5.47	12.81
C50/60 + DIN 5.8	3.26	1M24	3.46	120.35	709.40	2.91	4.62	1M22	0.52	1.16	5.47	12.81
C50/60 + DIN 8.8	2.00	1M20	2.41	134.12	723.17	1.41	3.95	1M16	0.52	1.16	5.47	12.81
C55/67 + S400	3.33	1Ø25	4.91	170.78	818.74	3.02	4.91	1Ø20	0.52	1.16	5.47	12.81
C55/67 + S500	2.64	1Ø20	3.14	136.52	784.48	2.14	4.56	1Ø20	0.52	1.16	5.47	12.81
C55/67 + DIN 4.8	4.22	1M27	4.52	125.77	773.73	4.19	5.51	1M27	0.52	1.16	5.47	12.81
C55/67 + DIN 5.8	3.33	1M24	3.46	120.35	768.30	3.02	4.91	1M24	0.52	1.16	5.47	12.81
C55/67 + DIN 8.8	2.05	1M20	2.41	134.12	782.08	1.43	4.11	1Ø16	0.52	1.16	5.47	12.81
C60/75 + S400	3.49	1Ø25	4.91	170.78	877.64	3.18	5.25	1Ø25	0.52	1.16	5.47	12.81
C60/75 + S500	2.77	1Ø20	3.14	136.52	843.38	2.24	4.85	1Ø20	0.52	1.16	5.47	12.81
C60/75 + DIN 4.8	4.42	1M27	4.52	125.77	832.63	4.43	5.59	1M27	0.52	1.16	5.47	12.81
C60/75 + DIN 5.8	3.46	1M24	3.46	120.35	827.21	3.18	5.25	1M24	0.52	1.16	5.47	12.81
C60/75 + DIN 8.8	2.15	1M20	2.41	134.12	840.98	1.49	4.35	1M16	0.52	1.16	5.47	12.81
C70/85 + S400	3.65	1Ø25	4.91	170.78	994.45	3.57	5.98	1Ø25	0.52	1.16	5.47	12.81
C70/85 + S500	2.89	1Ø20	3.14	136.52	961.19	2.50	5.47	1Ø20	0.52	1.16	5.47	12.81
C70/85 + DIN 4.8	4.62	1M30	5.52	153.60	978.27	5.00	6.40	1M30	0.52	1.16	5.47	12.81
C70/85 + DIN 5.8	3.65	1M27	4.52	157.22	981.89	3.57	5.98	1M27	0.52	1.16	5.47	12.81
C70/85 + DIN 8.8	2.24	1M20	2.41	134.12	958.79	1.65	4.89	1M18	0.52	1.16	5.47	12.81

Table A1. *Cont.*

Materials (Concrete + Steel)	$A_{s,min}$			Tensile Strength	Compressive Strength	Flexural Strength			M_{Ed}			
	$A_{s,req}$	$Ø_{eff}$	$A_{s,eff}$	$N_{pl,Rd}$	$N_{pl,Rd}$	$A_{s,satur}$	$M_{pl,Rd}$	$Ø_{satur}$	ULS(B) E_{wcm}	ULS(B) $E_{wcm,storm}$	ULS(B) $E_{wcm,maxstorm}$	ALS E_{wcme}
	cm^2	mm	cm^2	kN	kN	cm^2	kNm	mm	kNm	kNm	kNm	kNm
C80/95 + S400	3.81	1Ø25	4.91	170.78	1113.26	3.99	6.75	1Ø25	0.52	1.16	5.47	12.81
C80/95 + S500	3.02	1Ø20	3.14	136.52	1079.00	2.79	6.18	1Ø20	0.52	1.16	5.47	12.81
C80/95 + DIN 4.8	4.82	1M30	5.52	153.60	1096.08	5.62	7.24	1M33	0.52	1.16	5.47	12.81
C80/95 + DIN 5.8	3.81	1M27	4.52	157.22	1099.70	3.99	6.75	1M27	0.52	1.16	5.47	12.81
C80/95 + DIN 8.8	2.34	1M20	2.41	134.12	1076.60	1.84	5.49	1M18	0.52	1.16	5.47	12.81
C90/105 + S400	3.97	1Ø25	4.91	170.78	1231.07	4.49	7.59	1Ø25	0.52	1.16	5.47	12.81
C90/105 + S500	3.14	1Ø20	3.14	136.52	1196.81	3.14	6.95	1Ø20	0.52	1.16	5.47	12.81
C90/105 + DIN 4.8	5.02	1M30	5.52	153.60	1213.89	6.32	8.15	1M33	0.52	1.16	5.47	12.81
C90/105 + DIN 5.8	3.97	1M27	4.52	157.22	1217.50	4.49	7.59	1M27	0.52	1.16	5.47	12.81
C90/105 + DIN 8.8	2.44	1M22	2.99	166.40	1226.69	2.07	6.16	1M20	0.52	1.16	5.47	12.81

Colour legend: $A_{s,satur} \geq A_{s,eff}$; $Ø_{satur} \geq Ø_{eff}$; $A_{s,satur} << A_{s,eff}$; $Ø_{eff} = Ø_{satur}$; $M_{Rd} > M_{Ed,ALS\_Ewcme}$; $M_{Rd} < M_{Ed,ALS\_Ewcme}$.

Appendix B. CS Cross-Section

Table A2. Results of the parametric analysis of CS alternatives.

Materials (Concrete + Steel)	Steel Tube		Steel Ratio [1]	Compressive Strength	Tensile Strength	Flexural Strength	M_{Ed}	Reef Unit Weight	Reef Unit Cost
	D_{ext}	t_w	0.2 < d < 0.9	$N_{pl,Rd,comp}$	$N_{pl,Rd,trac}$	M_{Rd}	ALS_E_{wcme}		
	[mm]	[mm]	[-]	[kN]	[kN]	[kN.m]	[kN.m]	[kg]	[€]
C30/37 S235	88.9	6	0.20	493.24	210.32	13.30	12.81	167.76	52.24
C30/37 S275	76.1	6.3	0.21	500.79	215.71	12.71	12.81	153.79	48.38
	88.9	6	0.23	529.04	230.19	14.77	12.81	167.76	54.47
C30/37 S355	76.1	5	0.22	515.20	227.30	13.28	12.81	153.93	42.55
	88.9	4	0.21	507.89	219.57	13.90	12.81	168.03	40.85
C30/37 S420	60.3	5	0.20	497.91	207.13	11.57	12.81	139.72	37.81
	76.1	4	0.21	509.91	219.73	13.09	12.81	154.05	39.23
	88.9	3.2	0.20	501.36	210.76	13.59	12.81	168.14	35.77
C35/45 S275	88.9	6	0.21	576.19	246.12	15.54	12.81	167.76	55.07
C35/45 S355	76.1	6	0.22	600.28	266.95	15.45	12.81	153.82	50.19
	88.9	5	0.22	600.78	269.61	16.66	12.81	167.89	49.76
C35/45 S420	76.1	5	0.22	604.80	268.92	15.64	12.81	153.93	47.91
	88.9	4	0.21	596.15	259.77	16.37	12.81	168.03	46.02
C45/55 S355	88.9	6	0.21	742.10	317.72	20.04	12.81	167.76	58.56
C45/55 S420	76.1	6	0.20	744.39	315.82	19.32	12.81	153.82	56.00
	88.9	5	0.21	747.33	318.98	20.24	12.81	167.89	55.51

[1] steel ratio for perforated section; colour legend: $M_{Rd} > M_{Ed,ALS\_Ewcme}$; $M_{Rd} \approx M_{Ed,ALS\_Ewcme}$; $M_{Rd} < M_{Ed,ALS\_Ewcme}$.

References

1. Ng, K.; Phillips, M.R.; Calado, H.; Borges, P.; Veloso-gomes, F. Seeking harmony in coastal development for small islands: Exploring multifunctional artificial reefs for São Miguel Island, the Azores. *Appl. Geogr.* **2013**, *44*, 99–111. [CrossRef]
2. López, I.; Tinoco, H.; Aragonés, L.; García-barba, J. The multifunctional artificial reef and its role in the defence of the Mediterranean coast. *Sci. Total Environ.* **2016**, *550*, 910–923. [CrossRef] [PubMed]
3. Dafforn, K.A.; Glasby, T.M.; Airoldi, L.; Rivero, N.K.; Mayer-pinto, M.; Johnston, E.L. Marine urbanization: An ecological framework for designing multifunctional artificial structures. *Front. Ecol. Environ.* **2015**, *13*, 82–90. [CrossRef]
4. Lee, M.O.; Otake, S.; Kim, J.K. Transition of artificial reefs (ARs) research and its prospects. *Ocean Coast. Manag.* **2018**, *154*, 55–65. [CrossRef]

5. Pörtner, H.O.; Roberts, D.C.; Masson-Delmotte, V.P.; Zhai, P.; Tignor, E.; Poloczanska, E.; Mintenbeck, K.; Nicolai, M.; Okem, A.; Petzold, J.; et al. *The Ocean and Cryosphere in a Changing Climate*; Cambridge University Press: Cambridge, UK, 2019.
6. Bracho-Villavicencio, C.; Matthews-Cascon, H.; Rossi, S. Artificial Reefs around the World: A Review of the State of the Art and a Meta-Analysis of Its Effectiveness for the Restoration of Marine Ecosystems. *Environments* **2023**, *10*, 121. [CrossRef]
7. Kim, Y.C. *Handbook of Coastal and Ocean Engineering*; World Scientific Publishing Company: Singapore, 2010.
8. Ergin, S.; Soares, C.G. Sustainable Development and Innovations in Marine Technologies. *Sustain. Dev. Innov. Mar. Technol.* **2022**, *9*, 474. [CrossRef]
9. Carral, L.; Alvarez-feal, J.C.; Tarrio-Saavedra, J. Social interest in developing a green modular arti fi cial reef structure in concrete for the ecosystems of the Galician rías. *J. Clean. Prod.* **2018**, *172*, 1881–1898. [CrossRef]
10. Ramos, J.; Lino, P.G.; Himes-cornell, A.; Santos, M.N. Local fishermen's perceptions of the usefulness of artificial reef ecosystem services in Portugal. *PeerJ* **2019**, *6*, e6206. [CrossRef]
11. Cruz, F.; Pereira, E.; Valente, I.B.; Tiago, M.; Maslov, D.; Pinheiro, M. *Structural Design of an Innovative Multifunctional Artificial Reef*; Oceans 2018 MTS/IEEE Charleston; IEEE: Charleston, SC, USA, 2018; pp. 1–7. [CrossRef]
12. Boakes, Z.; Hall, A.E.; Ampou, E.E.; Jones, G.C.A.; Suryaputra, I.G.N.A.; Mahyuni, L.P.; Prasetijo, R.; Stafford, R. Coral reef conservation in Bali in light of international best practice, a literature review. *J. Nat. Conserv.* **2022**, *67*, 126190. [CrossRef]
13. Li, Y.; Ren, N.; Li, X.; Ou, J. Hydrodynamic Analysis of a Novel Modular Floating Structure System Integrated with Floating Artificial Reefs and Wave Energy Converters. *J. Mar. Sci. Eng.* **2022**, *10*, 1091. [CrossRef]
14. De Almeida, J.L. REEFS: An artificial reef for wave energy harnessing and shore protection—A new concept towards multipurpose sustainable solutions. *Renew. Energy* **2017**, *114*, 817–829. [CrossRef]
15. Bohnsack, J.A.; Sutherland, D.L. Artificial reef research: A review with recommendations for future priorities. *Bull. Mar. Sci.* **1985**, *37*, 11–39.
16. Kirkbride-Smith, A.E. The Economic, Social and Conservation Benefits of Recreation-orientated Artificial Reefs. Ph.D. Thesis, University of Hull, Hull, UK, 2014.
17. Ten Voorde, M. Contribution to the design of Multi-Functional Artificial Reefs. Ph.D. Thesis, University of Coimbra, Coimbra, Portugal, 2009.
18. Vieira, B.F.V.; Pinho, J.L.S.; Barros, J.A.O.; Antunes do Carmo, J.S. Hydrodynamics and morphodynamics performance assessment of three coastal protection structures. *J. Mar. Sci. Eng.* **2020**, *8*, 175. [CrossRef]
19. Vieira, B. Engineering with Nature: An Innovative Solution for Coastal Erosion Protection. Ph.D. Thesis, University of Minho, Largo do Paço, Braga, 2023.
20. Fabi, G.; Spagnolo, A.; Bellan-santini, D.; Charbonnel, E.; Çiçek, B.A.; García, J.J.G.; Jensen, A.C.; Kallianiotis, A.; Santos, M.N. Overview on artificial reefs in Europe. *Brazilian J. Oceanogr.* **2011**, *59*, 155–166. [CrossRef]
21. Silva, J.; Rosental, I.; Love, M. Overview and trends of ecological and socioeconomic research on artificial reefs. *Mar. Environ. Res.* **2019**, *145*, 81–96. [CrossRef]
22. Dini, E.; Nuzzolese, M.; Cafaro, F.; Lernia, A.; Gadaleta, A.; Sollecito, F. Subsidence of Artificial Reefs with Bamboo Foundations on a Soft Seabed. *Appl. Sci.* **2024**, *14*, 720. [CrossRef]
23. Wang, F.; Feng, Y.; Liu, S.; Chen, Y.; Ding, J. Artificial Fish Reef Site Evaluation Based on Multi-Source High-Resolution Acoustic Images. *J. Mar. Sci. Eng.* **2025**, *13*, 309. [CrossRef]
24. Broomfield, J.P. *Chapter 9 Galvanized Steel Reinforcement in Concrete: A Consultant's Perspective*; Elsevier Science: Amsterdam, The Netherlands, 2004.
25. Reeds, K.A.; Smith, J.A.; Suthers, I.M.; Johnston, E.L. An ecological halo surrounding a large offshore artificial reef: Sediments, infauna, and fish foraging. *Mar. Environ. Res.* **2018**, *141*, 30–38. [CrossRef]
26. Mendonça, A.; Neves, M.G.; Mendonça, A.; Fortes, C.J.; Carmo, J.S.A.D.; Moura, T. Hydrodynamics around an Artificial Surfing Reef at Leirosa, Portugal. *J. Waterw. Port. Coast. Ocean. Eng.* **2012**, *138*, 226–235. [CrossRef]
27. Soon, C.; Ng, L.; Chou, L.M. Artificial reefs as a reef restoration strategy in sediment—Affected environments: Insights from long—Term monitoring. *Aquat. Conserv. Mar. Freshw. Ecosyst.* **2017**, *27*, 976–985. [CrossRef]
28. Cruz, F.; Valente, I.; Almeida, J.; Neves, A.; Pereira, E. Experimental characterization of bond between concrete and HDG steel tubes for mixed steel—Concrete structures. *Mater. Struct.* **2024**, *57*, 89. [CrossRef]
29. UNEP-MAPUNEP-MAP. *Guidelines for the Placement at Sea of Matter for Purpose Other Than the Mere Disposal (Construction of Artificial Reefs)*; UNEP-MAP: Athens, Greece, 2005.
30. Fabi, G.; Scarcella, G.; Spagnolo, A.; Stephen, A.B.; Charbonnel, E.; Goutayer, J.J.; Haddad, N.; Altan, L.; Michel, T. Pratical guidelines for the use of artificial reefs in the Mediterranean and the Black sea. *Rome Stud. Rev.* **2015**, *96*, 1.
31. Great Barrier Reef Marine Park Authority. *Guidelines for the Management of Artificial Reefs in Great Barrier Reef Marine Park*; Great Barrier Reef Marine Park Authority: Townsville City, QLD, Australia, 2011; pp. 1–6.
32. Florida, S.; Reef, C.; Industry, M. *Guidelines and Management Practices for Artificial Reef Siting, Use, Construction, and Anchoring in Southeast Florida*; Florida Department of Environmental Protection: Tallahassee, FL, USA, 2011.

33. Vogler, V. A New Framework for Artificial Coral Reef Design. *J. Digit. Landsc. Archit.* **2022**, *2022*, 524–537. [CrossRef]
34. Vogler, V. A Framework for Artificial Coral Reef Design: Integrating Computational Modelling and High Precision Monitoring Strategies for Artificial Coral Reefs—An Ecosystem-Aware Approach. Ph.D. Thesis, Bauhaus-Universität Weimar, Faculty of Architecture and Urban Planning, Weimar, Germany, 2021. Available online: https://www.db-thueringen.de/receive/dbt_mods_00061874 (accessed on 18 May 2025).
35. Carral, L.; Lamas, M.I.; Barros, J.J.C.; López, I.; Carballo, R. Proposed Conceptual Framework to Design Artificial Reefs Based on Particular Ecosystem Ecology Traits. *Biology* **2022**, *11*, 680. [CrossRef]
36. Carral, L.; Alvarez-Feal, C.; Jesús Rodríguez-Guerreiro, M.; Vargas, A.; Arean, N.; Carballo, R. Methodology for positioning a group of green artificial reef based on a database management system, applied in the estuary of Ares-Betanzos (Nw Iberian Peninsula). *J. Clean. Prod.* **2019**, *233*, 1047–1060. [CrossRef]
37. Antunes do Carmo, J.S.; Neves, M.G.; tenVoorde, M. Designing a multifunctional artificial reef: Studies on the influence of parameters with most influence in the vertical plane. *J. Coast. Conserv.* **2011**, *15*, 99–112. [CrossRef]
38. Jackson, A.; Tomlinson, R.; Corbett, B.; Strauss, D. Long term performance of a submerged coastal control structure: A case study of the Narrowneck multifunctional artificial reef. *Coast. Eng. Proc.* **2012**, *1*, 1–13. [CrossRef]
39. Charbonnel, E.; Serre, C.; Ruitton, S.; Harmelin, J.G.; Jensen, A. Effects of increased habitat complexity on fish assemblages associated with large artificial reef units (French Mediterranean coast). *ICES J. Mar. Sci.* **2002**, *59*, 208–213. [CrossRef]
40. Pondella, D.J.; Claisse, J.T.; Williams, C.M. Theory, practice, and design criteria for utilizing artificial reefs to increase production of marine fishes. *Front. Mar. Sci.* **2022**, *9*, 983253. [CrossRef]
41. Stolk, P.; Markwell, K.; Jenkins, J.M. Artificial reefs as recreational scuba diving resources: A critical review of research. *J. Sustain. Tour.* **2007**, *15*, 331–350. [CrossRef]
42. London Convention and Protocol/UNEP. *London Convention and Protocol/UNEP Guidelines for the Placement of Artificial Reefs*; International Maritime Organization: London, UK, 2009; 100p, ISBN 978-92-807-3022-7. Available online: https://wwwcdn.imo.org/localresources/en/OurWork/Environment/Documents/London_convention_UNEP_Low-res-Artificial%20Reefs.pdf (accessed on 18 May 2025).
43. OSPAR Commission. *OSPAR Guidelines on Artificial Reefs in Relation to Living Marine Resources*; OSPAR Commission: London, UK, 2013; Volume 1.
44. Maslov, D.; Johnson, J.; Pereira, E.; Duarte, D.; Miranda, T.; Lima, M.; Cruz, F.; Valente, I.; Pinheiro, M. Experimental Testing and CFD Modelling for Prototype Design of Innovative Artificial Reef Structures. In Proceedings of the OCEANS 2019, Marseille, France, 17–20 June 2019; pp. 1–7. [CrossRef]
45. Maslov, D.; Pereira, E.; Duarte, D.; Miranda, T.; Ferreira, V.; Tieppo, M.; Cruz, F.; Johnson, J. Numerical analysis of the flow field and cross section design implications in a multifunctional artificial reef. *Ocean. Eng.* **2023**, *272*, 113817. [CrossRef]
46. Androulakis, D.N.; Dounas, C.G.; Banks, A.C.; Magoulas, A.N.; Margaris, D.P. An assessment of computational fluid dynamics as a tool to aid the design of the HCMR-artificial-reefsTM diving oasis in the Underwater Biotechnological Park of Crete. *Sustainability* **2020**, *12*, 4847. [CrossRef]
47. Jiang, Z.; Liang, Z.; Zhu, L.; Guo, Z.; Tang, Y. Effect of hole diameter of rotary-shaped artificial reef on flow field. *Ocean Eng.* **2020**, *197*, 106917. [CrossRef]
48. Liu, Y.; Dong, G.; Zhao, Y. *The Investigation of the Hydrodynamics of an Artificial Reef*; INTECH Open Access Publisher: London, UK, 2007.
49. Galdo, M.I.L.; Guerreiro, M.J.R.; Vigo, J.L.; Rodriguez, I.A.; Lorenzo, R.V.; Couce, J.C.C.; Couce, L.C. Definition of an Artificial Reef Unit through Hydrodynamic and Structural (CFD and FEM) Models—Application to the Ares-Betanzos Estuary. *J. Mar. Sci. Eng.* **2022**, *10*, 230. [CrossRef]
50. Xue, D.; Wang, C.; Huang, T.; Pan, Y.; Zhang, N.; Zhang, L. Flow field effects and physical stability of pyramidal artificial reef with different slope angles. *Ocean. Eng.* **2023**, *283*, 115059. [CrossRef]
51. Jang, S.; Jeong, J.; Lee, S.; Kim, D. Identifying Hydraulic Characteristics Related to Fishery Activities Using Numerical Analysis and an Automatic Identification System of a Fishing Vessel. *J. Mar. Sci. Eng.* **2022**, *10*, 1619. [CrossRef]
52. Shu, A.; Qin, J.; Rubinato, M.; Sun, T.; Wang, M.; Wang, S.; Wang, L.; Zhu, J.; Zhu, F. An Experimental Investigation of Turbulence Features Induced by Typical Artificial M-Shaped Unit Reefs. *Appl. Sci.* **2021**, *11*, 21. [CrossRef]
53. Cruz, F.; Pereira, E.; Lima, M.; Johnson, J.; Duarte, D.; Valente, I.B.; Miranda, T.; Maslov, D.; Pinheiro, M. Experimental Characterization of the Scour of Innovative Artificial Reef Prototypes Using Hydraulic Flume and Photogrammetry. In Proceedings of the OCEANS 2019, Marseille, France, 17–20 June 2019; pp. 1–10. [CrossRef]
54. Galdo, L.; Carballo, R. Numerical and Experimental Analysis of the Velocity Field Inside an Artificial Reef. Application to the Ares-Betanzos Estuary. *J. Mar. Sci. Eng.* **2022**, *10*, 1827. [CrossRef]
55. Kertorahardjo, S.P.; Vincent, H.; Duta, P.H. Structure Analysis on Artificial Reefs. *Appl. Mech. Mater.* **2018**, *874*, 27–34. [CrossRef]
56. Jin, L.; Du, L.; Zhou, Z.; Bao, X. Peak Acceleration Distribution on the Surface of Reef Islands under the Action of Vertical Ground Motion. *Appl. Sci.* **2024**, *14*, 2515. [CrossRef]

57. Maslov, D.; Cruz, F.; Pinheiro, M.; Miranda, T.; Valente, I.B.; Ferreira, V.; Pereira, E. Functional Conception of Biomimetic Artificial Reefs Using Parametric Design and Modular Construction. *J. Mar. Sci. Eng.* **2024**, *12*, 1682. [CrossRef]
58. Gomes, A.C.; Pinho, J.L.d.S.; Granja, H.P. Caracterização Da Morfodinâmica De Praias Sob Influência De Cilindros Geossintéticos: Caso De Estudo Da Praia De Ofir, Portugal. *Rev. Eletrônica Gestão e Tecnol. Ambient.* **2020**, *8*, 5. [CrossRef]
59. Loureiro, E. *Indicadores Geomorfológicos e Sedimentológicos na Avaliação da Tendência Evolutiva da Zona Costeira (Aplicação ao Concelho de Esposende)*; Universidade do Minho: Braga, Portugal, 2006.
60. Mendes, R.P.G. *Energia das Ondas Desenvolvimento de uma Tecnologia de Geração (Gerador Tubular)*; Universidade da Beira Interior: Covilhã, Portugal, 2011.
61. Vieira, F.V. Extreme wave value analysis under uncertainty of climate change scenarios off Iberian Peninsula coast. *Ocean. Eng.* **2021**, *229*, 15. [CrossRef]
62. Sorensen, R.M. *Basic Coastal Engineering*, 3rd ed.; Springer Science & Business Media: Berlin/Heidelberg, Germany, 2005.
63. Petroleum Institute (API). *API RECOMMENDED PRACTICE 2A-WSD: Planning, Designing, and Constructing Fixed Offshore Platforms—Working Stress Design*, 22nd ed.; API Publishing Services: Chicago, IL, USA, 2014; pp. 1–324.
64. Chakrabarti, S.K. *Handbook of Offshore Engineering*, 1st ed.; Elsevier: Amsterdam, The Netherlands, 2005; Volume I.
65. *DNV-RP-C205: Environmental Conditions and Environmental Loads*; Det Norske Veritas (DNV): Høvik, Norway, 2010; pp. 1–124.
66. *ISO 19903:2006*; Petroleum and Natural Gas Industries—Fixed Steel Offshore Structures. International Organization for Standardization: Geneva, Switzerland, 2006.
67. *EN 1992-1-1*; Eurocode 2: Design of Concrete Structures—Part 1-1: General Rules and Rules for Buildings. European Committee for Standardization (CEN): Plzen, Czech Republic, 2004; p. 225.
68. *EN 1990*; Eurocode 0: Basis of Structural Design. European Committee for Standardization (CEN): Plzen, Czech Republic, 2013; pp. 92–126. [CrossRef]
69. *NP EN 206-1*; ConcretePart 1: Specification, Performance, Production and Conformity. European Committee for Standardization (CEN): Brussels, Belgium, 2005; pp. 1–29.
70. *EN 1994-1-1*; Eurocode 4: Design of Composite Steel and Concrete Structures—Part 1-1: General Rules and Rules for Buildings. European Committee for Standardization (CEN): Brussels, Belgium, 2007; pp. 1–9.
71. Cruz, F.; Valente, I.B.; Monteiro, F.; Maslov, D.; Miranda, T.; Pereira, E.B. Development and characterization of structural connections for modular artificial reefs. *Ocean. Eng.* **2025**, *320*, 120311. [CrossRef]
72. Maslov, D.; Pereira, E.; Miranda, T.; Valente, I.B.; Cruz, F.; Pinheiro, M. Innovative monitoring strategies for multifunctional artificial reefs. In Proceedings of the OCEANS 2018 MTS/IEEE Charleston, Charleston, SC, USA, 22–25 October 2018; pp. 1–8.

applied sciences

MDPI

Article

Investigation of Lateral and Longitudinal Deformation of Submarine Nuclear Power Plant Water-Intake Tunnel on Non-Uniform Soft Soil during Earthquake

Jie Zhao [1,*], Bo Qian [1], Changjiang Gan [2], Jianshan Wang [1] and Yanli Peng [1]

1 School of Architectural Engineering, Dalian University, Dalian 116622, China; qian_jingbo@163.com (B.Q.); 13970381575@163.com (J.W.); 18783535782@163.com (Y.P.)
2 Gansu Comprehensive Railway Engineering Contracting Co., Ltd., Lanzhou 730030, China; gan420210@163.com
* Correspondence: zhaojie_gd@163.com

Abstract: The safety-grade water-intake immersed tunnel plays a vital role in the nuclear power cooling system, and its seismic safety is crucial. This paper employs the response displacement method and dynamic time-history analysis using the finite element software ANSYS to construct a beam–spring model and a 3D finite element model of a shield tunnel and foundation. It also develops equivalent linear dynamic constitutive and viscoelastic boundary element subprograms. This study focuses on the weak joint sections of immersed tunnels, conducting a seismic performance analysis under extreme safety earthquake conditions (SL-2). The results indicate that the joint stiffness of immersed tunnels and the increase in seismic peak values do not affect the trend of joint opening variation with longitudinal position. The change in joint opening is primarily located where the thickness of the cover layer changes abruptly or where the soil hardness is unevenly distributed. The joint opening is mainly influenced by seismic forces when considering static and dynamic superposition. When the stiffness of the joint GINA water stop exceeds a certain value, the correlation between stiffness change and joint compression–tension variation gradually weakens. This research can provide a reference for the seismic design of similar projects.

Keywords: immersed tunnels; seismic analysis; response displacement method; time-history analysis method; joint opening

1. Introduction

In recent years, China's intensified investment in new energy construction has spurred the rapid development of nuclear power as a viable alternative. Currently, most nuclear power plants in China are situated in coastal areas, utilizing seawater once-through cooling systems that require substantial water intake. Immersed tunnels are widely employed in these systems due to their minimal sea surface area requirements, eco-friendliness, cost efficiency, and straightforward construction. Given China's susceptibility to earthquakes, many plant sites are located in regions with high seismic activity. Earthquakes present a significant threat to the safety of nuclear power facilities, making seismic performance a critical concern in the construction of these projects. Currently, a well-developed research system exists for seismic studies in underground engineering. Seismic response data have been analyzed using scaled-down tunnel models designed for indoor shaking table tests. Yan et al. [1] proposed a design method for a compartmental particle damper suitable for immersed tunnels. They constructed a 1:60 scale model of an immersed tunnel and performed shaking table tests before and after installing the particle dampers. Cheng et al. [2] based their research on the Hong Kong–Zhuhai–Macao Bridge immersed tunnel project, conducting shaking table model tests on immersed tunnel systems under three site conditions (dry sand, saturated sand, and underwater saturated sand) with models at a 1:30 scale ratio. Zhang et al. [3] investigated the dynamic response characteristics of immersed tunnel structures and their joints using shaking table tests, considering longitudinal traveling wave excitation with varying apparent velocities. Yuan et al. [4] performed compression–shear tests on joints at a 1:10 geometric scale, examining their mechanical properties under low cyclic loading. Okamoto et al. [5] conducted underwater shaking table tests on small-scale immersed tunnel models to measure seismic response data. Chen et al. [6], through shaking table tests, explored the effect of unsaturated soil foundations on internal force distribution in immersed tunnels. Yang et al. [7] analyzed the displacement behavior of immersed tunnels under seismic activity, providing valuable references for seismic studies. Li et al. [8] performed shaking table tests under multi-dimensional seismic waves, focusing on the tunnel–soil system to study the dynamic response of tunnels and surrounding soil. With the continuous advancement of finite element method theory and computer technology, numerical simulation has become a prevalent and cost-effective method for studying the dynamic responses of underwater structures. Peng [9] developed a mechanical model to examine the interaction between immersed tunnels, soil, and fluid. Using the Newmark algorithm, he investigated stress changes in immersed tunnel structures under various seismic excitations and water depth conditions. Chen et al. [10] considered the dynamic nonlinear properties of seabed soil and tunnel concrete, along with the coupling effects between seawater and the seabed, developing a finite element model to analyze the interactions among seawater, seabed, and tunnel, focusing on the seismic response of undersea tunnels. Cui et al. [11] utilized finite element software ADINA to create an integrated model of seawater, immersed tunnels, and the seabed, analyzing the seismic responses of immersed tunnel structures under various conditions, specifically with vertically incident seismic P-waves.

Immersed tunnels, characterized by long linear structures with numerous joints, are often situated in complex marine environments. These joints, critical components connecting tunnel segments, possess significantly lower stiffness than the precast concrete segments. During seismic activities or uneven foundation settlement, the joints are subjected to considerable shear forces, necessitating the design of shear keys to resist these forces. The joints often represent the most vulnerable points in the structure, enduring substantial stress and requiring special attention in seismic analysis to ensure the tunnel's overall integrity and safety. Xiao et al. [12] employed a piecewise linear method to address the material nonlinearities of the joints, investigating their response under quasi-static loads through three-dimensional refined modeling. Zhang et al. [13] used a simplified model of nonlinear spring elements for immersed tunnel joints, complemented by three-dimensional finite element modeling for seismic analysis, enabling calculations of deformation and internal force responses under seismic loading. Yu et al. [14] developed a multi-scale seismic response analysis model, exploring characteristics such as stress on tunnel segments, joint deformation, and shear key forces, along with the influence of key factors. He et al. [15] used the finite element method combined with visco-spring artificial boundaries to formulate a three-dimensional planar SV wave incidence method, investigating the mechanical properties of shear keys in the Hong Kong–Zhuhai–Macao Bridge immersed tunnel joints. Cui et al. [16] combined shaking table tests and finite element simulations to examine the impact of tunnel joints and overlying water on the seismic response of immersed tunnels. Bai et al. [17] used ABAQUS finite element software to create a three-dimensional model for seismic dynamic analysis of part of the Hong Kong–Zhuhai–Macao Bridge immersed tunnel project. Immersed tunnels on the seabed frequently experience significant internal forces due to temperature-induced loads and uneven foundation settlement. Excessive internal forces can lead to joint cracks, leakage, and, in severe cases, misalignment of the joints. To mitigate leakage and accommodate deformations in complex environments, adding water stops at the joint areas is essential [18–20]. The most widely used water stops are the GINA and OMEGA types. The GINA water stop is primarily utilized for anti-deformation and waterproofing functions, while the OMEGA water stop [21–23], acting as a secondary defense line, connects immersed tunnels, providing additional waterproofing and effectively addressing deformations due to temperature and water pressure, concrete shrinkage, and stratum movements. Liu et al. [24] determined the minimum compression necessary to ensure the watertightness of GINA water stops using the average water pressure method and the contact stress method, establishing a safety warning level based on the remaining compression. Zhang et al. [25] investigated the material properties of GINA rubber water stops, the force and deformation mechanisms of vertical steel shear keys in joints under shear forces, and the impact of different axial water pressures on the vertical shear characteristics of the joints.

Given that China's undersea tunnels have not yet undergone strong seismic testing, the seismic stability of underwater immersed tunnels under ultimate seismic ground motion (SL-2) presents a significant technical challenge and key issue for seawater-intake projects at coastal nuclear power plants. This study focuses on the Changjiang Nuclear Power Units 3 and 4's discharge immersed tunnel project in Hainan. Utilizing the response displacement method and the time-history analysis method, the finite element software ANSYS (15.0) [26] was employed to conduct a seismic performance analysis of the safety level of seawater-intake immersed tunnels. This study involved establishing transverse and longitudinal beam–spring models and a three-dimensional soil–structure finite element model. Our research highlights the distribution of interlayer displacement within the immersed tunnel sections under seismic conditions, investigates the variations in compression–tension of GINA water stops and joint openings, and analyzes the primary factors affecting maximum deformation.

2. Methods and Principles

2.1. Response Displacement Method

2.1.1. Lateral Response Deformation Method

The reaction displacement method is applicable for performing seismic analysis of underground pipe culverts. During the analysis, the interaction between soil layers and the tunnel is modeled by arranging stratum springs around the immersed tunnel. Meanwhile, the seismic load is imparted to the structure through the foundation springs, as illustrated in Figure 1.

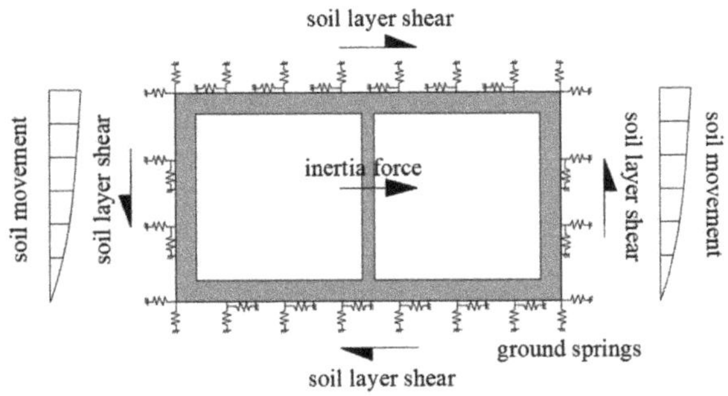

Figure 1. Calculation diagram of response displacement method.

2.1.2. Longitudinal Response Displacement Method

In this study, a beam–spring combined model is utilized to simulate the precast immersed tunnel segments and the surrounding soil. Precast immersed tunnels are modeled with beam elements, while the joint areas of the immersed tunnels are represented using spring elements. In this model, each node is equipped with soil springs in two directions, axial and transverse, along the tunnel axis to simulate the foundation soil layers, thereby forming a system for the analysis of the immersed tunnel–joint–foundation, as illustrated in Figure 2.

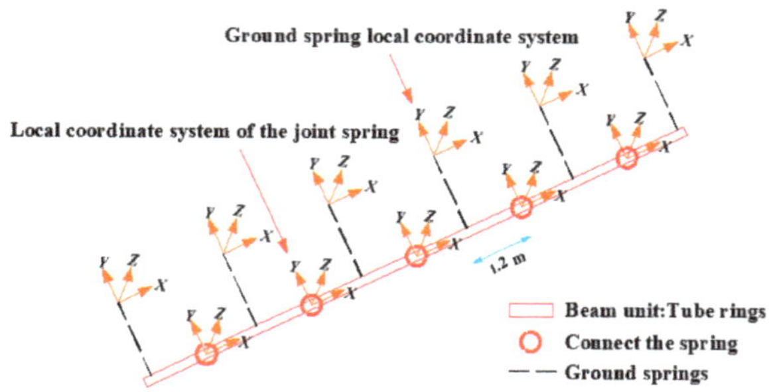

Figure 2. Longitudinal seismic response analysis model.

Utilizing the theories mentioned above, this study employs SuperFLUSH/2D (ver 6.0) software to perform free-field nonlinear dynamical analysis, thereby obtaining the relative displacements of the surrounding soil under seismic conditions. These displacements are subsequently applied as static loads to the restrained ends of the ground springs at the responsive segments of the analysis model for calculation.

2.2. *Equivalent Linear Dynamic Time-History Analysis Method*

The "Standard for Seismic Design of Nuclear Power Plants" in China [27] and the American ASCE 4-98 [28] advocate for the use of an equivalent linearization technique to characterize the nonlinear characteristics of non-rock foundations. By transforming nonlinear issues into linear problems, this approach greatly improves computational efficiency and is extensively used in the seismic analysis of nuclear power structures. Cao et al. [29] verified the feasibility of the equivalent linear method in finite element software using ANSYS's secondary development function. The key to applying the equivalent linearization technique is the calculation of the effective shear strain of the soil. First, an analysis is performed with a predefined initial effective shear strain derived from material characteristics. Subsequently, the new soil equivalent shear modulus and damping ratio are calculated from the G-γ and D-γ curves. This specific calculation process is delineated in Figure 3. In the plot, D represents the damping ratio, G represents the dynamic shear modulus ratio, and γ represents the equivalent shear strain. Based on practical engineering experience, generally, three to five iterations are typically sufficient to achieve satisfactory convergence results.

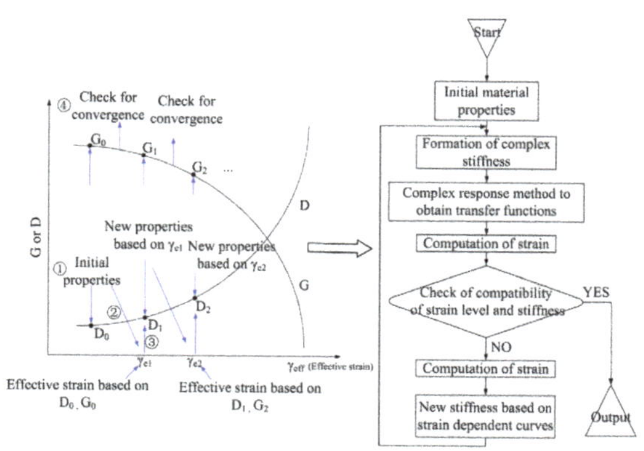

Figure 3. Schematic diagram of equivalent linear method.

2.2.1. Viscoelastic Artificial Boundary

The consideration of dynamic artificial boundaries is a key aspect in the study of the seismic performance of underwater immersed tunnels. The fundamental principle of the viscoelastic artificial boundary involves creating a new physical element by combining dampers and springs in parallel, which can be positioned at the foundation boundary to simulate the radiation damping effect of a semi-infinite foundation. The seismic waves encountered at the model boundaries are transformed into equivalent loads for each node at the boundary. When the scattered waves propagating towards the boundary are absorbed by the viscoelastic artificial boundary, the respective nodes at the boundary will exhibit free-field motion in response to the seismic load. A schematic diagram illustrating the boundary conditions is presented in Figure 4. In order to facilitate the analysis of three-dimensional seismic responses of complex structures such as submarine immersed tube tunnels, a secondary development based on Fortran was carried out using the UPFs provided by ANSYS. This involved writing a subroutine for equivalent viscoelastic boundary elements embedded in the software. The tangential and normal spring stiffness, denoted as K, and damping coefficient, denoted as C, are calculated using Equations (1) and (2) [30], respectively.

$$K = \alpha \frac{G}{r} \sum_{i=1}^{I} A_i \tag{1}$$

$$C = \rho c \sum_{i=1}^{I} A_i \tag{2}$$

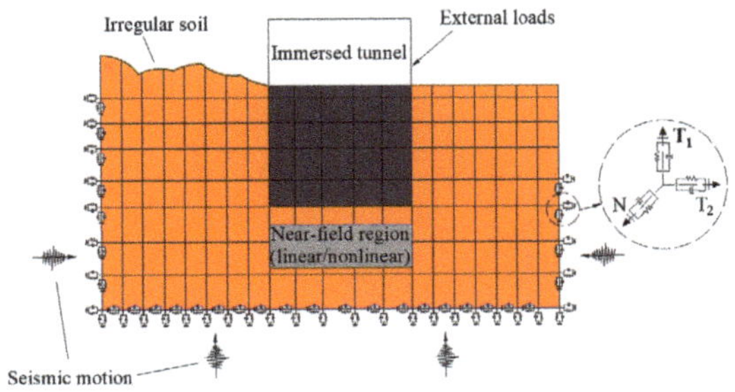

Figure 4. Schematic diagram of artificial boundary model.

In the formula, G represents the shear modulus, ρ denotes the mass density, r is the distance from the scattering source to the nodes on the artificial boundary, and c is the wave speed. The parameter α stands for the boundary parameters of different directional boundaries, while A represents the control area of the nodes at the outer boundary of the foundation region.

2.2.2. Hydrodynamic Pressure

It is widely acknowledged that seismic activity significantly influences the hydrodynamic pressure on the water-intake pipe culverts of nuclear power plants, markedly affecting the structure's dynamic response. Consequently, this pressure is a crucial dynamic load in the design of water-intake tunnels. Relevant regulations are specified in American standards (ASCE 4-98) [28]. In seismic response analysis, it is essential to consider both convective and impulsive effects arising from the oscillation of water flow within the immersed tunnel due to seismic activity. This oscillation induces pressure on the tunnel walls, necessitating the consideration of both horizontal and vertical components of fluid motion. The Housner spring–mass system [31], used to calculate hydrodynamic pressure, is illustrated in Figure 5.

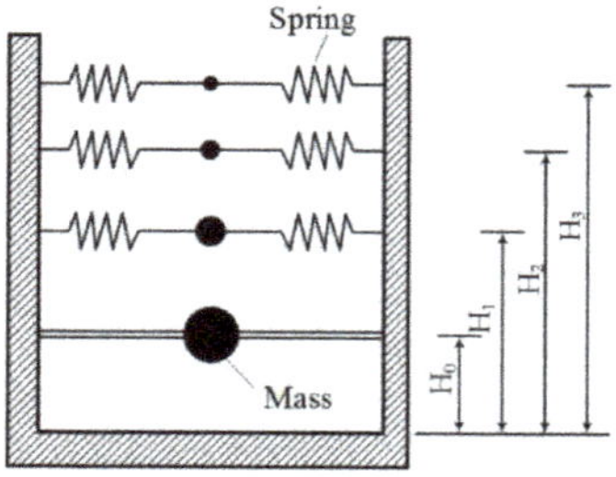

Figure 5. Housner system for hydrodynamic pressure.

3. Engineering Example

3.1. Project Parameters

The Changjiang Nuclear Power Plant, situated in Changjiang County, Hainan Province, features precast drainage immersed tunnels that extend for a total length of 3798.5 m, with a single-hole cross-sectional dimension of 6.0 × 6.0 m. Each tunnel segment measures 45 m in length, with the middle partition walls having a thickness of 40 cm and the remaining partition walls being 60 cm thick. The top of the immersed tunnels is overlaid with a rock layer at least 1.5 m thick. The tube's bottom features a fully paved crushed-stone foundation. The cross-section of the immersed tunnels is depicted in Figure 6. The tubes are connected via a flat joining method. At the joints between each pair of tunnel sections, GINA water stops are arranged around the outer wall and the mid-partition wall.

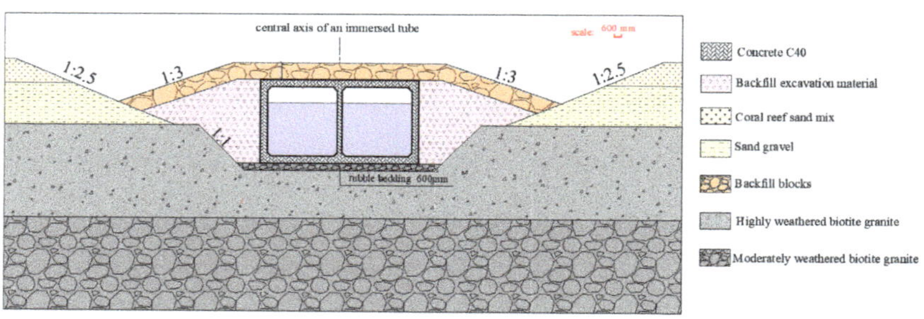

Figure 6. Cross-section view of immersed tunnel.

3.1.1. Soil Parameters

According to the geotechnical engineering survey report for the Changjiang Nuclear Power Plant site in Hainan, the site's strata primarily consist of intrusive rocks and the Quaternary system. The Quaternary system, predominantly distributed across the coastal plains and plateau geomorphic units of the plant area, mainly comprises marine and residual deposits, with a maximum thickness of 41.20 m as revealed by this survey. The site's intrusive rocks predominantly consist of biotite granite. Existing geological data indicate the presence of small areas of lamprophyre, granite porphyry, and other vein rocks. However, due to their limited distribution and minimal impact on the foundation, these formations are not considered in this study. For research convenience, rock bodies sharing similar mechanical properties and indistinct boundaries are amalgamated into a single layer.

By considering the submerged tunnel and stiffer rock foundation as elastic materials, the physical parameters are presented in Table 1. The remaining soil materials must account for their nonlinear characteristics, and the aforementioned equivalent linear method is utilized for dynamic analysis calculations. The necessary equivalent linear parameters are illustrated in Figure 7.

Table 1. Immersed tunnel and soil material parameters.

Material	Static Elastic Modulus (E/GPa)	Dynamic Elastic Modulus (E_d/GPa)	Poisson Ratio (μ)	Dynamic Poisson Ratio (μ_d)
Concrete C40	32.50	42.25	0.20	0.20
Backfill blocks	0.26	0.52	0.33	0.45
Backfill excavation material	0.26	0.52	0.33	0.45
Moderately weathered biotite granite	34.10	16.80	0.20	0.34

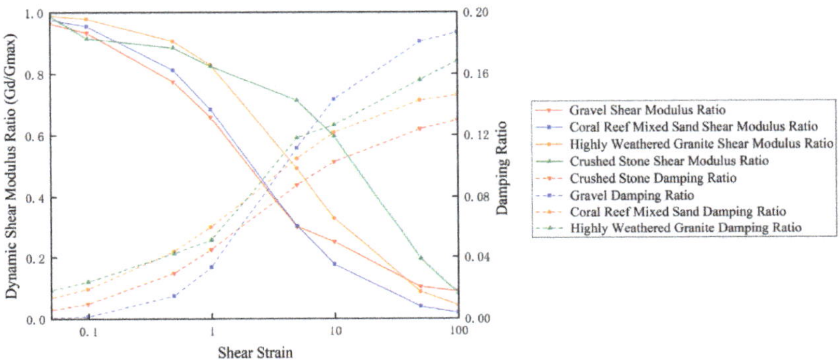

Figure 7. Dynamic shear modulus ratio and damping ratio of soil samples.

3.1.2. Seismic Parameters

This seismic analysis is based on seismic waves originating from the Changjiang Nuclear Power Plant site in Hainan. Figure 8 depicts the seismic waves and frequency spectra in three dimensions. Considering the SL-2 limit for safety seismic motion, the peak accelerations are 0.3 g in the X and Y directions and 0.2 g in the Z direction. This seismic wave lasts for 30 s, with significant fluctuations occurring between 3 and 14 s. The peak times in these directions occur at 13.59 s, 11.67 s, and 6.13 s, respectively.

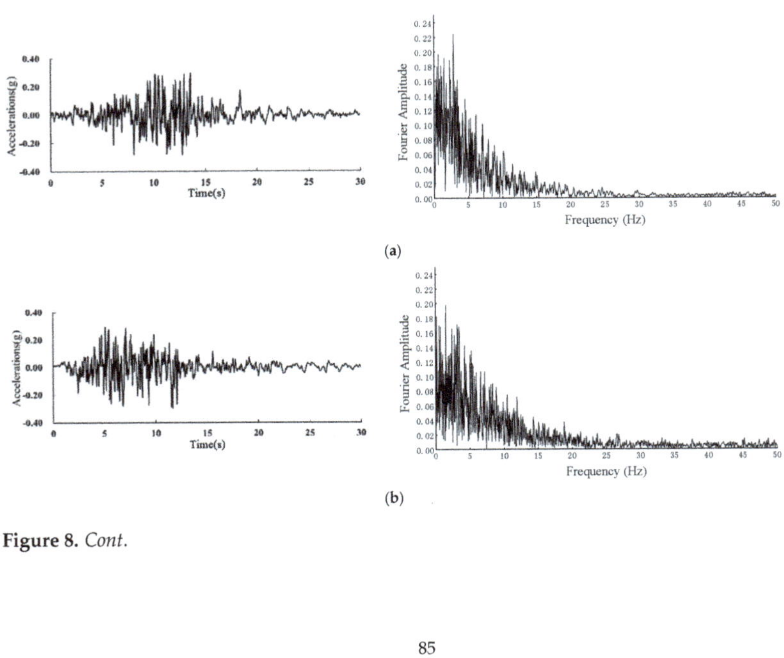

Figure 8. *Cont.*

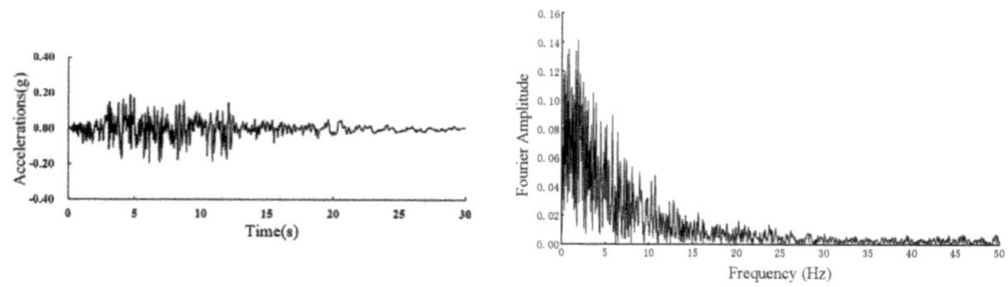

(c)

Figure 8. Time-history curve of seismic wave: (**a**) X-direction seismic wave and frequency spectrum; (**b**) Y-direction seismic wave and frequency spectrum; (**c**) Z-direction seismic wave and frequency spectrum.

3.2. Numerical Analysis Based on the Response Displacement Method

Due to the relative weakness of joint areas compared to the strength of immersed sections, seismic damage generally initiates at the joints. Compared to highway and railway tunnels, analyzing the seismic control indicators for the joints of nuclear power plant water-intake immersed tunnels, which demand higher water tightness, is particularly important. In this section, based on the response displacement method, a beam–spring model is established to study the interlayer displacement angle and joint opening under extreme seismic conditions in immersed tunnels.

3.2.1. Calculation Model

(1) Lateral Model of Immersed Tunnel

In order to study the stability of each wall of the immersed tunnel under earthquake conditions, the beam element (BEAM188) is used to simulate its walls, while the spring element (COMBIN14) is adopted to simulate the effect of surrounding soil on the tunnel's structure. Based on this, the lateral seismic response model is established, as shown in Figure 9.

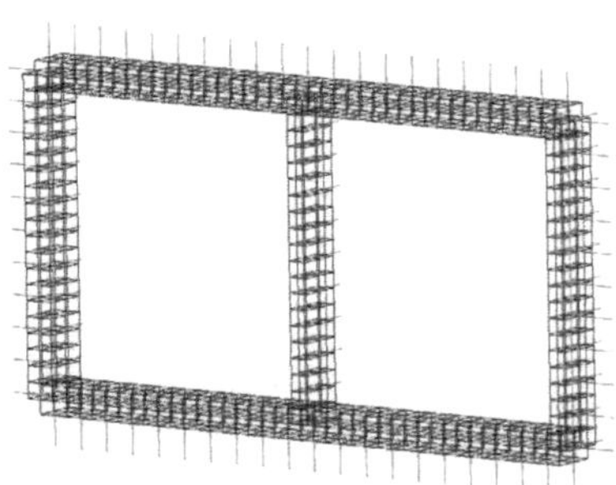

Figure 9. Transverse beam–spring model.

(2) Longitudinal Analysis Model of Immersed Tunnel

To investigate the deformation patterns of joints in the immersed tunnel under seismic conditions, 46 segments of the immersed tube in a representative sea area are selected for analysis, with a total length of 2070 m. The beam element (BEAM188) is chosen to simulate the immersed tunnel joints, while the spring element (COMBIN14) is used to simulate the rotating spring, axial spring, and shear spring, respectively, of the joints. Moreover, COMBIN14 is used to convey the forced displacement effect of the deformation in the foundation strata on the tunnel under earthquake conditions. The beam–spring model system formed is shown in Figure 10.

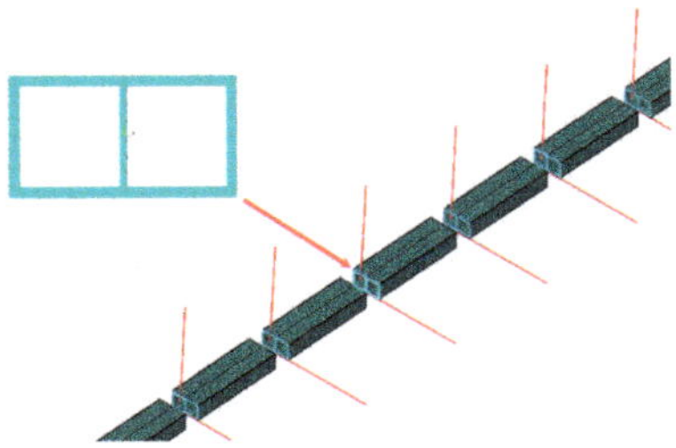

Figure 10. Longitudinal beam–spring model.

(3) Free-Field Analysis Model

The nonlinear material properties of soil are crucial in dynamic analysis. Consequently, a two-dimensional free-field model was established using the SuperFLUSH/2D program for equivalent linear analysis. The finite element model, depicted in Figure 11, incorporates viscous boundaries at the base and sides. Ground motion is introduced as an equivalent nodal force, enabling the determination of relative soil layer displacement along the immersed tunnel's entire domain through nonlinear seismic response analysis.

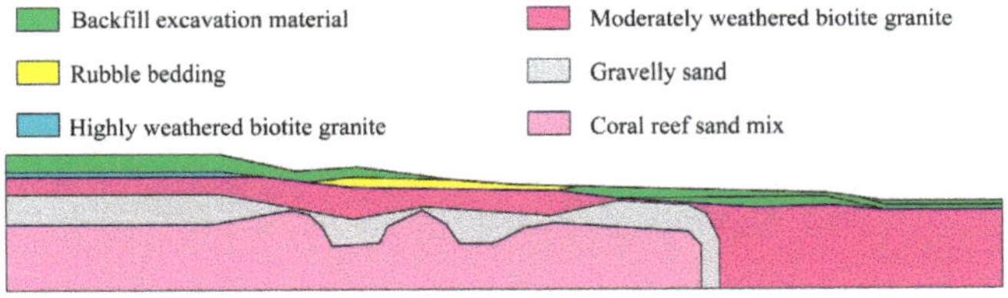

Figure 11. Longitudinal finite element model of the entire domain of immersed tunnel.

Relative displacement values of the surface, as well as the top and bottom of immersed tunnels under various ground motion conditions, are extracted and shown in Figure 12. These displacements are then applied to the ends of soil springs in the corresponding soil layers of the longitudinal beam–spring model. This approach allows for further analysis of the opening and relative angle of the immersed tunnel joints.

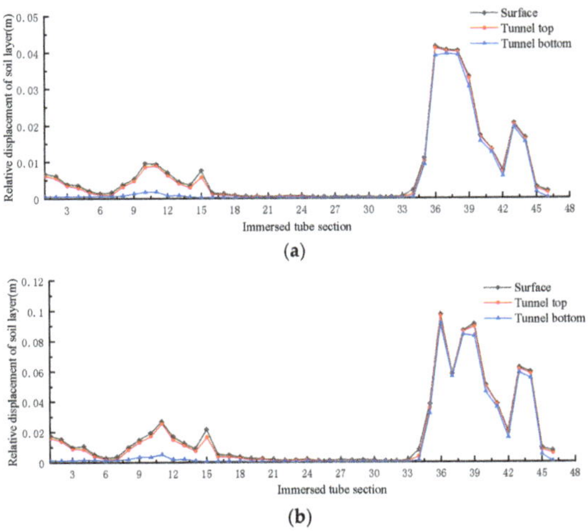

Figure 12. Relative displacement of soil layer in the entire domain of immersed tunnel: (**a**) working conditions under peak ground acceleration of 0.15 g; (**b**) working conditions under peak ground acceleration of 0.30 g.

3.2.2. Calculating Working Conditions

Investigations into nuclear power plant engineering reveal that the design value for the SL-2 level horizontal peak ground acceleration in China's coastal nuclear power plants typically ranges between 0.15 g and 0.30 g. For nuclear immersed tunnel conditions, it is essential to consider factors such as tunnel joint stiffness and varying peak ground accelerations. The seismic safety of immersed tunnels under diverse conditions must also be evaluated. This study simulates multiple scenarios to assess these aspects. Specifically, for the response displacement method in joint seismic performance analysis, peak accelerations of 0.15 g and 0.30 g and water stop stiffness values of 51 and 62 were selected, resulting in four distinct working conditions. Detailed information is presented in Table 2.

Table 2. Working conditions of calculation.

Peak Acceleration	Type of Water Stop	Joint Stiffness Coefficient				Working Condition
		Tensile Stiffness 10^6 ($N \cdot m^{-1}$)	Compressive Stiffness 10^6 ($N \cdot m^{-1}$)	Longitudinal Flexural Stiffness 10^9 ($N \cdot m^{-1} \cdot rad$)	Transverse Flexural Stiffness 10^{10} ($N \cdot m^{-1} \cdot rad$)	
0.15 g	320-370-51	150	1342	252	150	1
	320-370-62	207	1845	345	207	2
0.30 g	320-370-51	150	1342	252	150	3
	320-370-62	207	1845	345	207	4

3.2.3. Result Analysis

(1) Interlayer Displacement Angle

The interlayer displacement angle can be a good measure of the stability and damage degree of the structure under earthquake conditions. Table 3 summarizes the maximum interlayer displacement angles of the left wall, middle partition wall, and right wall under conditions 1 and 3. By analyzing these data, it is concluded that, under working condition 1, the largest angle is the right wall, with a maximum value of 1/7028; under working condition 3, the largest one is the right wall, with a maximum value of 1/7039. Both of these values are lower than the limit required by the "Standard for Seismic Design of Nuclear Power Plants".

Table 3. Interlayer displacement angle.

Working Condition	Peak Ground Acceleration	Left Side Wall	Middle Partition Wall	Right Side Wall	Code Limit Value
1	0.15 g	1/13745	1/9188	1/7028	1/550
3	0.30 g	1/10823	1/8428	1/7039	1/550

(2) Opening value of joints

For studying the change law in the longitudinal opening value of the immersed tunnel joint, the curves for each joint along the tunnel, depicting their positions under four different working conditions, are obtained. They are shown in Figure 13. It is found that all the parts with large joint openings are located in sections where there are changes in the thickness of the foundation strata and mutations in the soft and hard soil layers, while the parts with small opening values correspond to the positions where the thickness of the foundation strata is uniform.

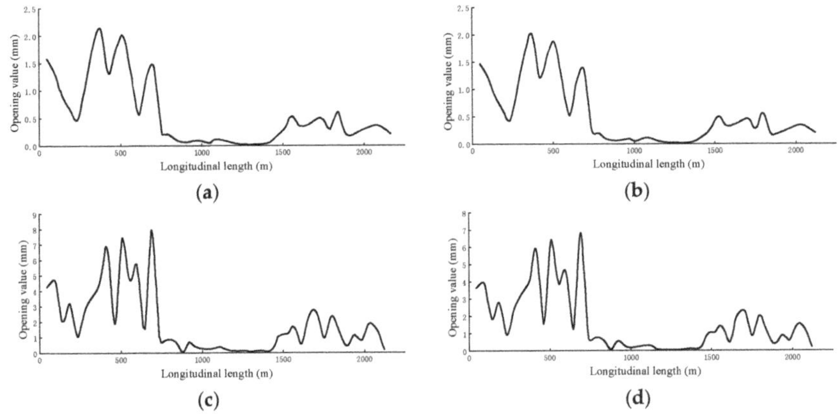

Figure 13. Distribution of joint opening values of immersed tunnel under different working conditions: (**a**) working condition 1, (**b**) working condition 2, (**c**) working condition 3, (**d**) working condition 4.

Compared with the change in the joint opening value under different working conditions, as the stiffness of the GINA waterproof strip increases, the longitudinal opening trend under different working conditions shows no significant changes with position, but the joint opening value decreases in all cases. For SL2 seismic peak accelerations of 0.15 g and 0.30 g, it was observed that the increase in seismic loading does not affect the trend of joint opening changes with position but results in an overall increase in joint opening.

3.3. *Numerical Analysis Based on Dynamic Time-History Method*

To conduct further research on the mechanical properties and seismic resistance of the immersed tunnel joints, based on the dynamic time-history method, the equivalent linear model and the viscoelastic boundary element are embedded in ANSYS by the subroutine of the UPFs, a secondary development program. Moreover, taking calculation time, calculation accuracy, storage cost, and other problems into consideration, a three-stage immersed tunnel with a relatively complex soil layer has been determined for modeling analysis. This study accurately simulates the tunnel joints and the effects of convection and impulse of the internal water flow, carrying out research on the interlayer displacement angle of the immersed tunnel, as well as the compression–tension variation and the opening value of its joints under ultimate seismic conditions. Lastly, the influence of joint stiffness on the seismic performance index has been analyzed.

3.3.1. Calculation Model

This study models three sections of an immersed tube, each 45 m in length, totaling 135 m. The immersed tunnels are simulated using eight-node shell elements (Shell281), GINA water stops with spring elements (Combine14), and soil and bedrock with eight-node solid elements (Solid181). Additionally, the convection and impulsive effects of moving water within the tunnel are modeled using mass point elements (Mass21) and spring elements (Combine14). This creates a comprehensive three-dimensional finite element model of the tunnel–joint–soil system. A viscoelastic boundary is set at the model's bottom and sides to simulate the radiation damping effect of an infinite foundation. The equivalent linear constitutive method describes the nonlinear characteristics of the soil mass. The soil layer model extends 50 m to both sides and 50 m below the tunnel center, with the tunnel buried at a depth of 13.2 m. The soil layers from top to bottom consist of coral reef mixed sand, medium sand, gravel cushion, highly weathered biotite granite, and moderately weathered biotite granite. In the mesh generation of the model, several key considerations are necessary. First, the balance between calculation cost and accuracy must be maintained. The mesh size should be controlled, adhering to the "Standard for Seismic Design of Nuclear Power Plants" for maximum mesh size in the foundation strata. The immersed tunnels, being central to this research, should have a mesh size between 0.2 and 2 m, as illustrated in Figure 14. For the foundation, the mesh height of soil near the tunnel should match that of the immersed pipe as closely as possible, while the mesh height of soil farther from the tunnel can be increased, with a maximum of 5 m. The overall finite element model is shown in Figure 15. The first joint of the tunnel is located in a highly weathered soil layer and a gravel sand layer, both relatively thick, resulting in weaker bearing capacity compared to the second joint.

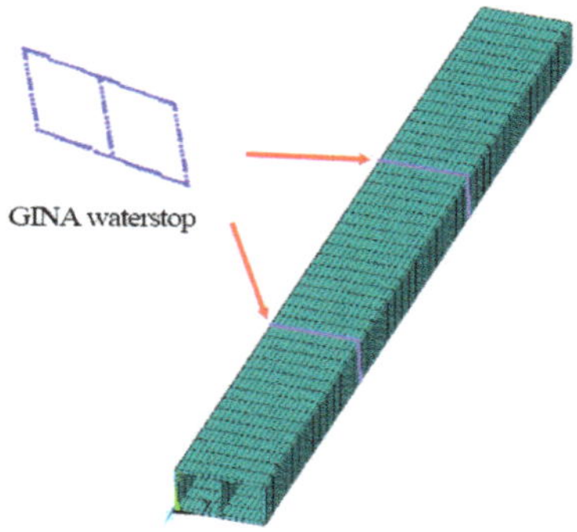

Figure 14. Immersed tunnel model and spring model.

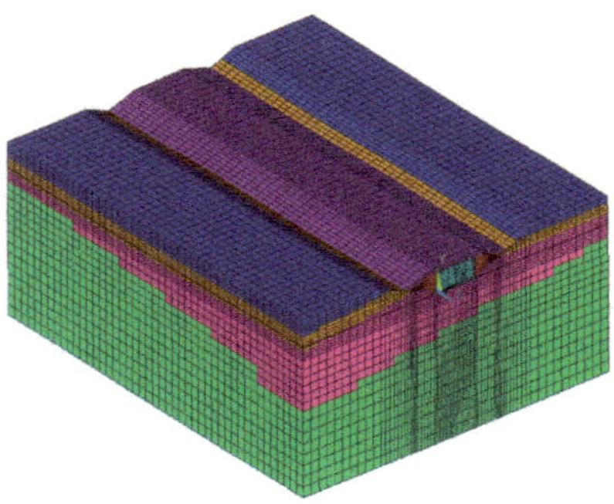

Figure 15. Immersed tunnel–joint–soil finite element model.

For convenience of expression, this paper has named the three sections of the immersed tube as follows: immersed tube No. 1, immersed tube No. 2, and immersed tube No. 3. Identically, the joints from left to right are named as follows: joint No. 1 and joint No. 2, which can be seen in Figure 16. Also, the walls of the immersed tunnel are named as follows: bottom plate 1, bottom plate 2, top plate 3, top plate 4, left side wall 5, middle partition wall 6, and right side wall 7, as shown in Figure 17.

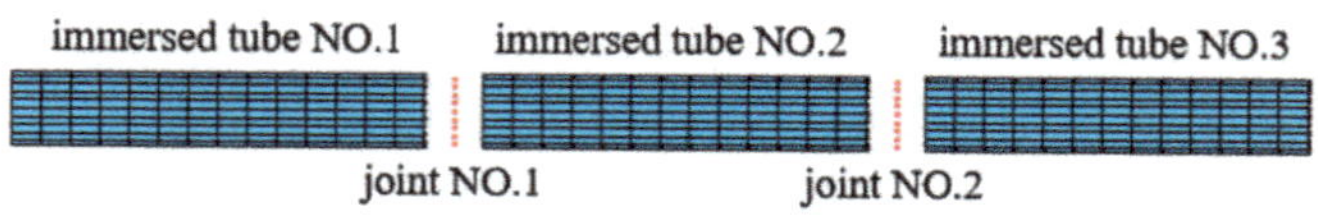

Figure 16. Numbers of each section of immersed tunnel.

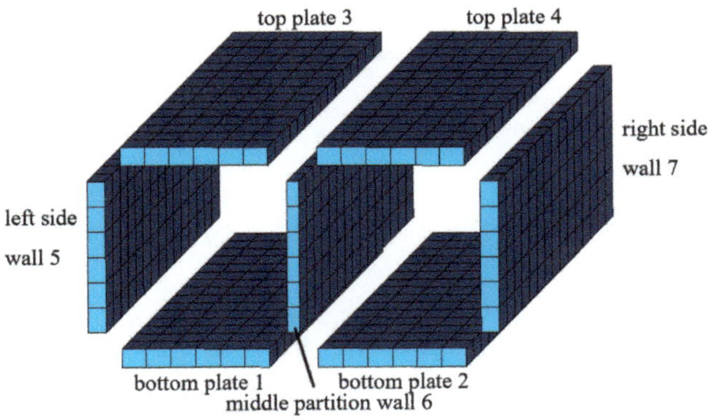

Figure 17. Locations of each wall of immersed tunnel.

3.3.2. Calculated Working Conditions

In this paper, a seismic load with a peak ground acceleration of 0.3 g is adopted to calculate the seismic response during the tunnel operating period. Additionally, hydrostatic pressure and hydrodynamic pressure, as well as high water levels, are taken into account. In addition, the combination of the action effects of the above loads is carried out according to the Code for Hydraulic Design of Nuclear Power Plants [32]. The joint GINA water stop is key to this study. Therefore, four water stops with different stiffness values and four working conditions are used in the calculation. The specific parameters of each condition are displayed in Table 4.

Table 4. Working conditions of calculation of seismic performance of immersed tunnel joints.

Peak Ground Acceleration	Type of Water Stop	Joint Stiffness Coefficient				Working Condition
		Tensile Stiffness 10^6 (N/m)	Compressive Stiffness 10^6 (N/m)	Longitudinal Flexural Stiffness 10^{11} (N/m·rad)	Transverse Flexural Stiffness 10^{10} (N/m·rad)	
SL2-0.3 g	320-370-51	150	1342	2.52	150	1
	320-370-62	207	1845	3.45	207	2

3.3.3. Analysis of Results

(1) Interlayer Displacement Angle

The interlayer displacement angles under dynamic and static conditions are superimposed, and the calculation results are shown in Table 5. From the table, the maximum displacement angle is 1/9429. Moreover, within the scope of the standard requirements, the interlayer displacement angle of the middle partition wall is always greater than that of the left and right side walls. From working condition 1 to 2, the stiffness of the joint water stop gradually increases while its interlayer displacement angle is gradually reduced, indicating that the interlayer displacement angle of the joint decreases with the increase in joint stiffness.

(2) Opening Value of Joints

Joints No. 1 and No. 2 of the immersed tunnel are located on a soil layer of varying thickness, so that the opening limits of the two joints are different under the same peak ground acceleration with identical stiffness. It can be seen from Figures 18 and 19 that the opening values of joints No. 1 and No. 2 under the two working conditions show a similar trend over time. The opening value is larger between 5 and 13 s, during which its changing value is more obvious. Hence, it can have a relatively serious impact on the joint water stop, which easily leads to the weakening of the water tightness of the immersed tunnel.

Table 5. Interlayer displacement angle under different working conditions.

Working Condition	Joint No. 1			Joint No. 2			Code Limit Value
	Left Side Wall 5	Middle Partition Wall 6	Right Side Wall 7	Left Side Wall 5	Middle Partition Wall 6	Right Side Wall 7	
Working condition 1	1/11579	1/10154	1/11000	1/10645	1/9429	1/10233	1/550
Working condition 2	1/15349	1/12692	1/14043	1/13469	1/11786	1/13018	

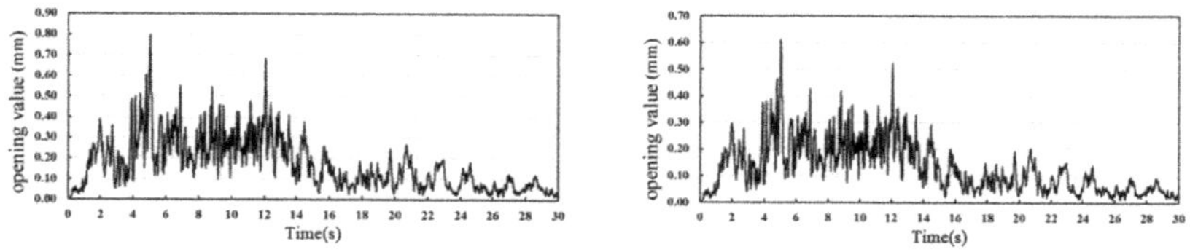

(a) (b)

Figure 18. Opening values of joint No. 1: (**a**) opening value under working condition 1; (**b**) opening value under working condition 2.

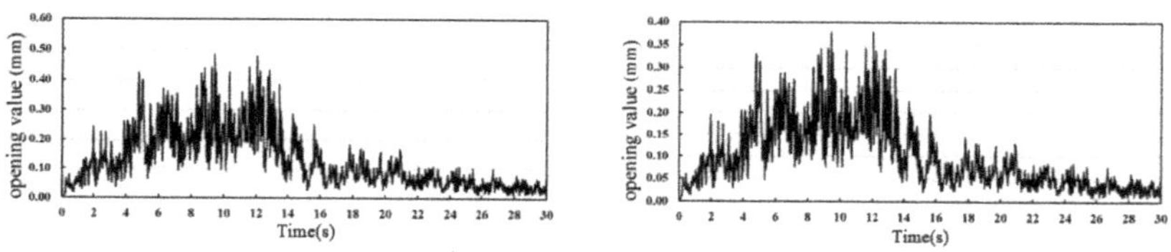

(a) (b)

Figure 19. Opening values of joint No. 2: (**a**) opening value under working condition 1; (**b**) opening value under working condition 2.

After the superposition of dynamic and static working conditions, the comprehensive opening value is obtained. Under working condition 1, the maximum opening value of joint No. 1 is 0.85 mm, while that of joint No. 2 is 0.49 mm. Under working condition 2, the former is 0.66 mm, while the latter is 0.38 mm. On the whole, the opening value under the dynamic condition is the main influencing factor, with that under the static condition having less effect. The opening values of joint water stops No. 1 and No. 2 under each working condition is shown in Table 6.

Table 6. Maximum opening value of joints under each working condition (unit: mm).

Joint	Working Condition	Type of Water Stop	Maximum Opening Value under Dynamic Condition	Maximum Opening Value under Static Condition	Comprehensive Opening Value
Joint No. 1	Condition 1	51	0.797	0.0496	0.8466
	Condition 2	62	0.611	0.0496	0.6606
Joint No. 2	Condition 1	51	0.486	0.0014	0.4874
	Condition 2	62	0.379	0.0014	0.3804

(3) Compression–Tension Value and Sensitivity Analysis of GINA Water Stop

The compression and tension values of GINA water stops at each joint are shown in Figures 20 and 21. Positive values indicate tension, while negative values indicate compression. An analysis of the change curves for joints No. 1 and No. 2 under the two working conditions reveals that joint No. 1 experiences the largest compression and tension, with maximum compression values of 0.55 mm and 0.73 mm and maximum tension values of 0.41 mm and 0.54 mm. Additionally, from working condition 1 to 2, the maximum compression values for joints No. 1 and 2 decrease sequentially, and their tension values significantly exceed their compression ones. This is attributed to the compressive stiffness of the joint water stop being greater than its tensile stiffness.

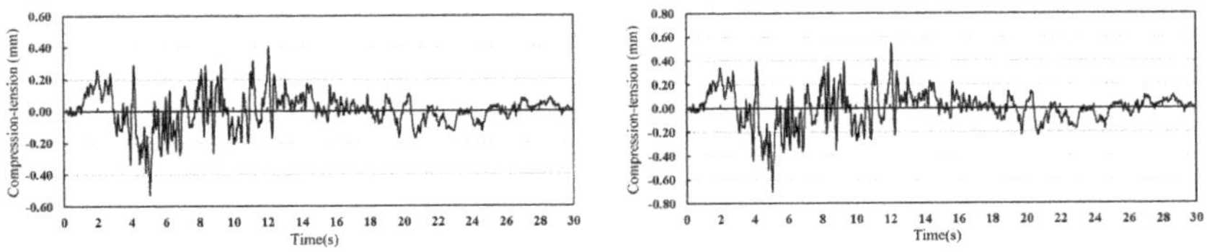

(a) (b)

Figure 20. Compression–tension values of joint No. 1: (**a**) compression–tension value under working condition 1; (**b**) compression–tension value under working condition 2.

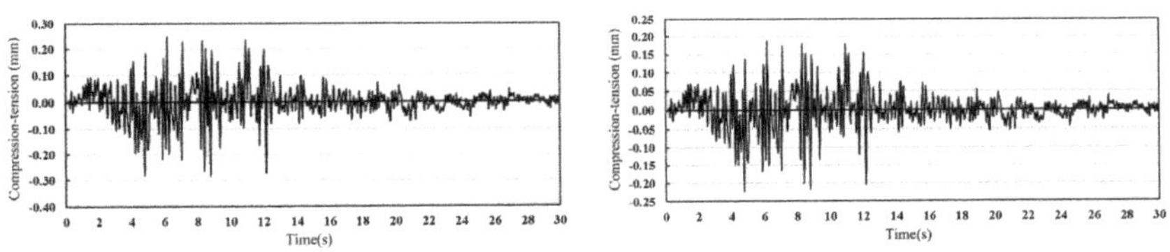

(a) (b)

Figure 21. Compression–tension values of joint No. 2: (**a**) compression–tension value under working condition 1; (**b**) compression–tension value under working condition 2.

So as to deeply analyze the response rule of the joint water stop of an immersed tunnel under seismic conditions, this paper also sets up two different water stop stiffness values as working conditions for calculation and discusses the impact of the different stiffness values of the GINA water stop on the compression and tension values of the immersed tunnel.

Figure 22 depicts the changes in the compressive stiffness of the maximum compression–tension values of the joint water stop under a peak ground acceleration of 0.3 g. When the compressive stiffness value of both joint No. 1 and joint 2 is 819 × 106 N/m, the maximum tension of joint No. 1 is far greater than that of joint No. 2. When the tensile stiffness is 92 × 106 N/m for both joints, the maximum tension value of joint No. 1 is also much greater than that of joint No. 2, which denotes that the thickness of the highly weathered biotite granite and medium-weathered biotite granite on the foundation where joints No. 1 and No. 2 are located has a large bearing on the maximum compression–tension value of the joint. Furthermore, with the increase in compressive and tensile stiffness of the GINA water stop, its maximum compression–tension value gradually decreases, the speed of which progressively slows down. Lastly, the maximum tension value of joint No. 1 is significantly greater than that of joint No. 2, implying that as the compressive and tensile stiffness values increase to a certain extent, the influence of the change in stiffness value on the value of the maximum tension of the joint gradually declines.

It is also worth noticing that there is no definite value for the ultimate compression–tension value of the immersed tunnel's joint stop belt by now, which mainly depends on factors including joint structure, the material of the water stop, construction errors, and so on. At present, among the seismic codes of underground engineering structures, only the "Code for Seismic Design of Highway Tunnel" [33] has made corresponding provisions on joint opening values for immersed tunnels, stipulating that the opening value should not exceed 40 mm under E2 earthquake conditions. Accordingly, the seismic calculation of the opening values in this paper is within the allowable range.

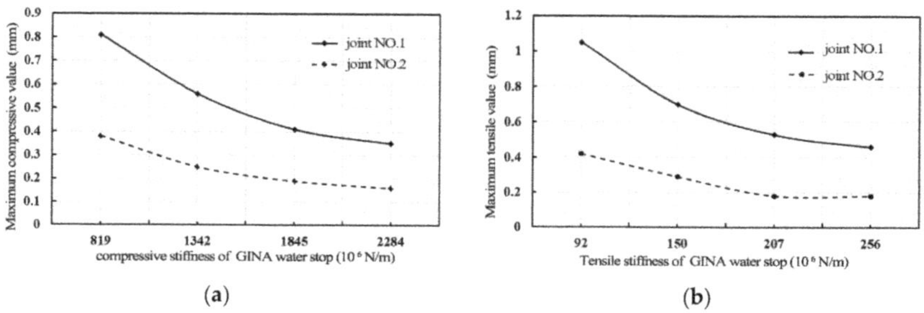

Figure 22. Relationship between the stiffness of GINA water stop and the maximum compression–tension value: (**a**) maximum compression value; (**b**) maximum tension value.

4. Conclusions

In this paper, the beam–spring model and the shield tunnel–foundation three-dimensional finite element model is established based on the water-intake project of an immersed tunnel in a nuclear power sea area. Additionally, we implemented secondary development in ANSYS to create equivalent linear dynamic constitutive and viscoelastic boundary element subroutines. Our research focused on the seismic performance of the immersed tunnel under ultimate seismic ground motion (SL-2), with particular attention to the seismic analysis of the tunnel joints. The primary findings are as follows:

(1) Using the response displacement method, we found that increases in joint stiffness and peak ground acceleration do not affect the variation trend of joint openings along the longitudinal position of the immersed tunnel. However, significant changes in joint openings are observed at locations where the overburden thickness changes abruptly or where there is an uneven distribution of soft and hard soil. Additionally, higher stiffness of the joint GINA water stop results in a smaller maximum opening value and a reduced maximum relative angle of the joint.

(2) Variations in soil thickness on either side of the foundation and tunnels significantly impact the maximum compression–tension value of the immersed tunnel joint water stop. As the stiffness of the GINA water stop increases, the compression–tension value gradually decreases, with diminishing reductions at higher stiffness levels. When the stiffness exceeds a certain threshold, its influence on the compression–tension value weakens.

(3) The deformation of the immersed tunnel is influenced by the surrounding soil layer's deformation. To mitigate forced displacement of the relatively rigid immersed tunnel due to uneven geological conditions, especially the adverse effects of ground motion, flexible joint structural measures should be adopted based on the actual foundation conditions. Additionally, it is essential to fully mobilize the supporting forces of the surrounding rocks to accommodate uneven soil layer deformation during earthquakes.

(4) The analysis results indicate that time-history analysis provides relatively quantitative and accurate results for seismic response analysis of structures such as immersed tunnels. Qualitative analyses can also be performed using the response displacement method in practical applications. Future research should focus on developing calculation programs for the generalized response displacement method.

Author Contributions: Resources, J.Z.; project administration, J.Z.; funding acquisition, J.Z.; methodology, J.Z., B.Q. and C.G.; conceptualization, J.Z., B.Q. and C.G.; writing—original draft, B.Q.; writing—review and editing, J.Z., B.Q., J.W. and Y.P.; software, C.G.; data curation, C.G.; visualization, J.W. and Y.P. All authors have read and agreed to the published version of the manuscript.

Funding: Key Project of the National Natural Science Foundation of China [51738010] and Dalian Science and Technology Innovation Fund Project [2021JJ13SN83].

Institutional Review Board Statement: Not applicable.

Informed Consent Statement: Not applicable.

Data Availability Statement: The raw data supporting the conclusions of this article will be made available by the authors on request.

Conflicts of Interest: Author Changjiang Gan was employed by the company Gansu Comprehensive Railway Engineering Contracting Co., Ltd. The remaining authors declare that the research was conducted in the absence of any commercial or financial relationships that could be construed as a potential conflict of interest.

References

1. Yan, W.; Xie, Z.; Zhang, X.; Wang, J.; Gao, X. Tests for compartmental particle damper's a seismic control in an immersed tunnel. *J. Vib. Shock* **2016**, *35*, 7–12.
2. Cheng, X.; Jing, L.; Cui, J.; Li, Y.; Dong, R. Research of shaking table model tests on immersed tunnels under different conditions. *J. Southwest Jiaotong Univ.* **2017**, *52*, 1113–1120.
3. Zhang, X.; Yan, W.; Chen, Y.; Chen, S.C.; Chen, H.J. Shaking tables tests for an immersed tunnel model considering wave passage effect. *J. Vib. Shock.* **2018**, *37*, 76–84.
4. Yuan, Y.; Yu, H.; Xiao, W.; Xu, G.; Ji, Z. Experimental failure analysis on concrete shear keys in immersion joint subjected to compression-shear loading. *Eng. Mech.* **2017**, *34*, 149–181.
5. Okamoto, S.; Tamura, C. Behaviour of Subaqueous Tunnels During Earthquakes. *Earthq. Eng. Struct. Dyn.* **1973**, *1*, 253–266. [CrossRef]
6. Chen, H.-J. Theoretical Analysis and Experimental Research on Seismic Performance of Immersed Tunnel and Its Joints. Ph.D. Thesis, Beijing University of Technology, Beijing, China, 2014.
7. Li, Y.; Tian, Y.; Zong, J. Shaking table test of a tunnel-soil system considering the influence of surface buildings under multi-dimensional seismic waves. *Chin. J. Rock Mech. Eng.* **2022**, *41*, 1453–1465.
8. Li, Y.; Tian, Y.; Zong, J. Shaking table test of soil and tunnel considering surface construction under multi-dimensional earthquake. *Chin. J. Rock Mech. Eng.* **2022**, *41*, 1–13.
9. Peng, H.; Meng, G.; Ding, Q.; Xu, F. The influence of hydrodynamic pressure on immersed tunnel under earthquake excitation. *J. Shanghai Jiaotong Univ.* **2008**, 1027–1031. [CrossRef]
10. Chen, W.; Lv, Z.; Xu, L.; Ruan, B.; Ma, J.; Chen, X. Seismic response of subsea tunnels considering seawater seabed coupling effect. *J. Eng. Geol.* **2021**, *29*, 1878–1886. [CrossRef]
11. Cui, J.; Zhou, P.; Li, Y.; Ouyang, Z. Earthquake dynamic response analysis of seabed under the action of immersed tunnel. *Earthq. Eng. Eng. Dyn.* **2016**, *4*, 96–102. [CrossRef]
12. Xiao, W.; Chai, R.; Yu, H.; Yuan, Y. Simulation method for nonlinear mechanical properties of immersed tunnel joint. *Mech. Eng.* **2014**, *36*, 757–763.
13. Zhang, X.; Zhao, G.; Ye, G.; Wang, J. Simplified method and three-dimensional finite element analysis of quake-proof for immersed tunnel joints. *Chin. J. Undergr. Space Eng.* **2011**, *7*, 1292–1297+1402.
14. Yu, H.; Song, Y.; Li, Y.; Zhang, S.; Xu, L. Multi-scale method and seismic response analysis of immersed tunnel. *J. Tongji Univ. Nat. Sci.* **2021**, *49*, 807–815.
15. He, C.; Xu, G.; Zhang, Z. Mechanical properties of shear keys of segment joints in immersed tunnels under seismic loading. *J. Northeast. Univ. Nat. Sci.* **2021**, *42*, 871–878.
16. Cui, J.; Lu, Y.; Qu, J.; Li, Y. Analysis of the Affecting Factors for Seismic Response of Immersed Tunnel. *J. Southwest Jiaotong Univ.* **2020**, *55*, 1224–1230.
17. Bai, L.; Zhao, X.; Du, X.; Yuan, Y.; Liu, H. Seismic Response of Joints of Immersed Tunnels. *J. Disaster Prev. Mitig. Eng.* **2015**, *35*, 153–159+165. [CrossRef]
18. Hu, Z. *Research on Action Mechanisms and Structural Properties of Segmental Joint Shear Keys on Immersed Tunnel*; Chang'an University: Xi'an, China, 2015.
19. Lin, W.; Liu, X. Analysis of GINA uneven compression during hydraulic connection of immersed tunnel element at curved plane design line. *China Harb. Constr.* **2016**, *36*, 51–53.
20. Liu, Y.; Wang, X.; Xu, Y. Technology for installation of GINA gasket. *China Harb. Eng.* **2016**, *36*, 66–68.
21. Bai, Y.; Lu, H. Damage Analysis and Repair Technology of OMEGA Gasket in Immersed Tunnel. *J. Railw. Eng. Soc.* **2016**, *33*, 87–92.
22. Chen, Y.; Zhang, B.; Wang, S.; Liu, Q.; Song, C. Research status and Engineering application of rubber water stop for immersed tunnel. *Spec. Purp. Rubber Prod.* **2012**, *33*, 60–64.
23. Zhang, W.; Yan, H. Development of New type Omega rubber water stop. *Spec. Purp. Rubber Prod.* **2014**, *35*, 50–52.
24. Liu, Z.; Huang, H. Performance Evaluation and Safety Pre-warning of GINA in Immersed Tube Tunnel. *Chin. J. Undergr. Space Eng.* **2009**, *5*, 347–353.

25. Zhang, Y.; Wang, H.; Guo, J.; Deng, X.; Luo, J. Vertical Compression Shear Mechanical Performance Study of Immersed Tunnel Joint. *Chin. J. Undergr. Space Eng.* **2016**, *12*, 24–31.
26. Wang, X. *ANASYS Structure Analysis and Application*; China Communication Press: Beijing, China, 2011.
27. *GB 50267-2019*; Standard for Seismic Design of Nuclear Power Plants. China Planning Press: Beijing, China, 2019.
28. *ASCE 4-98-2000*; Seismic Analysis of Safety-Related Nuclear Structures and Commentary. American Society of Civil Engineers: Reston, VA, USA, 2000.
29. Cao, J.; Tao, G.; Deng, Y. Research on dynamic response of ship cabin structure based on modified Housner model. *Water Transp. Eng.* **2021**, 117–122+155.
30. Guan, Q.; Fei, W.; Liu, J. Secondary development and application of equivalent linear viscoelastic model based on ANSYS. *Water Resour. Power* **2022**, *40*, 100–104.
31. Ma, S.; Chi, M.; Chen, H.; Chen, S. Implementation of viscoelastic artificial boundary in ABAQUS and comparative study of seismic input methods. *Chin. J. Rock Mech. Eng.* **2020**, *39*, 1445–1457.
32. *NB/T 25046-2015*; Code for Hydraulic Design of Nuclear Power Plants. China Planning Press: Beijing, China, 2015.
33. *JTGT 2232-2019*; Specifications for Seismic Design of Highway Tunnels. China Communication Press: Beijing, China, 2019.

applied sciences

MDPI

Article

Research on Seismic Performance of a Two-Story, Two-Span Underground Subway Station with Split Columns Based on the Quasi-Static Method

Zigang Xu [1,*], Zongyao Xia [1], Xiangbo Bu [2] and Runbo Han [3]

1 State Key Laboratory of Performance Monitoring and Protecting of Rail Transit Infrastructure, East China Jiaotong University, Nanchang 330013, China
2 Civil and Environmental Engineering Department, Technical University of Catalonia, 08034 Barcelona, Spain
3 College of Resources, Environment and Architectural Engineering, Chifeng University, Chifeng 024000, China; hanrunbo1987@126.com
* Correspondence: xuzigang1027@163.com

Abstract: The behavior of center columns in shallow-buried underground subway station structures resembles that of high-rise buildings. In both cases, these columns experience significant vertical loads during earthquake events and are susceptible to brittle failure due to inadequate deformation capacity. In this study, the design concept of split columns, commonly employed in high-rise structures, is adapted for application in a two-story, two-span subway station. Initially, a comparative analysis was conducted using quasi-static pushover analysis to assess the horizontal deformation characteristics of traditional and split columns under high axial loads. Subsequently, a comprehensive quasi-static pushover analysis model encompassing the soil–structure interaction was formulated. This model was employed to investigate differences in seismic performance between traditional and innovative underground structures, considering internal forces, deformation capacity, and plastic damage of crucial elements. The analysis results demonstrate that the incorporation of split columns in a two-story, two-span subway station enhances the overall seismic performance of the structure. This enhancement arises from the fact that split columns mitigate excessive shear forces while effectively utilizing their vertical support and horizontal deformation capacities.

Keywords: underground subway station; split column; quasi-static pushover analysis; vertical earthquake load; horizontal deformation capacity

1. Introduction

In the field of urban transportation, the integration of a high-capacity rapid public transportation system, centered around subway projects, assumes a pivotal role in the establishment of a contemporary and all-encompassing transportation network. Benefiting from the policy support, the urban rail transit sector in China is experiencing a rapid and comprehensive expansion, evidenced by substantial growth in both the length and quantity of operational routes. By the end of 2022, a total of 308 operational urban rail transit lines had been inaugurated across 55 cities in China, reflecting an annual rise of 8.8%. Impressively, the cumulative length of these operational lines had expanded to reach 10,287.45 km, marking a noteworthy year-on-year increase of 11.7%. This substantial mileage places China at the forefront globally in terms of total operational urban rail transit network length.

In 2016, China issued and implemented Seismic Ground Motion Parameters Zonation Map of China (GB18306-2015) [1], which completely eliminated the non-seismic zones. Most cities with a subway under construction in China are currently located in strong seismic activity zones, which means that underground structures in service, under construction, or proposed face seismic threats at any time. The historical records of seismic damage to structures demonstrate a noteworthy resemblance between underground constructions and surface buildings. When subjected to the effects of strong seismic events, underground structures are susceptible to varying degrees of damage and disruption [2]. For example, the Great Hanshin earthquake in 1995 caused the complete destruction of more than 30 columns in the Daikai subway station in Kobe City, resulting in the collapse of and damage to the overall structure [3]. In addition, the Wenchuan earthquake and the Chi-Chi earthquake caused damage to numerous urban underground structures and mountain tunnels [4,5]. In response to the complete collapse of the Daikai subway station in the Great Hanshin earthquake, scholars around the world have carried out research work from theoretical analysis, model tests, and numerical simulations [6–10]. Regarding the seismic damage mechanism of the Daikai station, many scholars believe that the overburden on the underground structure tends to lose its structural stability under the action of ground shaking, especially the inertial force of the vertical component effects of earthquakes. The incongruous deformation capacity of the sidewalls and center columns caused by the change in force distribution results in damage to the structure [11,12].

In order to protect underground structures from earthquake damage, the traditional seismic design generally aims to strengthen the structure itself. For example, in the structural design of the Daikai station, the safety factor of the center column is about 3. In addition, the seismic control techniques for underground structures can be divided into two main categories: one is used to install seismic damping materials around the structure; the other is used to install appurtenant devices to protect key components. Gao et al. [13] analyzed the effect of damping materials on the dynamic response of tunnels, and concluded that the smaller the ratio of elastic modulus between the damping layer and the surrounding rock, the more obvious the seismic damping effect. Zhao et al. [14] and Chen et al. [15] proposed a seismic isolation layer for underground structures and carried out theoretical analyses and experimental studies. Starting from the Daikai station, Ma et al. [16] and Xu et al. [17] studied the seismic damping effect of friction bearings at the top of the center column. Chen et al. [18,19] discussed the damping effect of the shear plate damper and lead-core rubber bearing installed on the center columns in subway stations and found that they could effectively reduce the structural internal forces. Du et al. [20] and Chen et al. [21] also proposed the use of a swaying column to mitigate the possible damage to underground structures. In addition, Yue et al. [22] and Ma et al. [23] strengthened the center column with carbon fiber and verified its effect by numerical simulation and experimental studies. Bu et al. [24,25] evaluated the seismic performances of underground structures with either a transverse traditional rigid layout or sliding interior columns via numerical analysis.

Similarly, the center columns in high-rise buildings generally need to bear large axial pressure. Such reinforced concrete columns have a relatively large shear span and are prone to shear damage. Li et al. [26,27] introduced the techniques of split columns, including the bearing capacity, seismic performance of split column frames and nodes, design, and construction recommendations. Hao et al. [28] conducted a series of experiments involving cyclic lateral loading on split columns and subsequently proposed a computational model to assess their load-carrying capacity. Li [29] used ABAQUS finite element software to analyze the effect of the separation ratio on the damage mode and load carrying capacity of split columns. In cases where the load-carrying capacity is comparable to that of monolithic columns, split columns exhibit a significant reduction in lateral stiffness and a notable increase in ductility, thereby demonstrating improved deformation and energy dissipation capabilities [30,31]. This feature of split columns is suitable for mitigating seismic hazards of underground structures. Du et al. [32] proposed a US patent for the application of split columns in underground structures, and Xu et al. [33] initially explored the seismic reduction effect of split columns in underground structures through integral dynamic time analysis.

In this study, taking a two-story and two-span subway station as a research case, the split column technology was adopted from the perspective of improving the seismic performance of underground structures. Based on the existing studies, the three-dimensional quasi-static pushover analysis models of the center column and soil–structure system were established using ABAQUS/CAE V 6.14 [34]. Additionally, the differences in seismic performance between the prototype and the new underground structure were compared from the perspective of the internal forces, deformation capacity, and plastic damage of critical members, which provide reference and guidance for the application of the split column technology in underground structures.

2. The Split Column Technology

The split columns are several independent unit columns divided using partition plates, and each unit column is independently reinforced. The structural diagram of split columns is shown in Figure 1. When cast-in-place reinforced concrete columns are used, partition plates are generally set between unit columns, and there is a bond between the unit column and the partition plates. In terms of assembled reinforced concrete columns, partition plates are not necessary, and frictional contact between unit columns is produced directly. Experimental and theoretical studies [27–31] have shown that when the split column technology is used in the ground buildings, although the flexural bearing capacity of the column is reduced, the shear bearing capacity remains basically unchanged, and the deformation capacity and ductility are significantly improved.

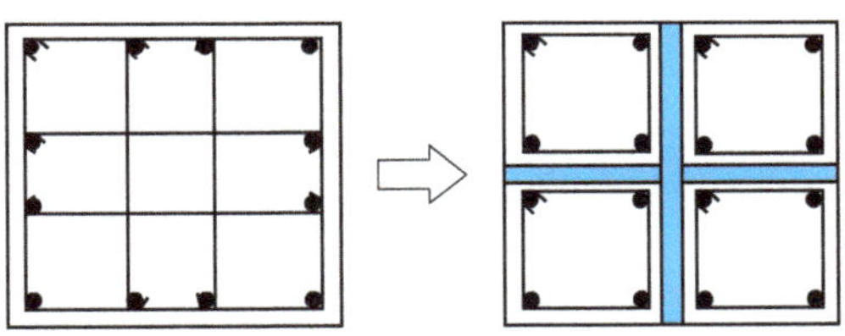

Figure 1. Structural diagram of split columns.

Taking the cast-in-place reinforced concrete column as an example, the loading process can be divided into three stages according to the degree of friction effect between the partition plate and the unit column. The strain distribution law of concrete in the cross section of the split column at different stress stages is shown in Figure 2, where w_1 is the width of the unit column and w_2 is the width of the partition plate. When the horizontal load on the split column is small, there is an effective bond between the partition plate and the concrete. The unit column on both sides of the partition plate is guaranteed to deform in a coordinated manner, and the strain distribution of the cross-section is similar to that of the monolithic column at that time. Furthermore, when the horizontal load increases to a certain degree, the bond between the partition plate and the concrete is gradually broken. The friction between the two restrains the deformation of the concrete on both sides of the partition plate, resulting in a smaller strain at the inner edge of the unit column. When the horizontal load continues to increase, the bond between the partition plate and the concrete is completely destroyed, and the friction between them is completely eliminated. The partition plate will no longer restrain the deformation of the concrete on both sides. Therefore, each unit column works independently, and the neutral axes of each unit columns move to their section center, respectively.

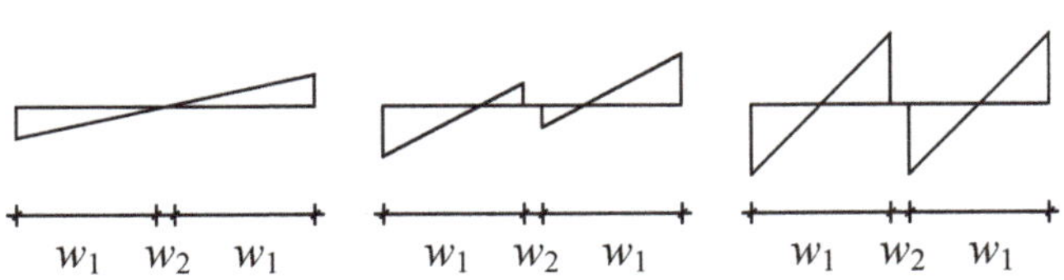

Figure 2. The strain distribution of the split column.

According to the distribution characteristics of strain within the cross section of the split column, the force of the split column is simplified as shown in Figure 3. Assuming that the height of the column is H, forces causing the bending deformation of the unit column are the horizontal load P and the shear force Q_f, where Q_f is the distributed shear force along the column height (i.e., the interaction friction between unit column and partition plate). When the bond between the unit column and the partition plate is intact or with large friction, the Q_f would be larger, and the loading condition of the column is closer to the monolithic one. When there is no friction between the unit column and the partition plate, Q_f is 0 with the unit columns working independently.

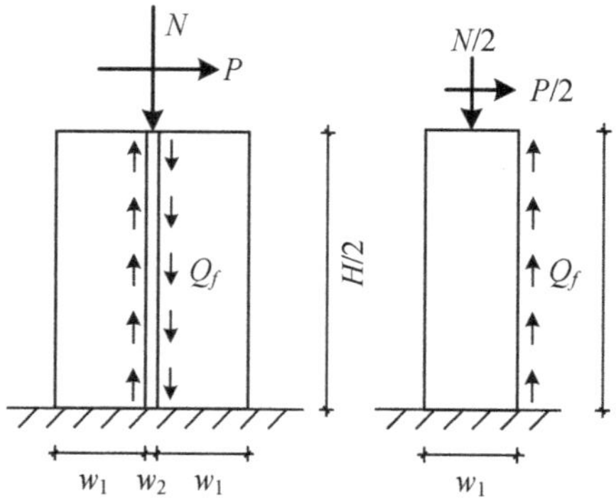

Figure 3. Analysis of the load condition of the split column.

3. The Overview of the Subway Station

The engineering case of the subway station is shown in Figure 4, which is a typical two-story and two-span rectangular frame structure. The dimensions of the subway station measure 18.0 m in width and 11.9 m in height. The burial depth of the station is 5 m. The thicknesses of sidewalls, the top, middle, and bottom slabs are 800 mm, 750 mm, 450 mm, and 800 mm, respectively, and the cross section of the center column is 800 mm × 800 mm. In the longitudinal direction of the station, the axis spacing of the center column is 6 m. The reinforcement ratios for the left and right sidewalls as well as the top and bottom slabs are 1.41%, while they are 1.26% and 2.69% for the middle slabs and columns, respectively. The reinforcement of the traditional monolithic center column is shown in Figure 5a. In addition, the split column applied in this station structure is the traditional center column equally separated into four-unit columns, and the new seismic reduction system is called split column structure. The split-column structure is identical to the prototype structure in all aspects, including member size and cross-sectional reinforcement rate, except for the cross-sectional form of the center column. As shown in Figure 5b, each unit column in the split column is equipped with eight reinforcement bars with the diameter of 26 mm, and the reinforcement rate of the whole section is 2.65%, which is comparable to that of the original center column.

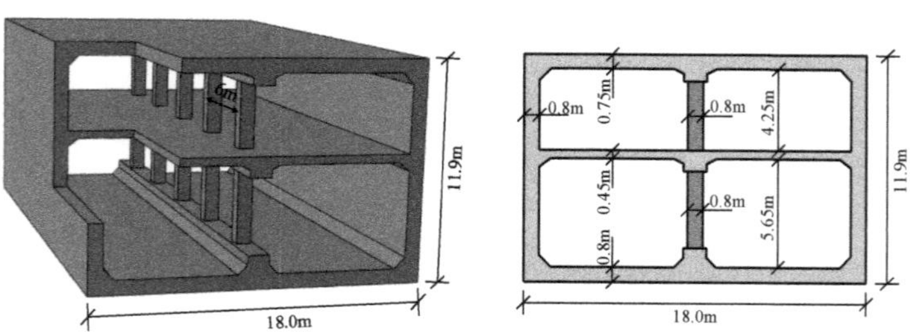

Figure 4. Three-dimensional diagram of an underground station.

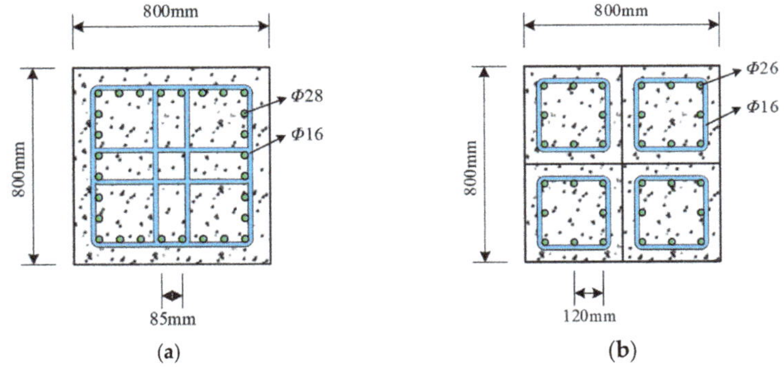

Figure 5. Three-dimensional diagram of an underground station. (**a**) Original center column reinforcement. (**b**) Split column reinforcement.

4. Finite Element Model

In this study, the seismic performance of the subway station structure is examined by using the quasi-static pushover analysis method. Considering the efficiency and the influence of calculation error, the length of the soil and structure model along the longitudinal direction of the station is 30 m in the soil–structure system. As shown in Figure 6, the total width of the soil–structure model is 140 m, which meets the requirement of more than seven times of the structural width. The model is fixed at the bottom, while the lateral boundaries are bundled, i.e., the soil nodes at the same height are bundled together for a consistent movement [35]. Both soil and structure are discretized by using C3D8R elements in ABAQUS, and the finite element mesh of soil is finer at the location near the station structure, in order to reduce the calculation error. Friction contact is set between the soil and the structure with kinetic friction coefficient taken as 0.4 [36,37]. In order to simplify the numerical model, the partition plate unit is not established in this model. The adjacent sides of each unit column are directly set as friction contact as well, and the kinetic friction coefficient is taken as 0.8 [38]. When assessing the internal force response of the central column, the analysis is conducted on the middle section of the subway station. The center column and section numbers are shown in Figure 6, where C1 and C2 denote the upper column and lower column, and S1 and S2 denote the bottom column sections, respectively.

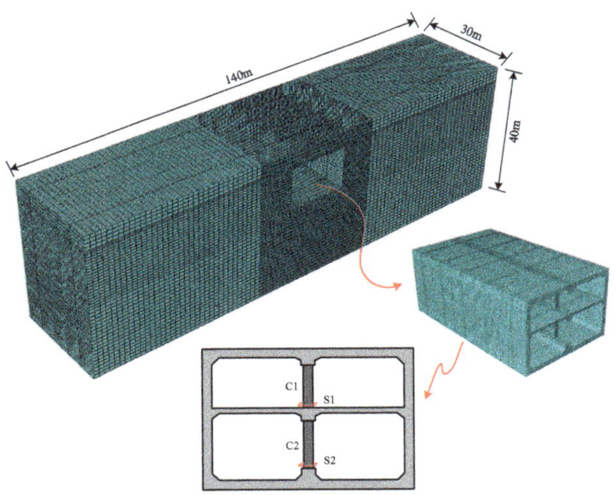

Figure 6. Finite element model for pushover analysis of the soil–structure system.

Concrete is simulated by creating solid elements and using the Concrete Damage Plasticity model proposed by Lee et al. [39] and Lubliner et al. [40], which describes the change in concrete stiffness in tension and compression by introducing a tensile damage factor and a compressive damage factor. Its material properties are defined in ABAQUS software as shown in Tables 1–3 and Figure 7 [41]. In addition, the reinforcement rebar is simulated by creating truss elements with an ideal elasto-plastic intrinsic model. The initial elastic modulus, Poisson's ratio, and yield stress is taken as 200 GPa, 0.3, and 335 MPa, respectively. The reinforcement is placed in the station concrete by embedding method, and the slip between the reinforcement and concrete is not considered in the numerical calculation process [42].

Table 1. Material properties of the concrete.

Parameters	Value	Parameters	Value
Density	2450 kg/m^3	Limited compressive yield stress	20.1 MPa
Elastic modulus	30 GPa	Initial tensile yield stress	2.4 MPa
Poisson's ratio	0.18	Compression stiffness recovery parameter	1
Dilation angle	36.31°	Tensile stiffness recovery parameter	0
Initial compressive yield stress	13 MPa		

Table 2. Compressive stress and damage factor versus plastic strain of the concrete.

Plastic Strain	Compressive Stress (MPa)	Compressive Damage Factor	Plastic Strain	Compressive Stress (MPa)	Compressive Damage Factor
0	14.64	0	2.4×10^{-3}	17.25	0.566
4.0×10^{-4}	17.33	0.113	3.6×10^{-3}	12.86	0.714
8×10^{-4}	19.44	0.246	5.0×10^{-3}	8.66	0.824
1.2×10^{-3}	20.10	0.341	7.5×10^{-3}	6.25	0.922
1.6×10^{-3}	20.18	0.427	1.0×10^{-2}	3.98	0.969
2.0×10^{-3}	18.72	0.501			

Table 3. Tensile stress and damage factor versus cracking displacement of the concrete.

Cracking Displacement (mm)	Tensile Stress (MPa)	Tensile Damage Factor	Cracking Displacement (mm)	Compression Stress (MPa)	Tensile Damage Factor
0	2.400	0	0.308	0.219	0.944
0.066	1.617	0.381	0.351	0.147	0.965
0.123	1.084	0.617	0.394	0.098	0.978
0.173	0.726	0.763	0.438	0.066	0.987
0.220	0.487	0.853	0.482	0.042	0.992

The site soil is divided into six layers, and the vertical distance from the soil surface to the bedrock is 40 m. The Davidenkov model is used for the site soil, which has been applied to the simulation of sandy soil liquefaction analysis of complex non-uniform sites, structural damage evaluation of large-scale subway stations, simulation of shield tunnels, and has shown excellent applicability [43–45]. The material parameters of each soil layer are shown in Table 4., and the shear modulus ratio varies with shear strain as follows:

$$\frac{G}{G_{\max}} = 1 - \left[\frac{(\gamma/\gamma_0)^{2B}}{1 + (\gamma/\gamma_0)^{2B}} \right]^A \tag{1}$$

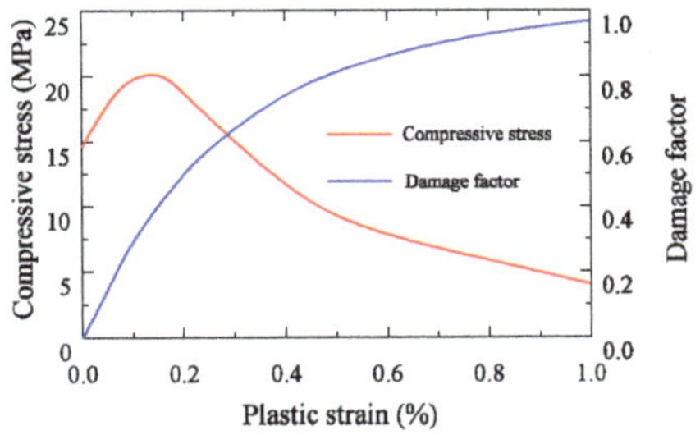

(a)

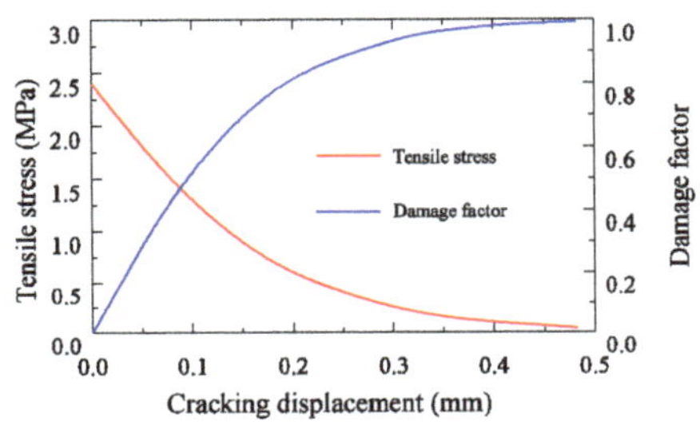

(b)

Figure 7. Material parameters of concrete. (**a**) Compressive behavior of concrete. (**b**) Tensile behavior of concrete.

Table 4. Parameters of soils.

Soil Layer	Thickness (m)	Density (kg/m^3)	Shear Wave Speed (m/s)	Poisson's Ratio	A	B	γ_0 (10^{-4})
Soil layer I	4	1900	200	0.3	1.02	0.35	4.0
Soil layer II	4	1950	260	0.3	1.05	0.34	3.5
Soil layer III	4	1980	310	0.3	1.10	0.35	3.8
Soil layer IV	8	1950	335	0.3	1.10	0.35	3.8
Soil layer V	10	2000	430	0.3	1.10	0.35	3.8
Soil layer VI	10	2100	520	0.3	1.20	0.35	2.5

In Equation (1), G and γ are the shear modulus and shear strain, respectively. G_{max} is the maximum shear modulus of the soil. A, B and γ_0 are the fitting parameters related to soil properties.

5. The Pushover Analysis of the Center Column

5.1. Analysis Methods

When subjected to seismic loading, the deformation mode of underground structures differs greatly from that of ground buildings. When a vertical member of a ground building (such as a bridge pier) is statically pushed over, a certain vertical pressure is first applied to achieve a predetermined axial pressure ratio, and then the member is loaded step by step through force control or displacement control until damaged. As shown in Figure 8 (left), the top of the member is often set without any restraints. For underground structures, the rotational degrees of freedom at the top of the center column are constrained to a large extent due to the restraining effect of the overlying soil and the top plate. Therefore, in order to truly reflect the deformation behavior of the center column during the earthquake, the top of the model is imposed with rotation constraints during pushover, and its loading pattern is shown in Figure 8 (right). In the pushover analysis of the center column, there are two analysis steps. Firstly, the axial pressure is applied to the top of the center column. The axial pressure considers two working conditions, one is considering only the self-weight of the soil and structure, and the other is considering both the self-weight and the vertical inertia force. In the second analysis step, the horizontal displacement is applied to the top of the center column. As shown in Table 5, the pushover analysis of the center column includes 16 cases.

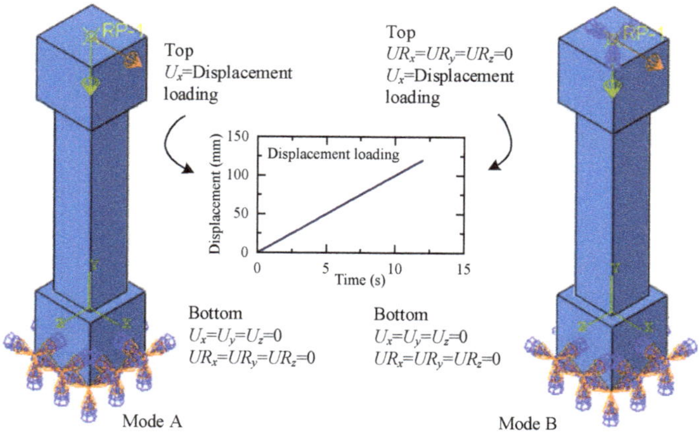

Figure 8. Pushover modes of center column.

Table 5. The pushover analysis of the working conditions of the center column.

Case	Column	Cross Section	Pushover Mode	Horizontal Load
UT-A-G	upper	traditional	Mode A	Gravity
UT-A-GE	upper	traditional	Mode A	Gravity and earthquake load
UT-B-G	upper	traditional	Mode B	Gravity
UT-B-GE	upper	traditional	Mode B	Gravity and earthquake load
US-A-G	upper	split	Mode A	Gravity
US-A-GE	upper	split	Mode A	Gravity and earthquake load
US-B-G	upper	split	Mode B	Gravity
US-B-GE	upper	split	Mode B	Gravity and earthquake load
LT-A-G	lower	traditional	Mode A	Gravity
LT-A-GE	lower	traditional	Mode A	Gravity and earthquake load
LT-B-G	lower	traditional	Mode B	Gravity
LT-B-GE	lower	traditional	Mode B	Gravity and earthquake load
LS-A-G	lower	split	Mode A	Gravity
LS-A-GE	lower	split	Mode A	Gravity and earthquake load
LS-B-G	lower	split	Mode B	Gravity
LS-B-GE	lower	split	Mode B	Gravity and earthquake load

5.2. Analysis Results

Before the pushover analysis of the center column, the axial pressure of the upper center column and lower center column need to be determined by static calculation, including the working conditions with only gravity force and with both gravity force and vertical seismic load. The calculation model under gravity conditions is consistent with the model shown in Figure 6. One of the calculation models only applies the gravity of the structure and soil, while the other model adds vertical inertial force. After preliminary calculations, the axial forces of the upper and lower center columns are about 5869 kN and 6669 kN, respectively, when the gravity load is considered. Considering the seismic condition, and with reference to the results of Xu et al. [46], it is assumed here that the axial force of the center column increases by 50% when the vertical seismic action is considered. The lateral force–drift ratio curves of the center column under different working conditions are shown in Figure 9.

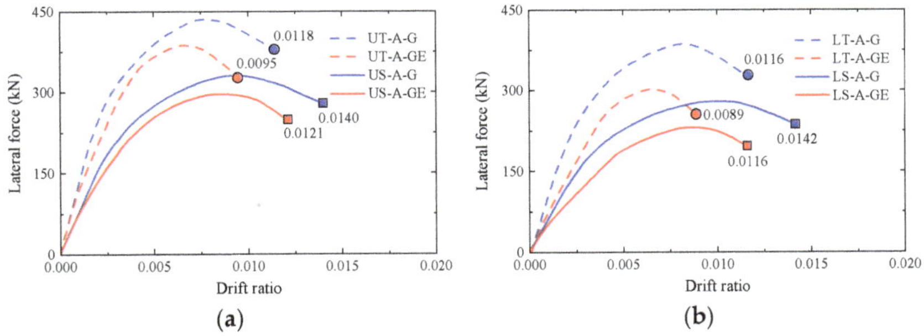

Figure 9. Lateral force–drift ratio curves of columns by Mode A. (**a**) Upper column. (**b**) Lower column.

From the lateral force–drift ratio curves, it can be seen that for the various working conditions listed in Table 5, the lateral force shows a trend of increasing and then decreasing with the increase in the lateral drift ratio. When the horizontal drift increases to a certain degree, the lateral thrust of the column reaches the peak, which does not mean that the column fails at this time. However, if the column has sufficient ductility performance, the column still has the ability to bear the horizontal load after the peak lateral thrust. Therefore, when the lateral force falls to 85% of the peak value, it is defined as the damage state; that is, the column no longer has the bearing capacity.

From Figure 9, it can be seen that the ultimate lateral drift ratio of the center column decreases when vertical inertia forces are considered for both traditional and split columns compared to the gravity load-only condition. For example, the ultimate lateral drift ratio of the lower center column is 0.0116 in the gravity load-only condition, and the ultimate lateral drift ratio of the lower center column is only 0.0089 after considering the vertical inertia force, which is about 23% lower. The vertical inertia effect of the overlying soil is one of the key factors affecting the seismic performance of the center column, which should be given sufficient attention in the seismic design process of underground structures, especially the shallow ones. When the split column is adopted, for the same level of axial load, the horizontal deformation capacity of both the upper and lower columns shows an increasing trend. For example, in working condition UT-A-GE, the ultimate lateral drift ratio of the upper column is only 0.0095, and after adopting the form of the split column section, the ultimate lateral drift rate of the upper column is increased to 0.0121, with an increase rate of about 27%.

Figure 10 further shows the lateral force–drift ratio curves of the upper and lower columns when Mode B pushover analysis is used. The comparison between Figures 9 and 10 shows that the lateral stiffness resistance of the center column is significantly increased when the top rotational freedom of the center column is constrained; moreover, the peak bearing capacity of the center column is much larger than that shown in Figure 9. Similarly, after considering both the gravity and vertical inertia forces, the deformation capacity of the traditional column is inferior, and the ultimate lateral drift ratio of the upper and lower columns are only 0.0064 and 0.0067. At this time, the lateral drift ratios of the split column when it is damaged are 0.0121 and 0.0109, respectively. Compared with the traditional center column, the ultimate lateral drift ratio of the split column is improved by more than 60%. In addition, when only the self-weight load is considered, the deformation capacity of the split column is stronger.

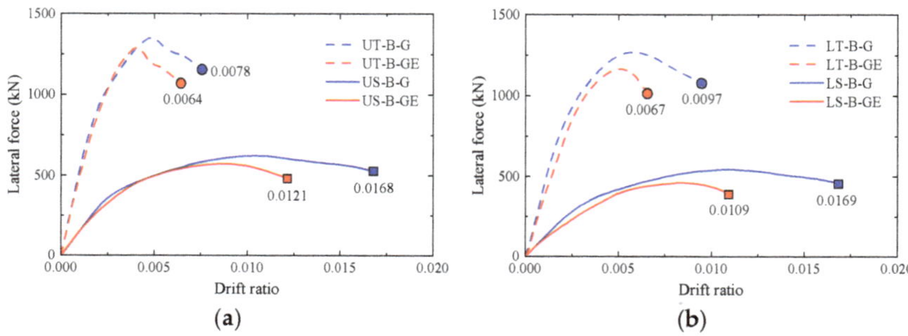

Figure 10. Lateral force–drift ratio curves of columns by Mode B. (**a**) Upper column. (**b**) Lower column.

The seismic design code for urban rail transit structures mentions that the underground structure is considered to reach a damage state when the inter-story drift angle reaches 1/250. Figures 11 and 12 show the concrete compressive and tensile damage graph for traditional and split columns under the combined action of gravity and vertical inertia forces at the lateral drift rate = 1/250 in both pushover modes. Among them, DAMAGEC represents compressive damage and DAMAGET represents tensile damage. It can be seen that the compression and tension damage to the concrete in split columns is much smaller than that of the concrete in monolithic column, which indicates that the split column still has a better load-bearing capacity at this time. In addition, by comparing the damage to the structure in the two pushover modes, it can also be found that the damage to the center column in the case of Mode A is mainly concentrated in the bottom of the column, while it is distributed at the ends of the column in the case of Mode B.

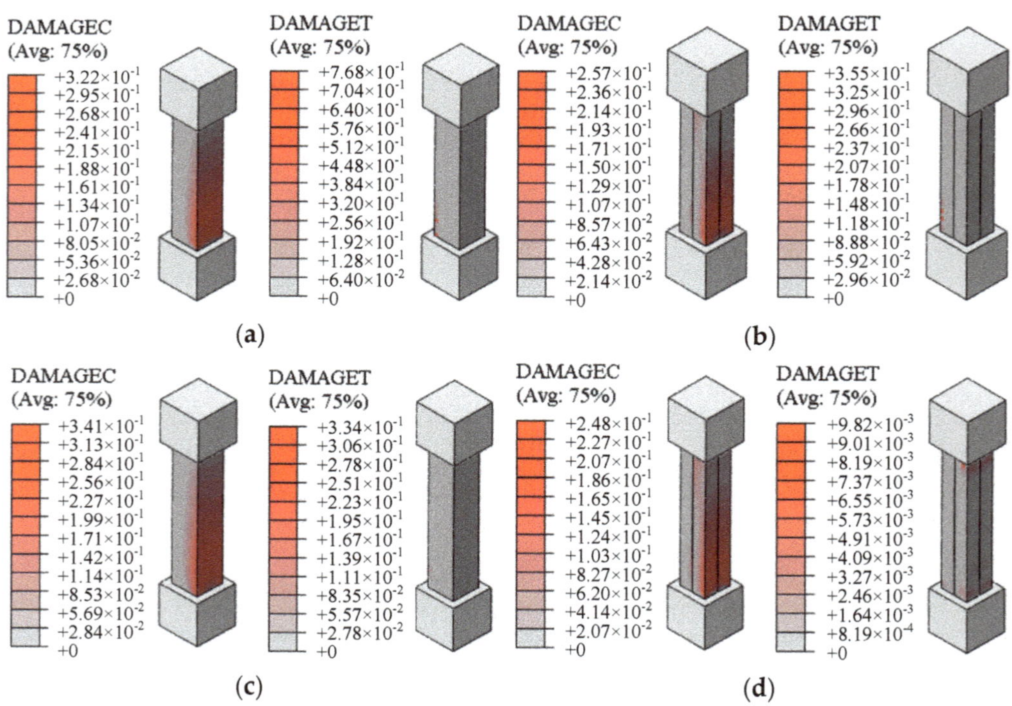

Figure 11. Concrete damage to the columns by Mode A. (**a**) UT-A-GE. (**b**) US-A-GE. (**c**) LT-A-GE. (**d**) LS-A-GE.

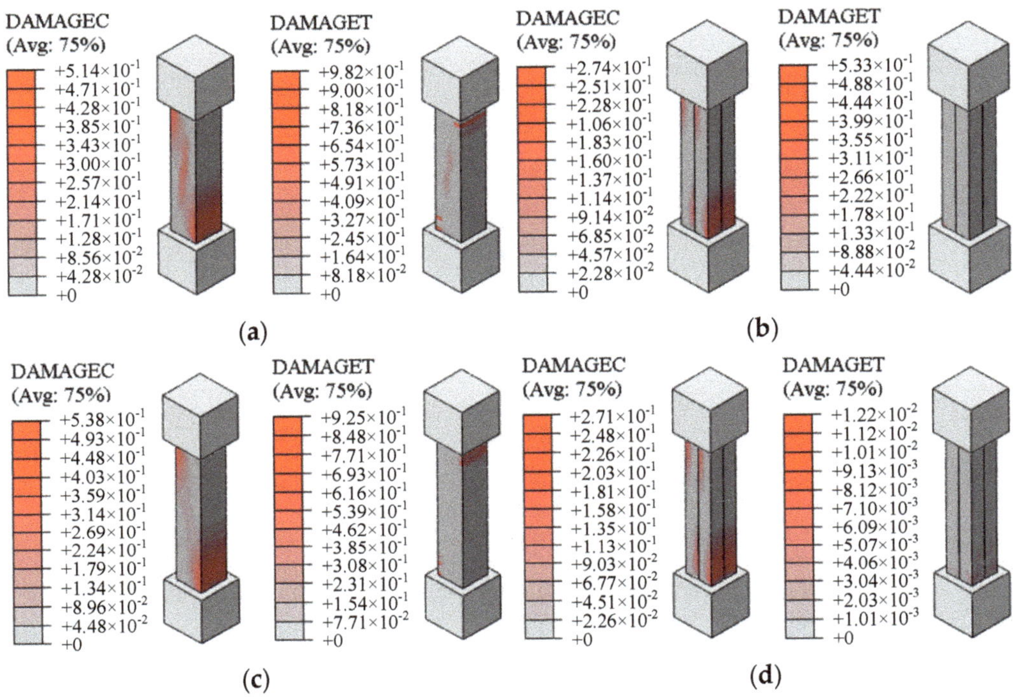

Figure 12. Concrete damage to the columns by Mode B. (**a**) UT-B-GE. (**b**) US-B-GE. (**c**) LT-B-GE. (**d**) LS-B-GE.

6. The Pushover Analysis of Soil–Structure System

6.1. Analysis Methods

The quasi-static pushover analysis is one of the most commonly used methods to study the seismic performance of underground structures. Liu et al. [47] established the pushover analysis method for underground structures based on their seismic response characteristics. In this method, gravity is first applied to the soil–structure system, and then horizontal inertial forces are applied to the system step by step until the structure is damaged, so as to obtain the seismic damage evolution process from the elastic to the elasto-plastic working state and finally to the failure. In the quasi-static pushover analysis of the soil–structure system, Jiang et al. [48] considered not only the gravity effect, but also the vertical seismic loads, so as to propose an improved pushover analysis method for underground structures. In the improved method, the loading conditions of the structure during earthquakes are simulated by applying horizontal and vertical inertia forces of certain distribution forms to the soil–structure system, respectively. The calculation models of the traditional and the improved pushover analysis method are shown in Figure 13 for Mode C and Mode D, respectively. The horizontal loads of Mode C and Mode D are the same in the form of inverted triangular distribution. The main difference between the two loading modes is the vertical loads, where Mode D includes vertical inertia forces in addition to gravity.

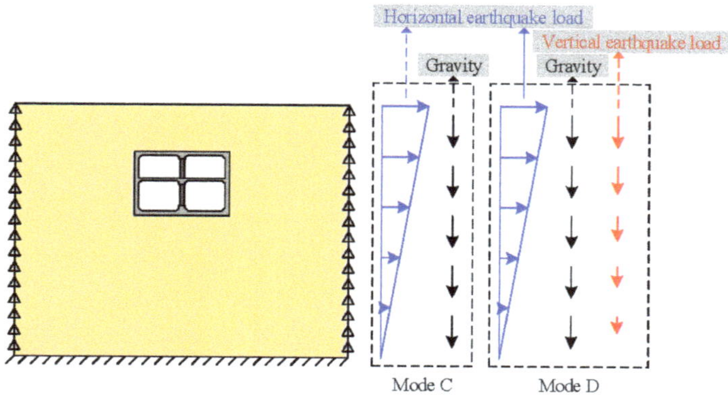

Figure 13. Pushover modes of the soil–structure system.

6.2. Analysis Results

Consistent with Section 5.2, it is assumed that the soil pressure of the overburden is increased by 50% considering the vertical earthquake effect [46]. The shear forces at the bottom of each center column of the structural mid-span under gravity are shown in Figure 14. From the figure, it can be seen that with the increase in applied horizontal acceleration, the lateral drift ratio of the center column is also increasing, and shear forces of each sections reach their peak with a lateral drift ratio of about 0.5%.

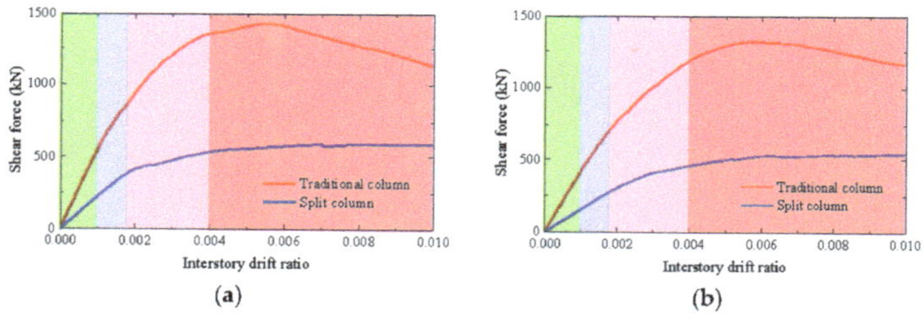

Figure 14. Lateral force–drift ratio curves of columns in the soil–structure system under gravity. (**a**) Upper column. (**b**) Lower column.

When the vertical load only considers the self-weight of soil and the structure, the shear force of traditional center column shows a gradual decrease after reaching the peak value. In contrast, although the peak shear force borne by the split column is only about 1/3 of that of the traditional one, its horizontal deformation capacity is excellent, and there is no obvious decreasing section. The split column can still bear the shear deformation even though the inter-story drift ratio has reached 1/100.

When the vertical loads consider both gravity and inertia forces, both the upper and the lower traditional columns show a significant decreasing trend after exceeding the peak shear force during the pushover process, and the decreasing trend in Figure 15 is more obvious than the decreasing trend in Figure 14. This step illustrates that the vertical inertia force of the overlying soil has an important influence on the seismic performance of underground structures. At this time, although the split column still needs to bear a large vertical pressure, its deformation capacity is outstanding and has good ductility performance.

Figure 16 presents the equivalent plastic strains of center column concrete under four typical cases of inter-story drift ratio, which are 1/1000, 1/550, 1/250, and 1/100 [49,50], corresponding to their respective color areas in Figure 16. Upon implementing the split column configuration, it is observed that the concrete's equivalent plastic strain throughout various phases consistently registers lower values in comparison to the traditional monolithic column. When the inter-story drift ratio reaches 1/100, the maximum equivalent plastic strain of concrete in the split column is about 0.25%. The zones of concrete damage are uniformly distributed around the top and bottom sections of split column, representing a prominent departure from the failure pattern exhibited by the prototype monolithic columns.

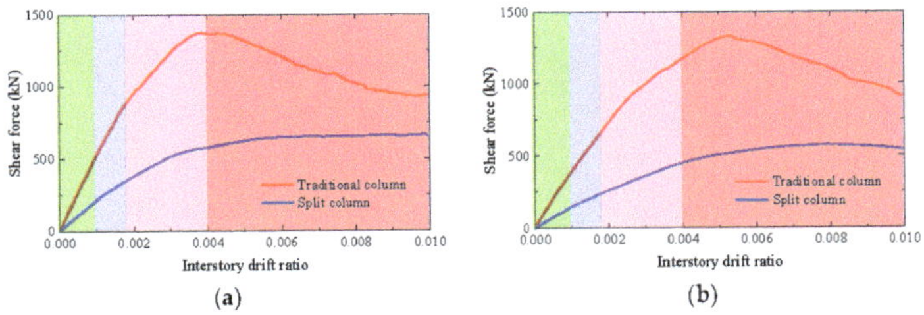

Figure 15. Lateral force–drift ratio curves of columns in the soil–structure system under gravity and vertical earthquake effect. (**a**) Upper column. (**b**) Lower column.

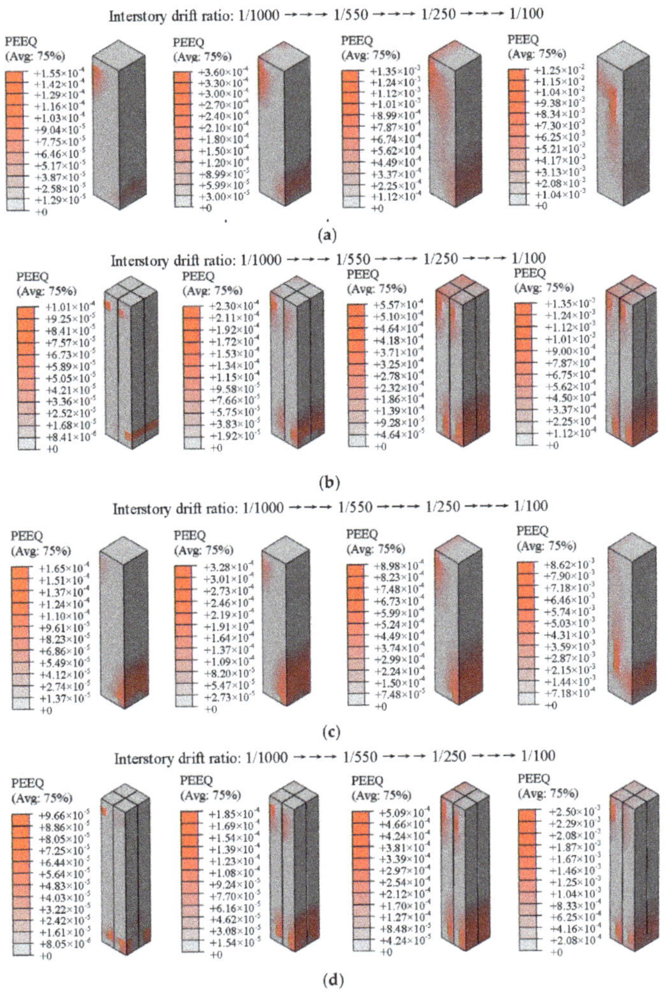

Figure 16. Equivalent plastic strain development of concrete in the center column. (**a**) Upper traditional column. (**b**) Upper split column. (**c**) Lower traditional column. (**d**) Lower split column.

In addition, comparing Figures 11, 12 and 16, it can also be found that the equivalent plastic strain distribution mode shown in Figure 12 is closer to that shown in Figure 16, which means that Mode B is preferred for the pushover analysis of underground structures. Specifically, the horizontal drift is applied while the rotation degrees of freedom at the top of the column are also restrained. This particular pushover methodology serves to accurately simulate the stress distribution within columns of underground structures, thereby yielding more authentic seismic performance curves.

7. Conclusions

This study draws inspiration from the seismic design concept of split columns in high-rise building structures, and sets up split columns in a two-story, two-span subway station to form a new structural system. Through quasi-static pushover analysis of the middle column and soil–structure system, the main conclusions obtained are as follows:

(1) When the cross-sectional area and the longitudinal reinforcement ratio is similar, the horizontal bearing capacity of split columns is reduced compared to traditional columns, and the stiffness of split columns is basically lower than that of traditional columns under various loading displacements.
(2) Compared to traditional columns, split column components have better displacement capacity, and the ultimate inter-story drift ratio of split columns under high axial compression ratio is 1.5-fold higher than that of traditional columns.
(3) Although the ability of split columns to share shear and bending moments is reduced compared to traditional columns, they still have higher vertical bearing capacity under larger horizontal deformations, and the damage to split columns is much smaller than that of traditional columns under the same inter-story displacement angle.
(4) The working mechanism is to avoid the central column bearing excessive shear and bending forces, while fully utilizing the vertical support capacity and horizontal deformation capacity of split columns. Compared to traditional columns, the construction steps of split columns are more complex. To promote the application of split columns in underground structures, it is necessary to strengthen both experimental research and construction technology research.

Author Contributions: Conceptualization, Z.X. (Zigang Xu) and Z.X. (Zongyao Xia); methodology, Z.X. (Zigang Xu) and R.H.; software, Z.X. (Zongyao Xia); validation, Z.X. (Zigang Xu) and X.B.; formal analysis, Z.X. (Zongyao Xia); data curation, Z.X. (Zigang Xu); writing—original draft preparation, Z.X. (Zongyao Xia); writing—review and editing, X.B.; visualization, Z.X. (Zongyao Xia); project administration, Z.X. (Zigang Xu); funding acquisition, Z.X. (Zigang Xu) and R.H. All authors have read and agreed to the published version of the manuscript.

Funding: This research is jointly funded by the National Natural Science Foundation of China (No. 52108453), the Natural Science Foundation of Jiangxi Province in China (No. 20212BAB214014 and No. 20232BAB204084), the China Postdoctoral Science Foundation Project (No. 2023M741159), and the Natural Science Foundation of Inner Mongolia Autonomous Region in China (No. 2023QN05012).

Institutional Review Board Statement: Not applicable.

Informed Consent Statement: Not applicable.

Data Availability Statement: Data are available in a publicly accessible repository.

Conflicts of Interest: The authors declare no conflicts of interest.

References

1. *GB18306-2015*; Seismic Ground Motion Parameters Zonation Map of China. Standardization Administration of China: Beijing, China, 2016. (In Chinese)
2. Du, X.L.; Li, Y.; Xu, C.S.; Lu, D.; Xu, Z.; Jin, L. Review on damage causes and disaster mechanism of Daikai subway station during 1995 Osaka-Kobe Earthquake. *Chin. J. Geotech. Eng.* **2018**, *40*, 223–236. (In Chinese)
3. Iida, H.; Hiroto, T.; Yoshida, N.; Iwafuji, M. Damage to Daikai subway station. *Soils Found* **1996**, *36*, 283–300. [CrossRef] [PubMed]

4. Wang, W.L.; Wang, T.T.; Su, J.J.; Lin, C.H.; Seng, C.R.; Huang, T.H. Assessment of damage in mountain tunnels due to the Taiwan Chi-Chi earthquake. *Tunn. Undergr. Space Technol.* **2001**, *16*, 133–150. [CrossRef]
5. Wang, Z.Z.; Gao, B.; Jiang, Y.J.; Yuan, S. Investigation and assessment on mountain tunnels and geotechnical damage after the Wenchuan earthquake. *Sci. China Ser. E Technol. Sci.* **2009**, *52*, 546–558. [CrossRef]
6. Kuesel, T.R. Earthquake design criteria for subways. *J. Struct. Div.* **1969**, *95*, 1213–1231. [CrossRef]
7. Liu, J.B.; Zhang, X.B. Practical seismic analysis of large underground structures: Theory and application. *Sci. China Technol. Sci.* **2018**, *61*, 1417–1425. [CrossRef]
8. Jiang, L.; Chen, J.; Li, J. Seismic response of underground utility tunnels: Shaking table testing and FEM analysis. *Earthq. Eng. Eng. Vib.* **2010**, *9*, 555–567. [CrossRef]
9. Ulgen, D.; Saglam, S.; Ozkan, M.Y. Dynamic response of a flexible rectangular underground structure in sand: Centrifuge modeling. *Bull. Earthq. Eng.* **2015**, *13*, 2547–2566.
10. Jiang, X.; Wang, F.; Yang, H.; Sun, G.; Niu, J. Dynamic Response of Shallow-Buried Small Spacing Tunnel with Asymmetrical Pressure: Shaking Table Testing and Numerical Simulation. *Geotech. Geol. Eng.* **2018**, *36*, 2037–2055. [CrossRef]
11. An, X.; Shawky, A.A.; Maekawa, K. The collapse mechanism of a subway station during the Great Hanshin earthquake. *Cem. Concr. Compos.* **1997**, *19*, 241–257. [CrossRef]
12. Huo, H.; Bobet, A.; Fernández, G.; Ramírez, J. Load transfer mechanisms between underground structure and surrounding ground: Evaluation of the failure of the Daikai station. *J. Geotech. Geoenviron. Eng.* **2005**, *131*, 1522–1533. [CrossRef]
13. Gao, F.; Shi, Y.C.; Yan, S.H.; Guan, B.S. Study of two shock absorption measures in tunnel. *Chin. J. Rock Mech. Eng.* **2005**, *24*, 222–229. (In Chinese) [CrossRef]
14. Zhao, W.; Chen, W.; Yang, D. Interaction between strengthening and isolation layers for tunnels in rock subjected to SH waves. *Tunn. Undergr. Space Technol.* **2018**, *79*, 121–133. [CrossRef]
15. Chen, Z.; Liang, S.; Shen, H.; He, C. Dynamic centrifuge tests on effects of isolation layer and cross-section dimensions on shield tunnels. *Soil Dyn. Earthq. Eng.* **2018**, *109*, 173–187. [CrossRef]
16. Ma, C.; Lu, D.; Du, X. Seismic performance upgrading for underground structures by introducing sliding isolation bearings. *Tunn. Undergr. Space Technol.* **2018**, *74*, 1–9. [CrossRef]
17. Xu, Z.; Du, X.; Xu, C.; Han, R. Numerical analyses of seismic performance of underground and aboveground structures with friction pendulum bearings. *Soil Dyn. Earthq. Eng.* **2020**, *130*, 105967. [CrossRef]
18. Chen, Z.; Chen, W.; Bian, G. Seismic performance upgrading for underground structures by introducing shear panel dampers. *Adv. Struct. Eng.* **2014**, *17*, 1343–1357. [CrossRef]
19. Chen, Z.; Zhao, H.; Lou, M. Seismic performance and optimal design of framed underground structures with lead-rubber bearings. *Struct. Eng. Mech.* **2016**, *58*, 259–276. [CrossRef]
20. Du, X.L.; Wang, Z.L.; Liu, H.T. Study of a seismic new system of underground frame structure based on toughness design. *Technol. Earthq. Disaster Prev.* **2018**, *13*, 493–501. (In Chinese)
21. Chen, Z.; Zhou, Y. Seismic performance of framed underground structures with self-centering energy-dissipation column base. *Adv. Struct. Eng.* **2019**, *22*, 2809–2822. [CrossRef]
22. Yue, C.; Zheng, Y.; Deng, S. Shaking table test study on seismic performance improvement for underground structures with center column enhancement. *J. Earthq. Tsunami* **2019**, *13*, 1950009. [CrossRef]
23. Ma, C.; Wang, Z.H.; Lu, D.C.; Du, X. Seismic damage evaluation of CFRP-strengthened columns in subway stations. *Chin. J. Geotech. Eng.* **2020**, *42*, 2249–2256. (In Chinese)
24. Bu, X.; Ledesma, A.; López-Almansa, F. Low lateral stiffness underground structures for improved seismic performance. *Appl. Kobe Daikai Stn. Eng. Struct.* **2023**, *291*, 116388.
25. Bu, X.; Xu, Z.; Ledesma, A.; López-Almansa, F. Seismic target drifts for underground railway stations. Cases of traditional rigid connections and sliding interior columns. *Tunn. Undergr. Space Technol.* **2024**, *144*, 105555. [CrossRef]
26. Li, Z.X. Theory and technology of split reinforced concrete columns. *Eng. Mech.* **2005**, *22*, 128–140. (In Chinese) [CrossRef]
27. Li, Z.X.; Hao, Y.C.; Zhang, J.Y. Research on seismic behavior of interior beam-column joints of reinforced concrete frames. *J. Build. Struct.* **2001**, *22*, 55–60. (In Chinese)
28. Hao, Y.C.; Hu, Q.C.; Xu, Y.F.; Kang, G.Y. Experimental study on bearing capacity of separated columns under horizontal cyclic loading. *J. Build. Struct.* **1999**, *20*, 18–25. (In Chinese)
29. Li, R.W. Experimental Study on the Mechanical Behavior of Split Columns with Different Separation Ratios. Master's Thesis, Yantai University, Yantai, China, 2017. (In Chinese).
30. Li, Z.X.; Hao, Y.C.; Zhou, B.; Kang, G.; Hu, Q.; Xu, Y. Model experimental study on seismic behavior of reinforced concrete frame with split columns. *J. Build. Struct.* **2003**, *24*, 1–10. (In Chinese)
31. Ma, Q.Y.; Zhao, G.Q.; Jiang, C.Y. Experimental research on seismic performance of central column in subway station. *Build. Struct.* **2021**, *51*, 77–85. (In Chinese)
32. Du, X.L.; Xu, Z.G.; Xu, C.S. n-Shape Prefabricated Connective Split Column for Underground Structures. U.S. Patent 10669719 B2, 18 December 2023.
33. Xu, C.S.; Wang, Y.X.S.; Du, X.L.; Xu, Z.; Zhang, Z. Seismic mitigation effects of split columns in underground station structures. *Chin. J. Geotech. Eng.* **2021**, *43*, 624–633. (In Chinese)

34. Hibbitt; Karlsson; Sorensen. *ABAQUS/Explicit: User's Manual*; Hibbitt, Karlsson and Sorenson Incorporated: Birmingham, AL, USA, 2001.
35. Kampitsis, A.E.; Sapountzakis, E.J.; Giannakos, S.K.; Gerolymos, N.A. Seismic soil-pile-structure kinematic and inertial interaction-A new beam approach. *Soil Dyn. Earthq. Eng.* **2013**, *55*, 211–224. [CrossRef]
36. Bu, X.; Ledesma, A.; López-Almansa, F. Novel seismic design solution for underground structures. Case study of a 2-story 3-bay subway station. *Soil Dyn. Earthq. Eng.* **2022**, *153*, 107087. [CrossRef]
37. Yang, J.; Zhuang, H.; Wang, W.; Zhou, Z.; Chen, G. Seismic performance and effective isolation of a large multilayered underground subway station. *Soil Dyn. Earthq. Eng.* **2021**, *142*, 106560.
38. Nie, X.; Liu, T. Experimental and numerical study on bending behavior of concrete-concrete composite beams. *China Civ. Eng. J.* **2022**, *55*, 14–25. (In Chinese)
39. Lee, J.; Fenves, G.L. Plastic-Damage Model for Cyclic Loading of Concrete Structures. *J. Eng. Mech.* **1998**, *124*, 892–900. [CrossRef]
40. Lubliner, J.; Oliver, J.; Oller, S.; Oñate, E. A plastic-damage model for concrete. *Int. J. Solids Struct.* **1989**, *25*, 299–326. [CrossRef]
41. Zhuang, H.; Hu, Z.; Wang, X.; Chen, G. Seismic responses of a large underground structure in liquefied soils by FEM numerical modelling. *Bull. Earthq. Eng.* **2015**, *13*, 3645–3668. [CrossRef]
42. Xu, Z.; Zhuang, H.; Xia, Z.; Yang, J.; Bu, X. Study on the effect of burial depth on seismic response and seismic intensity measure of underground structures. *Soil Dyn. Earthq. Eng.* **2023**, *166*, 107782. [CrossRef]
43. Jiang, J.; El Nggar, H.M.; Xu, C.; Zhong, Z.; Du, X. Effect of ground motion characteristics on seismic fragility of subway station. *Soil Dyn. Earthq. Eng.* **2021**, *143*, 106618. [CrossRef]
44. Jiang, J.; El Naggar, M.H.; Xu, C.; Du, X. Effect of parameters associated with soil-to-structure relative stiffness on seismic fragility curves of subway station. *Tunn. Undergr. Space Technol.* **2023**, *135*, 105057. [CrossRef]
45. Jiang, J.; El Naggar, M.H.; Du, X.; Du, X. Seismic fragility curves for shallow buried subway station using pushover-based method. *Tunn. Undergr. Space Technol.* **2023**, *135*, 105059. [CrossRef]
46. Xu, Z.; Du, X.; Xu, C.; Jiang, J.; Han, R. Simplified equivalent static methods for seismic analysis of shallow buried rectangular underground structures. *Soil Dyn. Earthq. Eng.* **2019**, *121*, 1–11. [CrossRef]
47. Liu, J.B.; Wang, W.H.; Dasgupta, G. Pushover analysis of underground structures: Method and application. *Sci. China Technol. Sci.* **2014**, *57*, 423–437. [CrossRef]
48. Jiang, J.; Xu, C.; El Naggar, H.M.; Du, X.; Xu, Z.; Assaf, J. Improved pushover method for seismic analysis of shallow buried underground rectangular frame structure. *Soil Dyn. Earthq. Eng.* **2021**, *140*, 106363. [CrossRef]
49. Du, X.; Jiang, J.; El Naggar, M.H.; Xu, C.; Xu, Z. Interstory drift ratio associated with performance objectives for shallow-buried multistory and span subway stations in inhomogeneous soil profiles. *Earthq. Eng. Struct. Dyn.* **2021**, *50*, 655–672. [CrossRef]
50. Zhong, Z.; Shen, Y.; Zhao, M.; Li, L.; Du, X. Seismic performance evaluation of two-story and three-span subway stations in different engineering sites. *J. Earthq. Eng.* **2022**, *26*, 7505–7535. [CrossRef]

applied sciences

MDPI

Article

Seismic Response and Mitigation Analysis of a Subway Station in the Site with Weak Interlayers

Zigang Xu [1,*], Chunyu Li [1], Zongyao Xia [1] and Runbo Han [2]

[1] State Key Laboratory of Performance Monitoring and Protecting of Rail Transit Infrastructure, East China Jiaotong University, Nanchang 330013, China
[2] School of Resources, Environment and Architectural Engineering, Chifeng University, Chifeng 024000, China
* Correspondence: xuzigang1027@163.com

Abstract: Challenges related to seismic performance and seismic mitigation are more pronounced in the presence of weak interlayers compared to typical layered soil conditions. This study focuses on a double-layer double-span rectangular frame subway station structure. A coupled static–dynamic finite element analysis model of the soil-structure system is established by using the finite element software ABAQUS/CAE V 6.14. The research investigates the influence of factors such as interlayer thickness, location, and strength on the seismic response of subway station structures. Furthermore, in order to evaluate the effectiveness of FPB in mitigating seismic effects in the weak interlayer ground, two different schemes are proposed in this paper. One is the structure without FPB and the other is the structure with FPB on the top of the central column. The findings reveal that weak interlayers exert a significant influence on the seismic response of subway station structures, especially when these lower-strength weak interlayers are located within the central portion of the subway station structure and exhibit considerable thickness. The FPB on the top of the central column can reduce the overall lateral stiffness of the subway station structure. This, in turn, results in a slight increase in the deformation of sidewall and inter-story displacement angles, accompanied by a marginal exacerbation of sidewall damage. However, the implementation of FPB effectively reduces the deformation of the central column and substantially mitigates the extent of damage to the central column.

Keywords: weak interlayer; subway station structure; friction pendulum bearing; numerical simulation; seismic response

1. Introduction

With rapid global development and exponential population growth, the pursuit of more extensive living spaces has made the development and utilization of underground areas a vital direction for advancement. Currently, underground structures have broad applications across various construction domains, with subway projects being notable examples within this realm. Rock formations and soil exert constraints on the vibrational deformations of underground structures. Hence, it was commonly believed that underground structures exhibited superior seismic performance compared to their above-ground counterparts. However, until the 1990s, there was a scarcity of seismic damage data related to underground structures, which resulted in a prolonged lack of emphasis on seismic analysis theories and design methods in this field [1–3]. Recent seismic disasters have highlighted the vulnerability of underground structures to earthquakes, with some experiencing severe damage [4–7]. For instance, during the Great Hanshin Earthquake in Japan, Daikai station suffered extensive destruction. The Great Hanshin Earthquake's powerful horizontal and vertical seismic forces caused the central column to fracture, resulting in the roof's collapse and creating a distinctive "M" shaped pattern of devastation [8]. This earthquake-induced damage has acted as a wake-up call for experts and scholars worldwide.

Complex site characteristics significantly impact the safety of underground structures and weak interlayers are a common variety among these complex site types. Consequently, the study of weak interlayer's impact on underground structures holds significant importance for seismic design. Seokwon et al. [9] employed a combination of scaled model testing and numerical analysis to investigate the influence of rock faults and weak surfaces on the stability of underground structures. Huang et al. [10] conducted model testing and numerical simulations, concluding that factors such as position, angle, thickness, and distance from the tunnel significantly affect tunnel stability. Some scholars [11–14] pointed out in their work that adverse soft soil conditions exacerbate the deformation of subway station structures. Huang et al. [15,16]. taking into account the influence of the soil-structure interaction and structural depth, conducted vulnerability assessments for circular tunnels located within soft interlayers in the Shanghai subway system.

Since the Great Hanshin Earthquake, seismic mitigation measures for underground structures have garnered significant research attention. Some engineers [17–20] have proposed the seismic mitigation scheme by installing an isolation layer between the rock and lining, resulting in the isolation layer reducing the deformation of the lining. Chen et al. [21] delved into the utilization of isolation layers for underground structures and conducted pertinent experimental studies. The results showed that isolation layers can absorb seismic energy and efficiently reduce structural bending moments. Several scholars [22–25] carried out research on Daikai stations, which suffered the most damage during the Great Hanshin Earthquake. Their results demonstrated that the central column of subway station structure is particularly susceptible to seismic forces and prone to shear failures under earthquake motion. Therefore, mitigating shear deformations of central column under seismic forces is a crucial step in enhancing the seismic performance of subways. Studies have investigated the seismic mitigation effects of various approaches, including rubber bearings on the top of the central column [26–28], shear plate dampers [29], sliding bearings [30–32], and split columns [33]. The results indicated that all these measures effectively reduce the shear deformations of the central column.

Rubber bearing is a commonly used form of seismic mitigation support. However, in comparison to rubber bearing, friction pendulum bearing (FPB) is cost-effective, highly durable, and possess self-centering capabilities. In 1987, Zayas et al. [34] introduced FPB at the University of California, Berkeley. The bearing operates on the principle of a pendulum, extending the structural natural period by the swaying motion of the upper support and slider. The swaying motion can effectively dissipate seismic energy. In practical engineering applications, FPB have become widely employed. Tesopelas et al. [35] explored the application of FPB in bridge structures and found that the bearing effectively reduces deformation and damage to bridge structures under seismic forces. Jangid et al. [36] studied the mitigation seismic performance of FPB under near-fault motion and derived the optimal range for the friction coefficient of the bearing. Mazza [37] conducted seismic analyses on reinforced concrete frames fitted with FPB and observed that a lower friction coefficient leads to greater story rotations and residual displacements. However, many studies on seismic mitigation mechanisms have been focused on above-ground structures and the research of seismic mitigation of underground structure started relatively late. Xu et al. [38] delved into the effectiveness of FPB in underground structures by drawing inspiration from the accomplishments in seismic mitigation research for above-ground structures.

Based on the research findings of several scholars, it is evident that weak interlayers have detrimental effects on subway station structure deformations and structural forces. The seismic mitigation effects of the FPB in subway station structure situated within the weak interlayer site is not clear and needs further study. This paper proposes a seismic mitigation scheme with FPB on the top of the central column for the subway station structure situated within the weak interlayer site. This study focuses on a two-storey two-span subway station structure, employing the finite element software ABAQUS to establish a two-dimensional model of the soil and subway station structure. The impact of weak interlayers on the seismic response of the subway station structure and the effectiveness of

FPB in improving the seismic performance of the subway station structure within the weak interlayer site is studied by time-history analysis.

2. Seismic Response of the Subway Station in the Site with Weak Interlayers

2.1. Engineering Profile

The structural model used in this study is a double-story double-span rectangular frame structure for the station, with cross-sectional dimensions and reinforcement details shown in Figure 1. The station's dimensions are as follows: a width of 17 m and a height of 12 m. The thickness of the top slab, middle slab, bottom slab, and side wall of the station are 0.75 m, 0.45 m, 0.8 m, and 0.8 m, respectively. The cross-sectional size of the central column is 0.7 m × 0.7 m, with a longitudinal spacing of 7 m. The site consists of 11 layers of soil, with a vertical distance of 40 m from the soil surface to the bedrock. The site has a width of 119 m and the structural depth is 7 m. Table 1 provides the relevant parameters for the soil in the normal layered site, including soil density (ρ), shear wave velocity (Vs), and Poisson's ratio (ν).

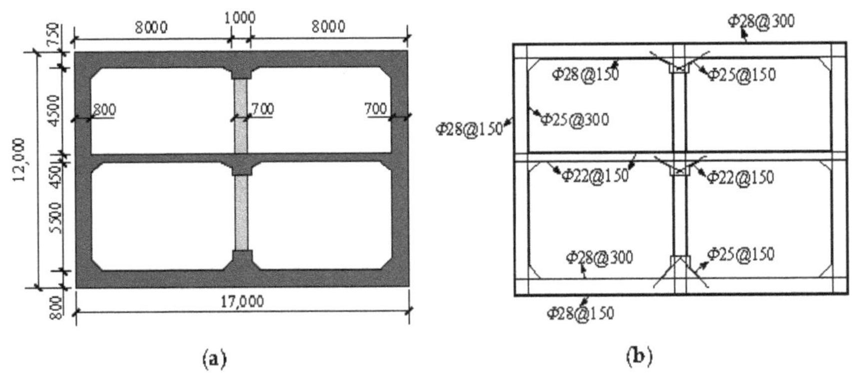

Figure 1. Size and reinforcement of the subway station. (**a**) Size of the subway station. (**b**) Reinforcement diagram of the subway station.

Table 1. Site soil parameters.

Layer No.	Thickness (m)	ρ (kg/m^3)	Vs (m/s)	ν
1	4	2000	170	0.3
2	4	2000	205	0.3
3	4	2000	240	0.3
4	2	2000	275	0.3
5	2	2000	275	0.3
6	4	2000	310	0.3
7	4	2000	345	0.3
8	4	2000	380	0.3
9	4	2000	415	0.3
10	4	2000	450	0.3
11	4	2000	485	0.3

In order to analyze the seismic response patterns of subway station structures situated within weak interlayer sites, three factors such as the thickness, location, and strength of the weak interlayer are discussed. The simplified diagram of the conditions is presented in Figure 2. By modifying the shear wave velocity of soil layer 4 (adjusted from 275 m/s to 80 m/s), a weak interlayer site is introduced at the midsection of the subway station structure. The thickness of the weak interlayer is varied (1, 1.5, 2, and 2.5 m) to study its impact on the seismic response of the subway station structure. Five different conditions are set up to discuss the effects of the interlayer's position relative to the subway station structure. To analyze the influence of the interlayer's strength on the seismic response of the subway station structure, assuming that the interlayer passes through the center of the structure, this paper designs five conditions with varying shear wave velocities for the weak interlayer (80, 90, 100, 110, and 120 m/s). The parameters for each condition are summarized in Table 2.

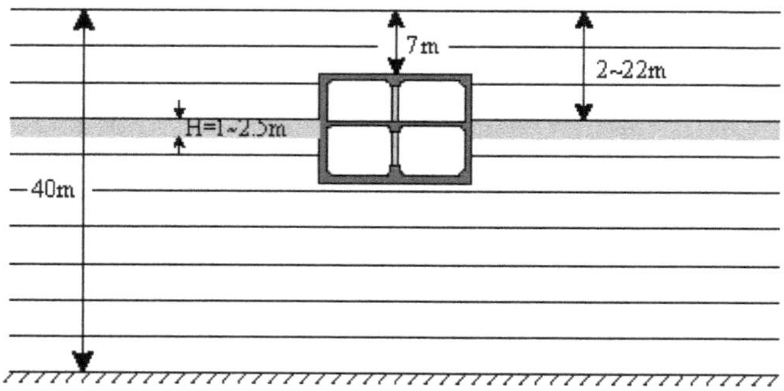

Figure 2. Schematic diagram of the soil-underground structure working condition.

Table 2. List of operating conditions.

Condition	Interlayer Position (m)	Relationship between Interlayer and Underground Structure Position	Thickness of Interlayer (m)	Shear Wave Velocity (m/s)
1	Layered soil site			
2	2	Upper	2	80
3	6	Top	2	80
4	12	Middle	2	80
5	19	Bottom	2	80
6	22	Lower	2	80
7	12.5	Middle	1	80
8	12.25	Middle	1.5	80
9	11.75	Middle	2.5	80
10	11.5	Middle	3	80
11	12	Middle	2	90
12	12	Middle	2	100
13	12	Middle	2	110
14	12	Middle	2	120

2.2. Finite Element Model

The numerical simulation method is a research method in the civil engineering field, which is verified by some studies [39,40]. In this study, a two-dimensional model of the soil and subway station structure is established by using the finite element software ABAQUS [41]. The soil model has a width of 119 m and a height of 40 m. The specific values for the soil parameters are listed in Table 1 and the elastic modulus can be calculated using Equations (1) and (2). Here, ρ represents the density of the soil layer, Vs stands for the shear wave velocity of the soil layer, and ν represents the Poisson's ratio of the soil layer.

$$G = \rho {V_s}^2 \tag{1}$$

$$E = 2(1+\nu)G \tag{2}$$

During the modeling process, the selection of soil layer damping deserves careful consideration as it is a pivotal factor. In this study, the Rayleigh damping model was chosen. Using the frequency analysis module in the ABAQUS software, we determined the first and second natural frequencies of the soil layers. Subsequently, the Rayleigh damping coefficients of the soil layers were calculated using Equations (3) and (4).

$$\alpha = \frac{2D \times w_1 \times w_2}{w_1 + w_2} \tag{3}$$

$$\beta = \frac{2D}{w_1 + w_2} \tag{4}$$

In the equation, where w_1 and w_2 represent the first and second natural frequencies of the soil layers, D is the damping ratio, set as 0.1 [42].

The concrete's constitutive model employed is the plastic damage model proposed by Jeeho Lee et al. [43], which is an improvement based on the model introduced by Lubliner et al.'s work [44]. This model is grounded in the principles of fracture energy within concrete and introduces two key variables: tensile damage and compressive damage factors. These factors are employed to represent the stiffness degradation in concrete under tension and compression, respectively. The C40 concrete is used for the main structure of the station, with a density of 2500 kg/m^3, an elastic modulus of 3.25×10^4 MPa, and a Poisson's ratio of 0.2. The other corresponding parameters are listed in Tables 3–5. The materials for the station's column include C50 concrete. The column is uniformly spaced along the alignment, with a 7 m interval. The material properties of the column are equivalently determined following established principles [38]. As a result of this equivalency, the column has a density of 250 kg/m^3, an elastic modulus of 3.45×10^3 MPa, a Poisson's ratio of 0.2, a peak compressive strength of 1.83 MPa, and a peak tensile strength of 0.13 MPa. Reinforcement is simulated using truss elements, employing an ideal elastic–plastic constitutive model. The initial elastic modulus is set at 200 GPa, with a Poisson's ratio of 0.3 and a yield stress of 335 MPa.

Table 3. Dynamic plastic–damage model parameters of concrete.

Model Parameters	Parameter Values	Model Parameters	
Initial yield compressive stress σ c/MPa	20.72	Tensile stiffness recovery coefficient/ωt	0
Ultimate compressive stressσ cu/MPa	29.6	Compression stiffness recovery coefficient/ωc	1
Initial yield tensile stressσ t0/MPa	2.95	Damage factor	dc, dt
Divergence angle ψ/(°)	38		

Table 4. Compressive stress and damage factor versus plastic strain.

Plastic strain (%)	0.0	0.13	0.19	0.26	0.31	0.42	0.53	0.78	1.48
Compressive stress (MPa)	29.6	26.5	21.53	17.32	14.18	10.15	7.79	5.25	2.32
Damage factor	0.0	0.37	0.49	0.59	0.65	0.74	0.79	0.85	0.93

Table 5. Tensile stress and damage factor versus cracking displacement.

Cracking strain (%)	0.0	0.011	0.018	0.058	0.107	0.205	0.399	1.56	3.07
Tensile stress (MPa)	3.42	2.7	2.17	1.05	0.68	0.43	0.27	0.11	0.07
Damage factor	0.0	0.34	0.49	0.78	0.86	0.93	0.96	0.98	0.99

The outer surface of the station is set to have frictional contact with the surrounding soil, in which the tangential friction coefficient is set as 0.4. A hard contact is established in the normal direction of the interface. The hard contact allows for separation after contact but prevents embedding. The soil model employs fixed constraints at the bottom; the lateral artificial boundary is implemented using a binding boundary approach. The binding boundary utilizes the MPC (Multi-Point Constraint) node degree of freedom coupling feature within ABAQUS, effectively binding the finite element model nodes of the soil to the boundary nodes at the same height, ensuring that these nodes share identical displacements throughout the analysis. Research by Kampitsis [45] has demonstrated that the equidistant displacement boundary condition model effectively simulates free-field motion. The acceleration time history curves at various elevations remain largely consistent and seismic waves do not radiate toward the soil boundaries, significantly enhancing the calculation accuracy. Mesh division is crucial for finite element models as it directly impacts the model's calculation accuracy. Uniform mesh dimensions of 0.5 m × 0.5 m are used for the entire soil. The station is divided into elements with a mesh size of 0.25 m. Both the soil and station are simulated using four-node plane strain elements, as shown in Figure 3 in the finite element model.

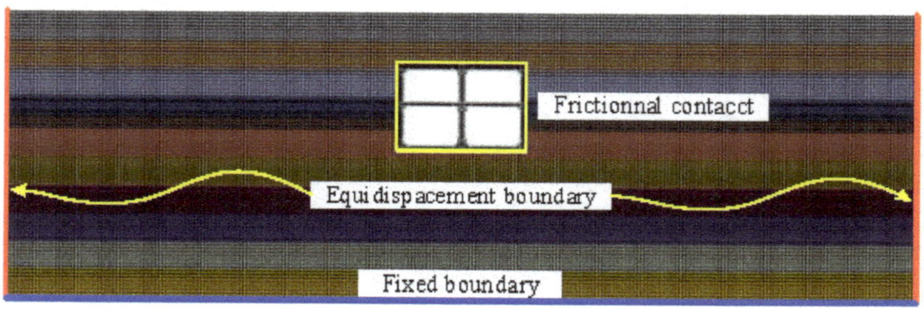

Figure 3. Finite element model.

The seismic input in this study considers only horizontal seismic motion. To prevent the idiosyncrasies of single-spectrum seismic waves from affecting the computational results, we have chosen to analyze three typical seismic motions: the Chi-Chi earthquake, Duzce earthquake, and Manjil earthquake. Furthermore, we have adjusted the amplitudes of these seismic waves, setting peak accelerations at 0.05 g, 0.1 g, 0.2 g, and 0.4 g. The acceleration time history curves for these seismic waves are depicted in Figure 4.

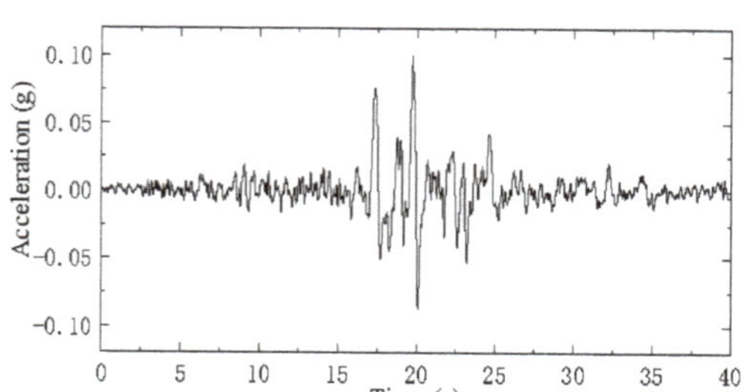

(**a**)

Figure 4. *Cont.*

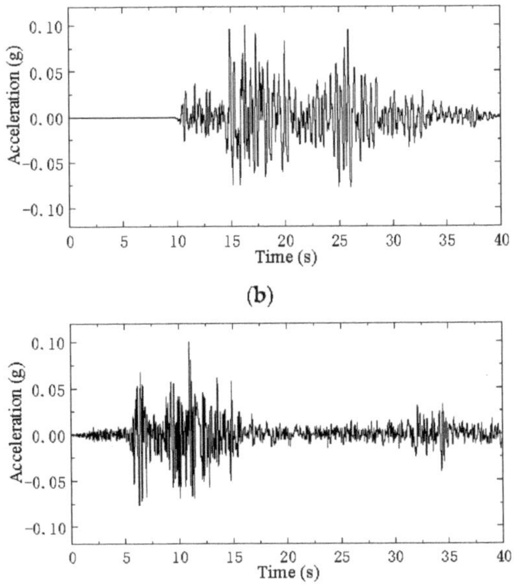

Figure 4. Input ground motion acceleration time history curve. (**a**) Chi-Chi earthquake. (**b**) Dzuce earthquake. (c) Manjil earthquake.

2.3. Computational Result

Under seismic loading, significant lateral displacement of the station's sidewall is a critical factor in the structural damage of the station. Simultaneously, damage to the station's central column often precedes that of the sidewall. The extraction of horizontal displacements for the sidewall and central column along with an evaluation of seismic damage to the station's structure are selected as analytical indicators. This paper uses these indicators to assess the impact of the weak interlayer on the station's seismic response. As shown in Figure 5, the absolute values of the relative horizontal displacement values at points A and B represent the horizontal displacement values for the traditional structure; the absolute values of the relative horizontal displacement values at points C, D and E, F represent the horizontal displacement values of the upper and lower central columns of the station, respectively.

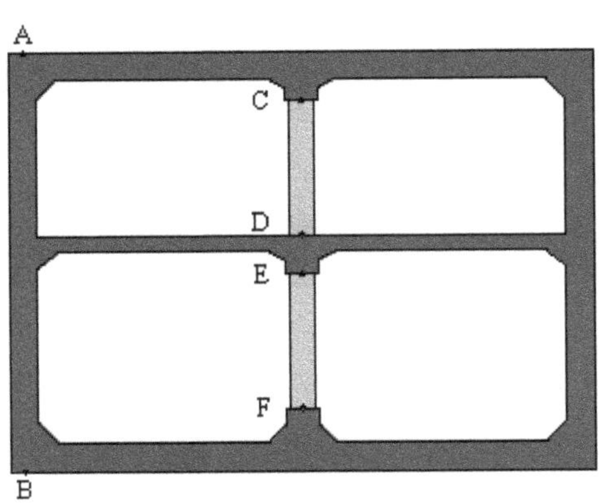

Figure 5. Schematic diagram of the section node.

(1) Influence of the position of weak interlayers

In this condition, a weak interlayer with a thickness of 2 m is assumed, positioned at various heights within the subway station structure, including the upper, top, middle, bottom, and lower sections. A layered site without an interlayer is established for comparison by analyzing these six different conditions to assess the influence of the interlayer's position on structural seismic response. Figures 6–8 present the maximum displacement values of the station's sidewall and central column when the interlayer is located at different positions within the station. Due to the length of the article, input accelerations only consider 0.2 g and 0.4 g in this context. From Figures 6–8, it becomes evident that when the weak interlayer is situated in the middle of the station, it has the most significant seismic impact. When the weak interlayer is at the top or bottom of the subway station structure, it substantially affects the subway station structure. However, when the weak interlayer is placed in the upper or lower sections of the station, its impact is similar to that of the layered site condition, or even less. The weak interlayer effectively separates the soil into upper and lower portions. When the weak interlayer is positioned in the upper or lower sections of the structure, it does not make direct contact with the subway station structure. Consequently, the forces generated by the deformation of the weak interlayer under seismic action do not directly affect the subway station structure, resulting in minimal impact. When the weak interlayer is located at the top or bottom of the station, only half of its thickness is in contact with the subway station structure. The deformation generated by the weak interlayer during seismic action is only partially transferred to the structure. Therefore, its impact on the underground structure is less pronounced compared to the condition where the interlayer is in the middle of the subway station structure.

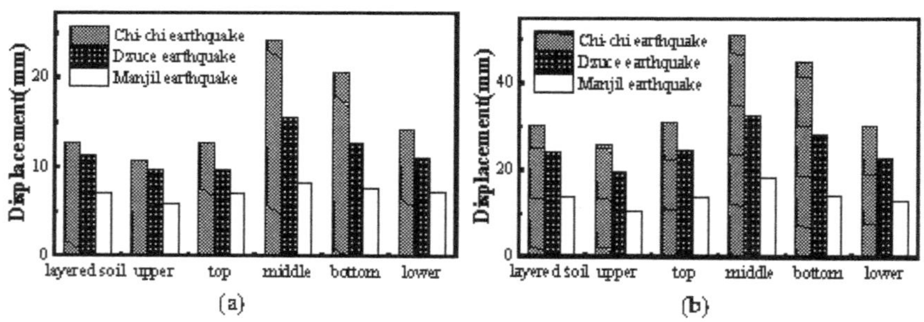

Figure 6. Maximum displacement of the station side wall at different positions. (**a**) 0.2 g. (**b**) 0.4 g.

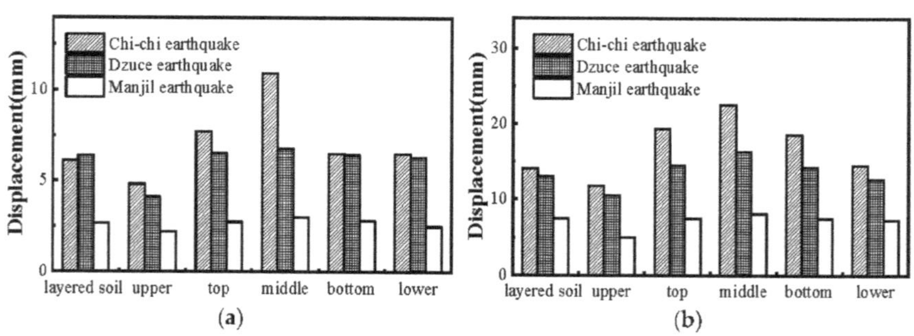

Figure 7. Maximum displacement of the station upper column at different positions. (**a**) 0.2 g. (**b**) 0.4 g.

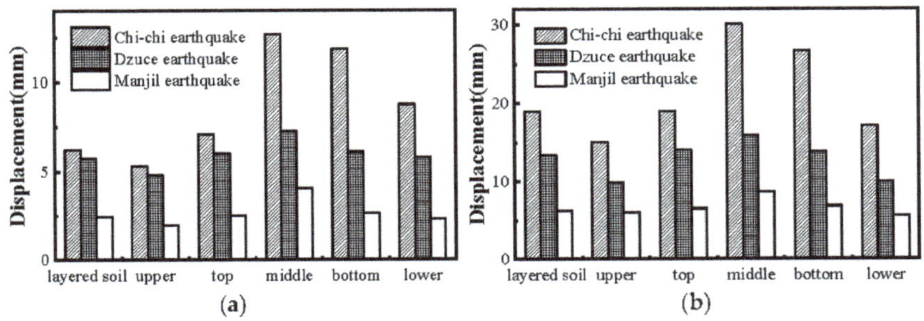

Figure 8. Maximum displacement of the station lower column at different positions. (**a**) 0.2 g. (**b**) 0.4 g.

Figures 9 and 10 depict the tensile damage of the subway station structure when the weak interlayer is positioned at different locations. When DAMAGET approaches 1, it indicates that the concrete has completely cracked under tension. Given the more pronounced seismic response patterns under the Chi-Chi earthquake, this section only illustrates subway station structure tensile damage under 0.2 g and 0.4 g Chi-Chi earthquakes. It can be seen from these illustrations that, compared to the subway station structure situated on a layered soil site, the presence of a weak interlayer at the station's midsection significantly exacerbates the tensile damage of the main subway station structure. The next most affected condition is when the weak interlayer is located at the station's bottom, resulting in noticeable increases in tensile damage to the station's bottom slab, lower-level central column, and middle plate. When the weak interlayer is positioned at the station's top, the weak interlayer increases the tensile damage to the upper central column and top slab. The influence on the subway station structure is nearly negligible when the weak interlayer is situated at the top or bottom of the station. These patterns become even more conspicuous when the input acceleration is set to 0.4 g.

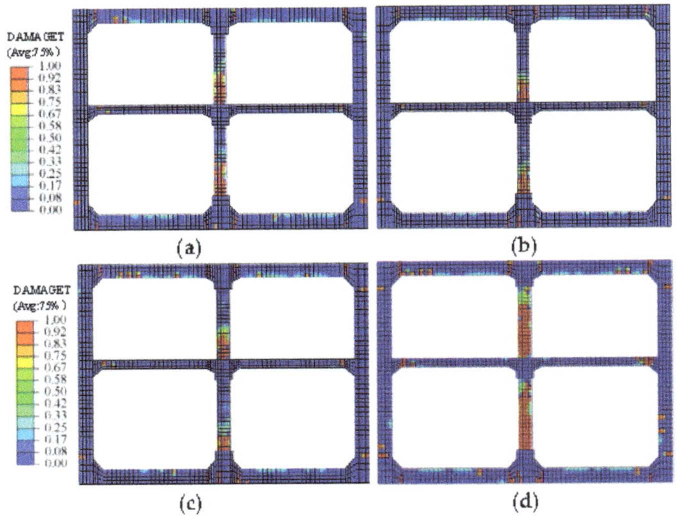

Figure 9. *Cont.*

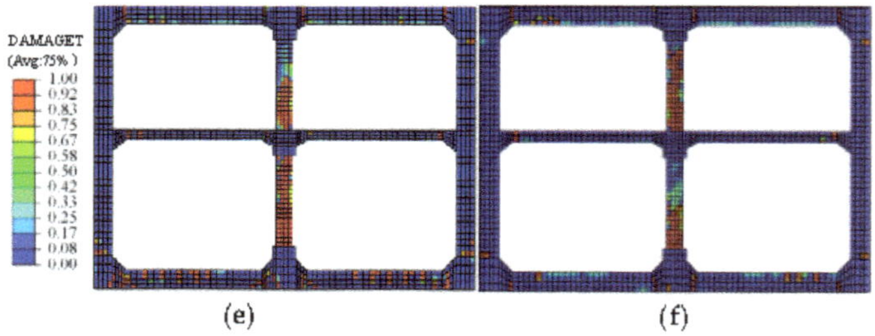

Figure 9. Tensile damage of the station at different positions under a 0.2 g Chi-Chi earthquake. (**a**) Layered soil site. (**b**) Upper. (**c**) Top. (**d**) Middle. (**e**) Bottom. (**f**) Lower.

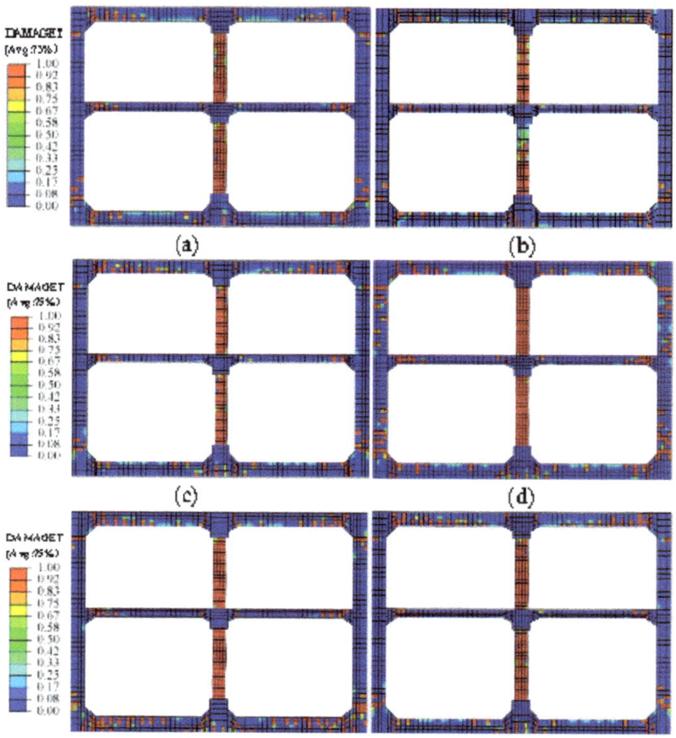

Figure 10. Tensile damage of the station at different positions under a 0.4 g Chi-Chi earthquake. (**a**) Layered soil site. (**b**) Upper. (**c**) Top. (**d**) Middle. (**e**) Bottom. (**f**) Lower.

(2) Influence of the Influence of the thickness of weak interlayers

Figures 11–13 present the maximum displacements of the structure's side wall and central column under various thicknesses of the weak interlayer, where H represents the interlayer's thickness. From these figures, it is evident that the maximum horizontal displacements of the station's side wall and central column in the weak interlayer site are greater than those in the layered soil site. This indicates that the weak interlayer has an adverse effect on structural horizontal deformations. It is important to note that the maximum horizontal displacement of the station's side wall and central column does not continuously increase with an increase in the thickness of the weak interlayer. Instead, it follows a pattern of decreasing and then increasing as the thickness reaches a certain level.

When the thickness of the weak interlayer is 3 m, it has the most significant impact on the horizontal deformation of the structure.

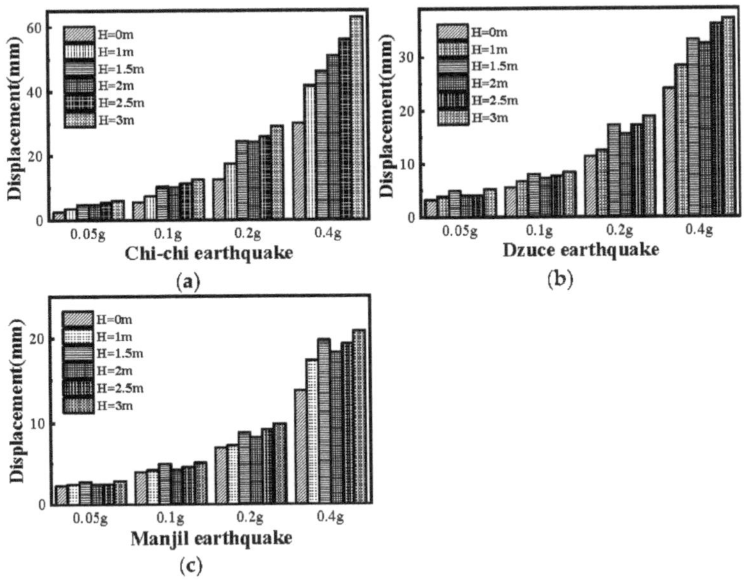

Figure 11. Maximum displacement of the station side wall. (**a**) Chi-Chi earthquake. (**b**) Dzuce earthquake. (**c**) Manjil earthquake.

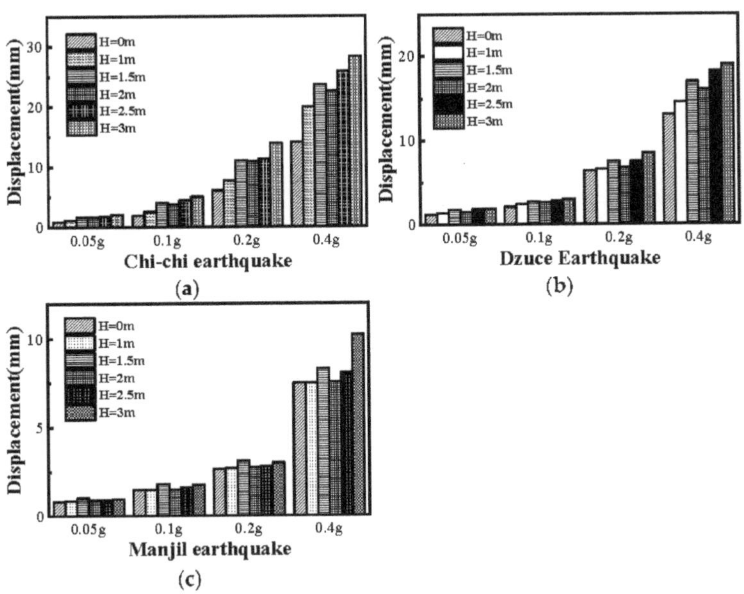

Figure 12. Maximum displacement of the station upper column. (**a**) Chi-Chi earthquake. (**b**) Dzuce earthquake. (**c**) Manjil earthquake.

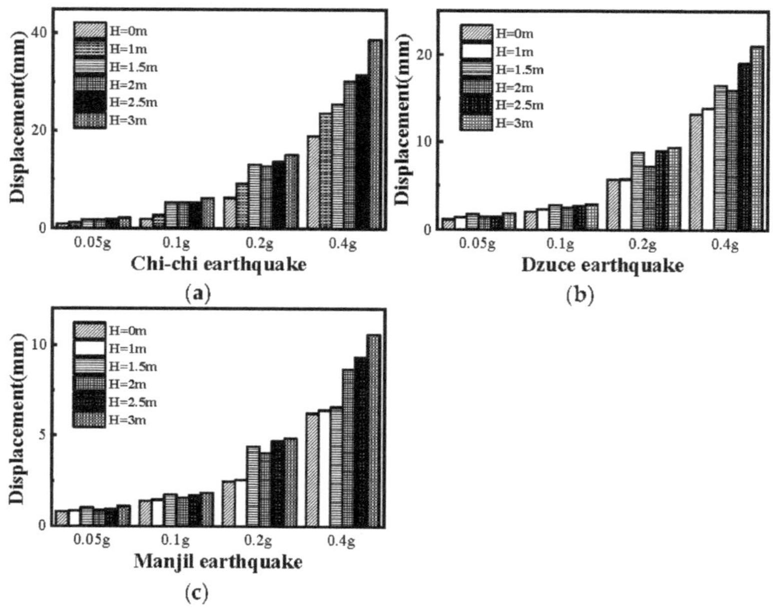

Figure 13. Maximum displacement of the station lower column. (**a**) Chi-Chi earthquake. (**b**) Dzuce earthquake. (**c**) Manjil earthquake.

To further analyze the impact of the thickness of the weak interlayer on the seismic response of subway station structure, Figures 14–17 present the damaged nephograms of the station's main structure under varying weak interlayer thicknesses. Observing these figures reveals that under different levels of seismic intensity, station structures located within the weak interlayer experience significantly more severe damage compared to those on layered soil sites. This difference becomes more pronounced as the seismic intensity increases. When the seismic input acceleration is 0.05 g, the subway station structure experiences minimal damage. As the thickness of the weak interlayer increases, only minor damage occurs to the top and bottom slabs of the station. However, the damage to the structural column, side wall, and middle slab intensifies as the thickness of the weak interlayer increases when the input acceleration is set to 0.1 g. In the case of a 3 m thick weak interlayer during the Chi-Chi earthquake, seismic tensile damage almost permeates the entire lower part of the middle column. Under strong seismic activity (with peak input accelerations of 0.2 g and 0.4 g), the influence of the weak interlayer's thickness becomes more pronounced. In comparison to layered soil sites, structures within the weak interlayer experience severe tensile damage at the connections between the middle plates and side walls on both sides. As the thickness of the weak interlayer increases, the extent of damage to the middle column, side wall, top slab, and bottom slab intensifies. The subway station structure is nearly destroyed under seismic activity with a 0.4 g acceleration, resulting in severe damage throughout the subway station structure. Particularly noteworthy is the pronounced damage to the station's side wall as the interlayer thickness increases. The computational results presented above underscore the adverse effect of the weak interlayer on subway station structure deformation and its exacerbation of structural seismic damage.

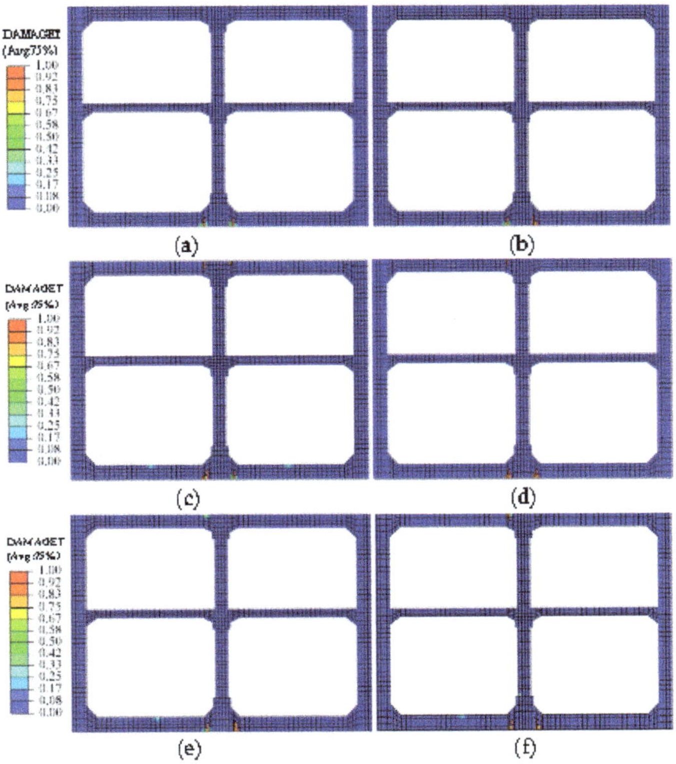

Figure 14. Tensile damage of the station at different positions under a 0.05 g Chi-Chi earthquake. (**a**) H = 0 m. (**b**) H = 1 m. (**c**) H = 1.5 m. (**d**) H = 2 m. (**e**) H = 2.5 m. (**f**) H = 3 m.

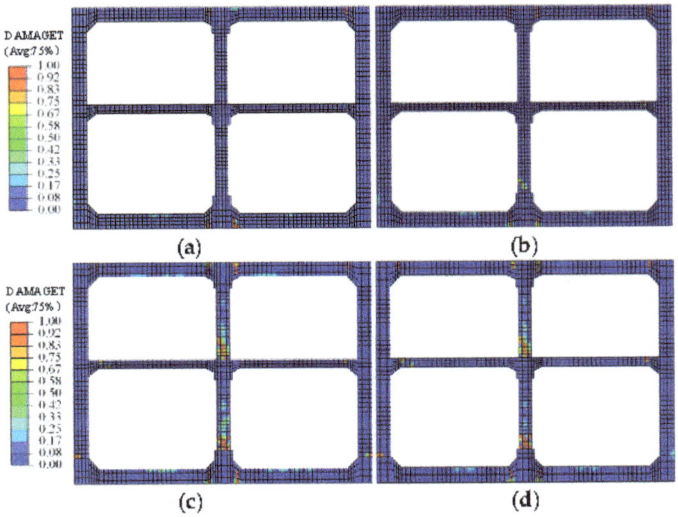

Figure 15. *Cont.*

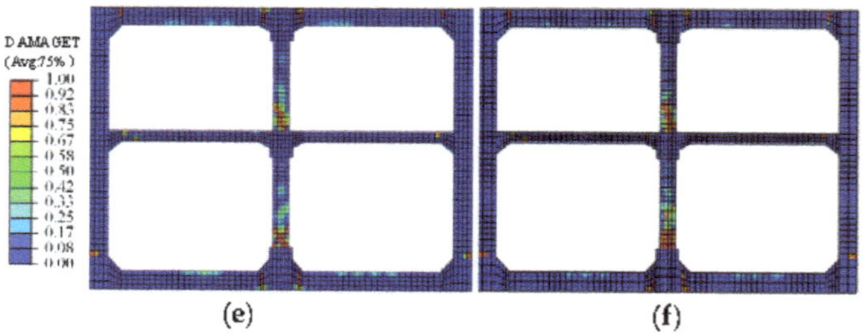

Figure 15. Tensile damage of the station at different positions under a 0.1 g Chi-Chi earthquake. (**a**) H = 0 m. (**b**) H = 1 m. (**c**) H = 1.5 m. (**d**) H = 2 m. (**e**) H = 2.5 m. (**f**) H = 3 m.

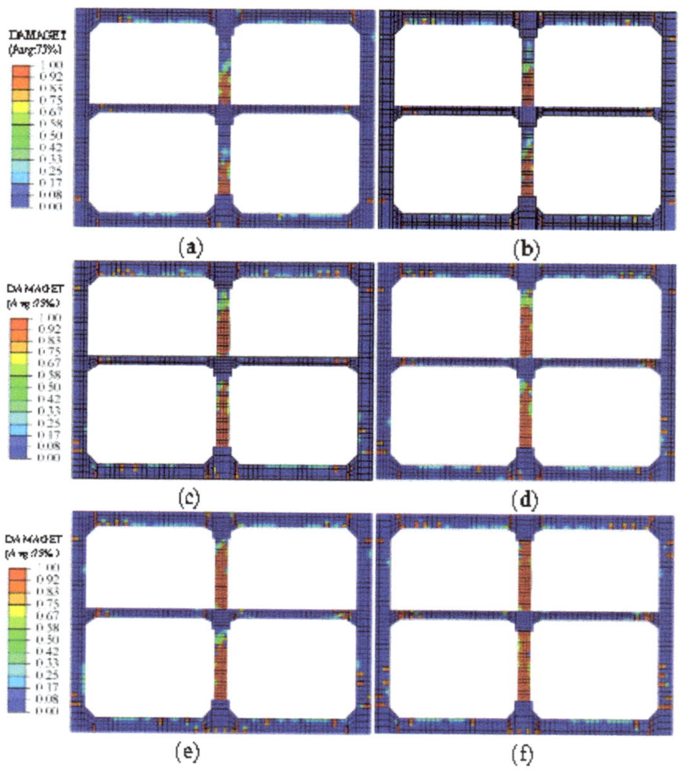

Figure 16. Tensile damage of the station at different positions under a 0.2 g Chi-Chi earthquake. (**a**) H = 0 m. (**b**) H = 1 m. (**c**) H = 1.5 m. (**d**) H = 2 m. (**e**) H = 2.5 m. (**f**) H = 3 m.

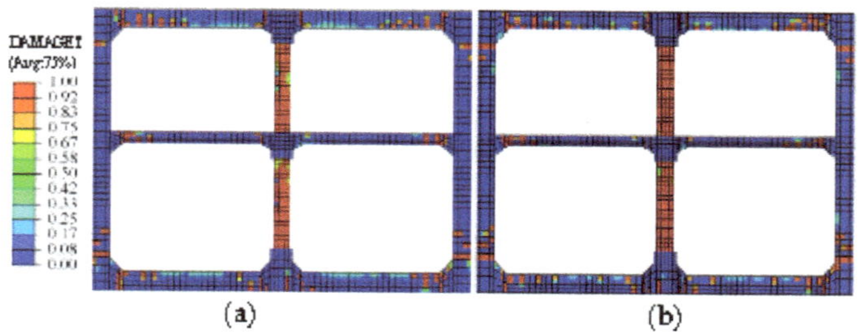

Figure 17. *Cont.*

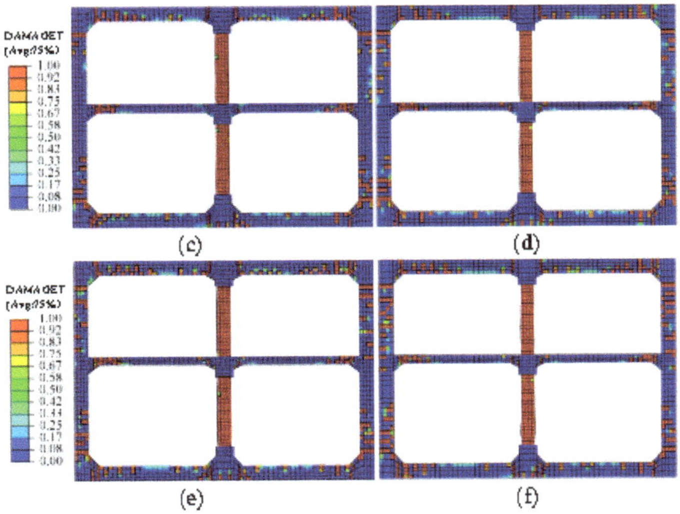

Figure 17. Tensile damage of the station at different positions under a 0.4 g Chi-Chi earthquake. (**a**) H = 0 m. (**b**) H = 1 m. (**c**) H = 1.5 m. (**d**) H = 2 m. (**e**) H = 2.5 m. (**f**) H = 3 m.

(3) Influence of the strength of the weak interlayer

A 2 m thick weak interlayer located in the middle of the subway station structure is assumed in this section. The aim is to investigate the impact of the interlayer's strength on the seismic response of the subway station structure by varying the interlayer's shear wave velocity. Figure 18 illustrates the maximum displacements of the station's side wall and middle columns under a 0.4 g earthquake. From this figure, it is evident that as the shear wave velocity increases, the maximum displacement of the structure's side wall and middle column decreases. In general, firmer soil layers correspond to higher shear wave velocities. Therefore, an increase in the stiffness of the weak interlayer can, to some extent, limit the deformation of the structure.

Figure 19 displays the seismic tensile damage nephograms of the subway station structure under various shear wave velocities of the weak interlayer conditions. The seismic input used here is the Chi-Chi earthquake, with a magnitude of 0.4 g. From the figure, it becomes apparent that as the shear wave velocity of the weak interlayer increases, there is a noticeable reduction in damage to the station's side wall. However, the impact on other parts of the subway station structure remains relatively minor. This highlights the positive effect of increased stiffness in the weak interlayer on the subway station structure's seismic resistance.

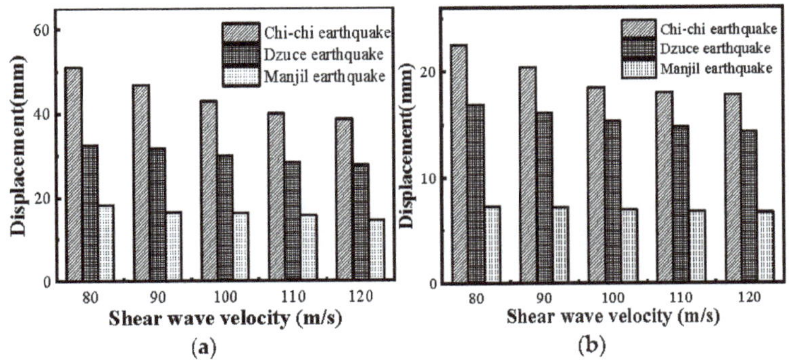

Figure 18. *Cont.*

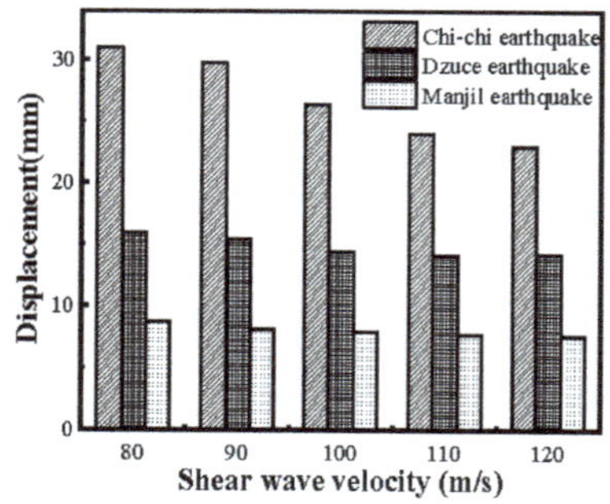

(c)

Figure 18. Maximum displacement of the station side wall and column. (**a**) Side wall. (**b**) Upper column. (**c**) Lower column.

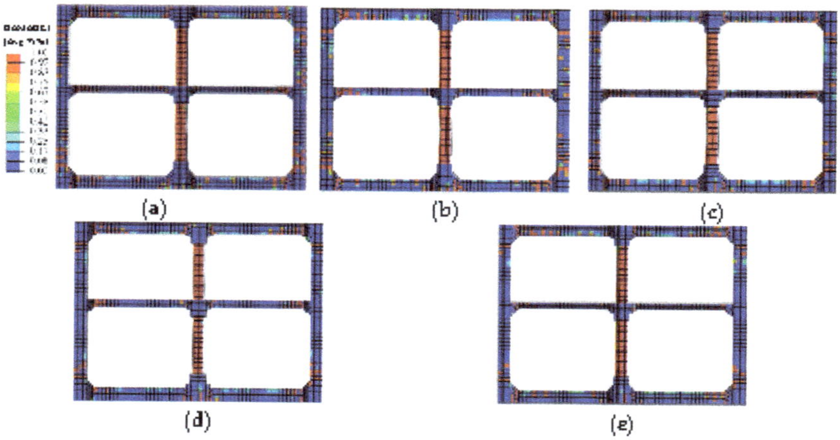

Figure 19. Tensile damage of the station with different shear wave velocities. (**a**) Shear wave velocity of 80 m/s. (**b**) Shear wave velocity of 90 m/s. (**c**) Shear wave velocity of 100 m/s. (**d**) Shear wave velocity of 110 m/s. (**e**) Shear wave velocity of 120 m/s.

3. Seismic Mitigation Effect of FPB in Weak Interlayer Site

The research in the previous section indicates that the weak interlayer has an adverse impact on subway station structure. This section proposes a seismic mitigation measure that involves installing FPB on the top of the central column in subway station structure situated within weak interlayer site. In order to investigate the seismic damping effect of FPB in ground with a weak interlayer, the structural dynamic time-history analysis of this proposed mitigation measure is conducted.

3.1. The Mechanical Model of FPB

FPB is typically categorized into single-sliding-surface and double-sliding-surface bearings. The single-sliding-surface FPB is the most commonly used type. The FPB primarily consists of upper support, a slider, and lower support. The main structural material of the bearing is typically metal, ensuring high reliability and easy quality assurance. The friction surface is coated with low-friction materials such as polytetrafluoroethylene (PTFE). The specific structural configuration of the bearing is depicted in Figure 20.

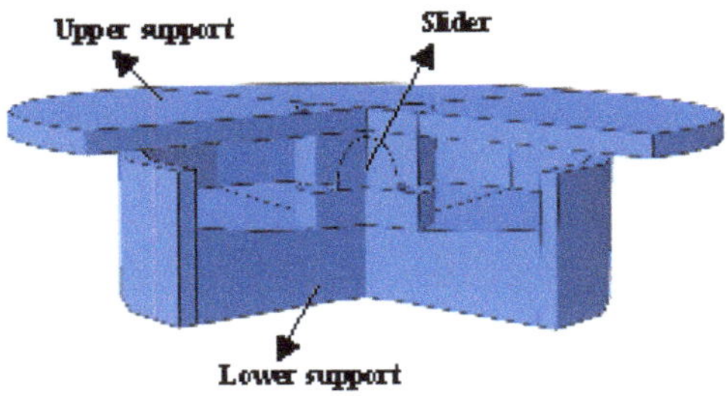

Figure 20. Section of the friction pendulum bearing.

Within the friction pendulum bearing, the slider, under the influence of seismic motion, experiences a vertical load denoted as W. The sliding friction force on the bottom surface of the slider is represented as f. The horizontal restoring force of the friction pendulum bearing is designated as F. The normal force on the sliding surface is labeled as N. The radius of the sliding surface is R and the horizontal sliding distance of the slider is D. The angular displacement of the slider relative to the axis of symmetry is represented as θ. As illustrated in Figure 21, when analyzing the forces acting on the bearing, the friction force can be expressed as follows, with μ denoting the friction coefficient of the sliding surface and the normal force being represented by N. The equilibrium of forces on the slider can be expressed as

$$F \cdot R\cos\theta = W \cdot D + f \cdot R \tag{5}$$

When the angular displacement of the slider is small, N is almost equal to W, and Equation (5) can be simplified to

$$F = \frac{W \cdot D}{R} + f = \frac{W \cdot D}{R} + \mu W \mathrm{sgn}(\theta) \tag{6}$$

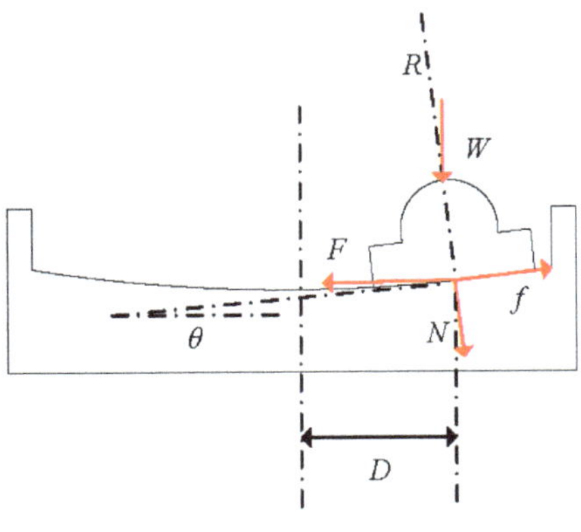

Figure 21. Force analysis diagram of the slide block.

3.2. Validation of Model Suitability

To ensure the model's validity, it is imperative to conduct finite element simulations on the FPB. Using the finite element software ABAQUS, a three-dimensional model of the friction pendulum bearings is constructed. The cross-section and finite element model of FPB are shown in Figures 22 and 23, respectively. The FPB was modeled by using an elastic constitutive model. The material of FPB is steel characterized by a density of 7850 kg/m^3, an elastic modulus of 2×10^5 MPa, and a Poisson's ratio of 0.3. The friction coefficient of the FPB is 0.1. The normal direction indicates a rigid contact. The contact between the slider's sliding surface and the lower support's sliding surface follows the same conditions.

The mesh employed consists of C3D8R (a linear reduced integration solid element with eight nodes) with a mesh size ranging from 10 to 50 mm.

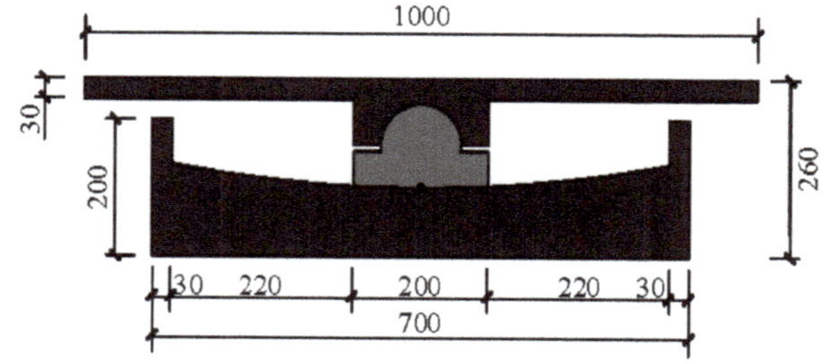

Figure 22. Cross section of the friction pendulum bearing.

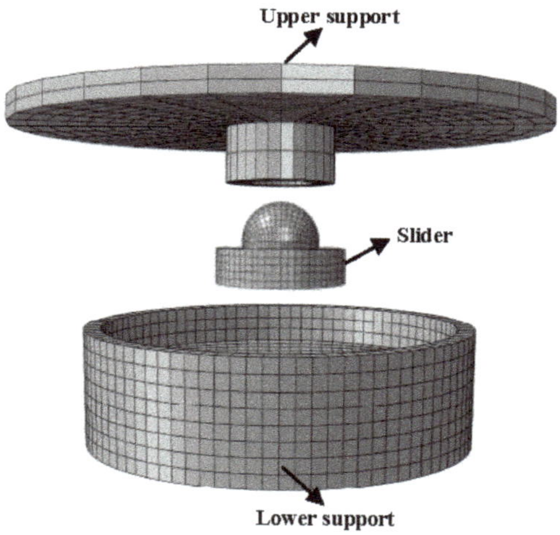

Figure 23. 3D detailed finite element model of the friction pendulum bearing.

Vertical concentrated forces of 100 kN and 200 kN were individually applied at the center of the upper support to obtain the hysteresis curve. The curve representing the restoring force–displacement relationship of the FPB is shown in Figure 24. The curve calculated using finite element software is compared with the curve obtained from Equation (6). The comparison reveals a close alignment between the two curves, demonstrating a consistent trend. This indicates that the finite element model of the FPB effectively simulates its mechanical characteristics.

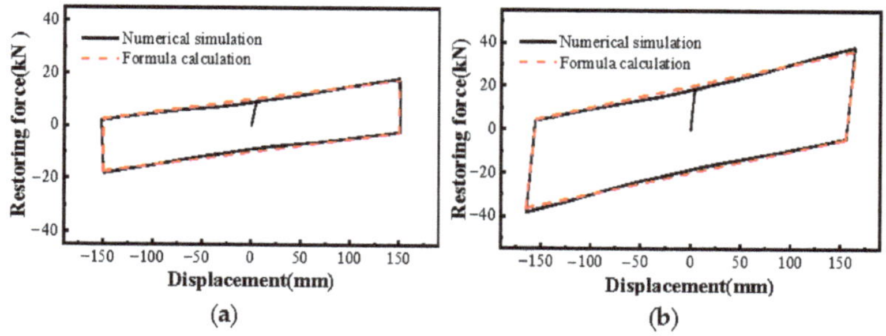

Figure 24. Verification of the finite element simulation results of the friction pendulum bearing. (**a**) 100 kN. (**b**) 200 kN.

3.3. The Seismic Mitigation Effect of FPB

The dimensions of the FPB used in this study are shown in Figure 22. The radius of the bottom plate sliding surface is 1000 mm and the friction coefficient of the FPB is assumed as 0.1. The upper support surface of the FPB is anchored to the bottom of the beam and the lower support surface of the lower support block is anchored to the top of the central column. The specific finite element model is depicted in Figure 25. The boundary condition of the soil layer and the contact between the soil and the station are consistent with the calculation model in Section 2. Here, the Chi-Chi earthquake is selected as the input earthquake, only considering the horizontal ground motion. Calculations are performed for the scenarios listed in Table 2. The seismic response of the subway station structure with and without the FPB is analyzed to assess the seismic mitigation effects of the FPB. In this section, the subway station structure without the FPB model is referred to as the original structure, while the subway station structure with the FPB is referred to as the seismic mitigation structure.

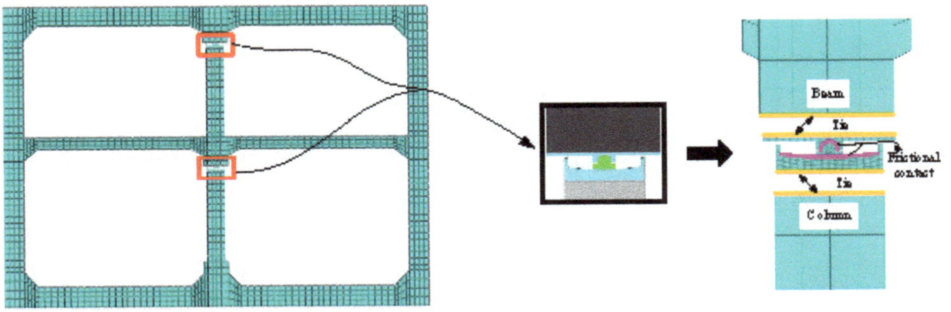

Figure 25. Finite element model of friction pendulum bearing.

Figures 26–28 present the differences in maximum displacements of the subway station structure's side wall and central column under various conditions. A positive difference in displacement indicates that the FPB can reduce the structure's displacement, while a negative difference suggests that the FPB may increase the structure's displacement. In some of the figures, red boxes are used to highlight and magnify specific areas, which are also shown within the same figure. From Figures 26–28, it is evident that the seismic mitigation structure has slightly increased the maximum displacement of the side wall compared to the original structure, but the increase is relatively small. On the other hand, the subway station structure with FPB has led to a reduction in the maximum displacement of the central columns. This reduction is relatively small under minor seismic forces (0.05 g, 0.1 g) but more significant under severe seismic conditions (0.2 g, 0.4 g); the maximum reduction could reach approximately 36 mm. The displacement values of the side wall of the subway station structure with FPB on the top of the central column are slightly bigger than that of the original structure, which is due to the decreased lateral stiffness of the structure with FPB on the top of the central column. The reduction in central column displacement is attributed to the change in boundary constraints on the top of the central column, transforming it from a fixed connection to an approximate hinge. This change substantially releases the horizontal displacement of the central column, resulting in the displacement of the top slab of the seismic mitigation structure not being transferred to the upper central column effectively.

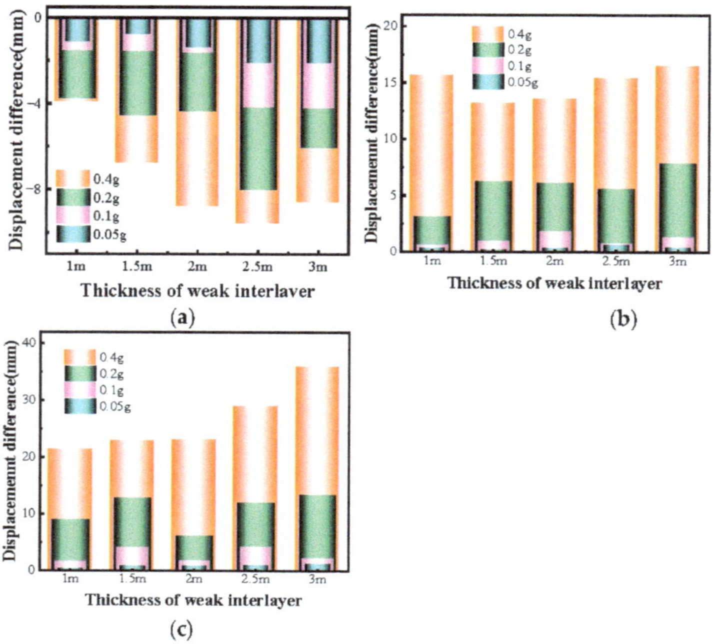

Figure 26. Difference in the maximum displacement of the side wall and column with different thicknesses. (**a**) Side wall. (**b**) Upper column. (**c**) Lower column.

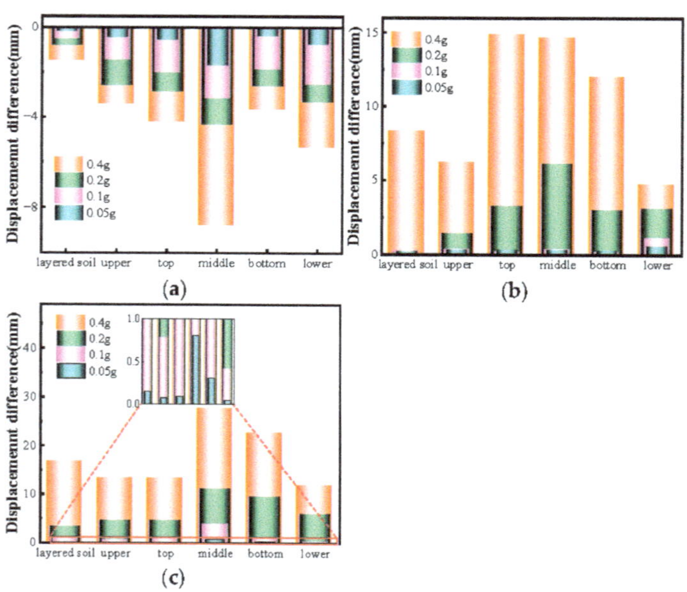

Figure 27. Difference in the maximum displacement of the side wall and column at different positions. (**a**) Side wall. (**b**) Upper column. (**c**) Lower column.

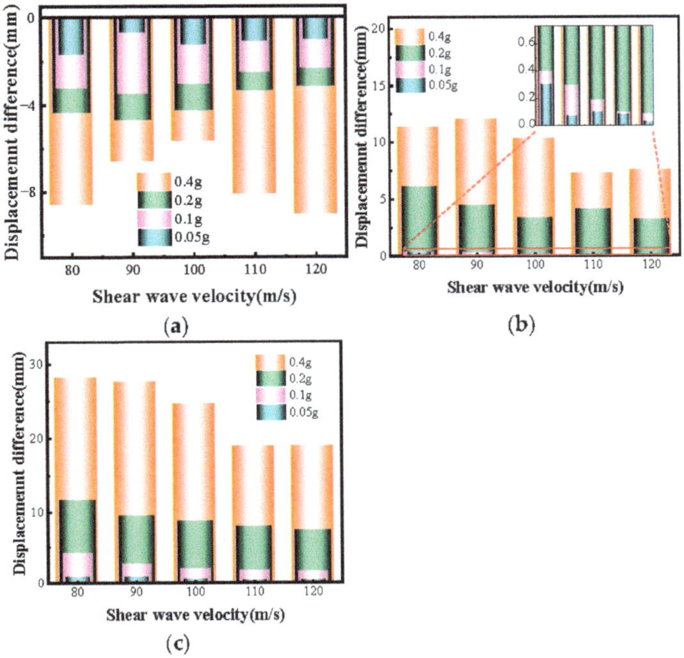

Figure 28. Difference in the maximum displacement of the side wall and column with different shear wave velocities. (**a**) Side wall. (**b**) Upper column. (**c**) Lower column.

In order to gain a deeper understanding of the deformation characteristics of the subway station structure, Figure 29 illustrates the inter-story displacements for various conditions. As shown in Figure 29, the inter-story displacements of the subway station's structural layers remain well below 1/250 under seismic forces, which meets the design standard. Importantly, the implementation of FPB on the top of the central column leads to an increase in inter-story displacements. The maximum inter-story displacement angle occurs at the lower level of the subway station structure. The increase in the inter-story displacement angle of the subway station structure is attributed to the release of horizontal displacement in the central column by the FPB, resulting in an overall reduction in the station's lateral stiffness.

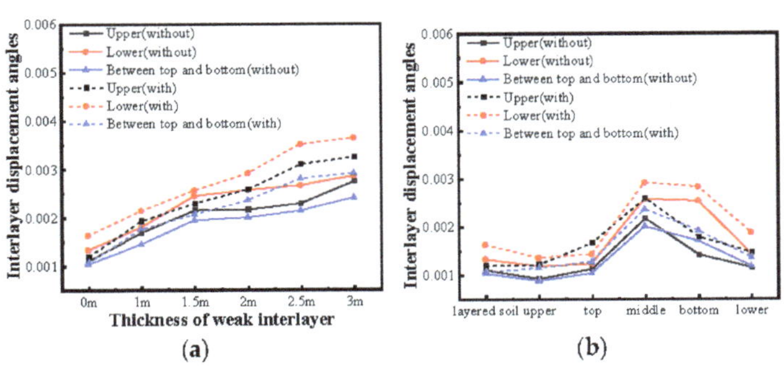

Figure 29. *Cont.*

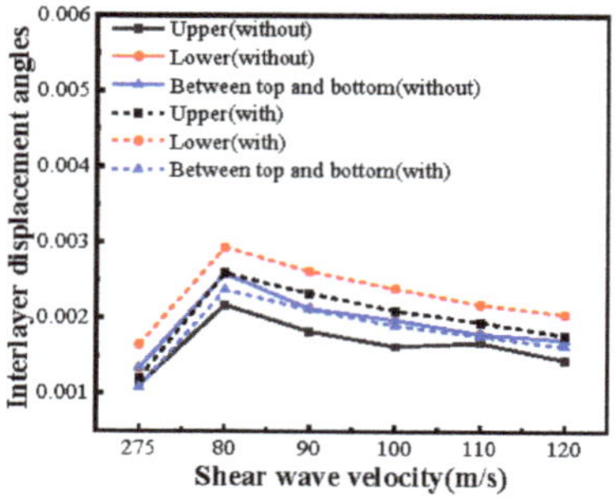

(c)

Figure 29. Interstorey displacement angle of the station under different conditions. (**a**) Thickness of the interlayer. (**b**) Positions. (**c**) Shear wave velocity.

Figure 30 compares tensile damage nephograms of the structure under two unfavorable conditions and in a layered soil site. From this illustration, it is evident that the central column of the original structure is nearly destroyed, while the seismic mitigation structure experiences only minor tensile damage. When the subway station structure is situated in a site with a weak interlayer, the implementation of FPB shifts the damage that occurs at the connection of the top and bottom slabs with the central column to the connection points between the top and bottom slabs with the side wall. During an earthquake, the damage to the central column typically occurs earlier than that to the side wall. The use of FPB can effectively mitigate the seismic damage to the central column, ensuring the protection of the subway station structure.

To provide a more intuitive illustration of the seismic damping effect of FPB in a site with a weak interlayer, the shear forces and bending moments on the bottom of the structural central column between the original and seismic mitigation structures are compared. In this context, the damping ratio is employed as the criterion for evaluating the damping effectiveness. The damping ratio can be defined using Equation (7) in this context.

$$\gamma = \frac{A_0 - A_R}{A_0} \tag{7}$$

In the equation, under the same conditions, A_0 represents the internal force values within the top and bottom cross-sections of the central columns of the original structure, while A_R represents the internal force values within the top and bottom cross-sections of the central columns of the seismic mitigation structure.

Figure 31 provides the damping ratio plots for structures under different site conditions. It can be seen from Figure 31 that the damping ratio of the seismic mitigation structure in the weak interlayer site is higher compared to that in the layered soil site. This implies that the seismic mitigation effect of FPB in weak interlayer sites surpasses their seismic mitigation effect in layered soil sites. The maximum damping ratio observed in the shear force of the central column could reach 94%. The central column of the subway station structure is the component most vulnerable to seismic forces and prone to shear failure under seismic motion. The seismic mitigation scheme with FPB on the top of the central column of the subway station in a site with a weak interlayer can effectively reduce the damage to the subway station structure, demonstrating excellent seismic mitigation effects. Therefore, when the station is situated on a site with a weak interlayer, the use of FPB for seismic design is worth considering.

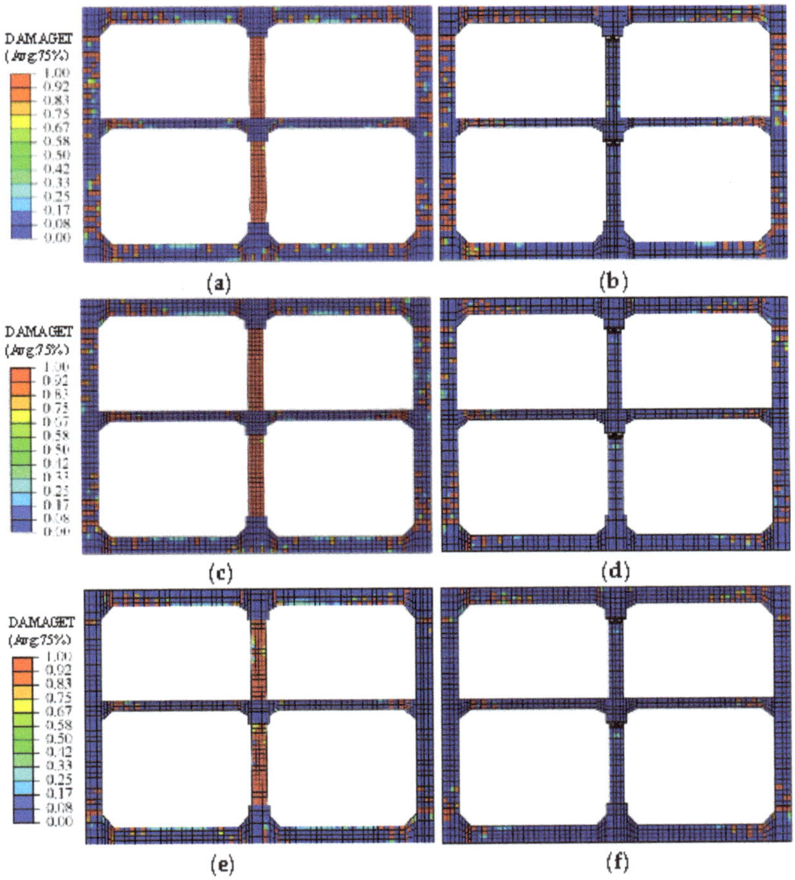

Figure 30. Tensile damage of the station under different conditions. (**a**) H= 3 m at 0.4 g (original structure). (**b**) H = 3 m at 0.4 g (shock absorption structure). (**c**) Middle at 0.4 g (original structure). (**d**) Middle at 0.4 g (shock absorption structure). (**e**) Layered soil site at 0.4 g (original structure). (**f**) layered soil site at 0.4 g (shock absorption structure).

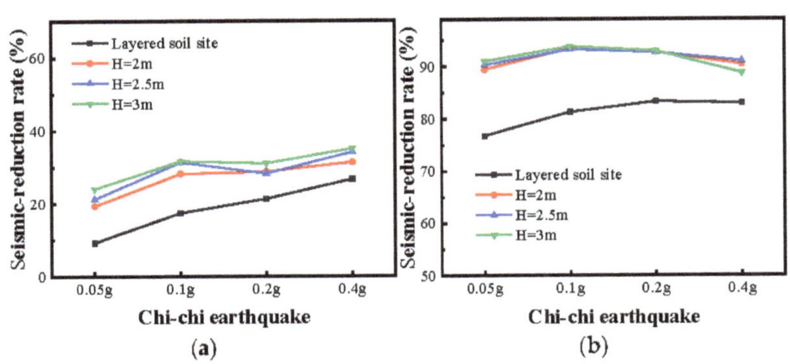

Figure 31. *Cont.*

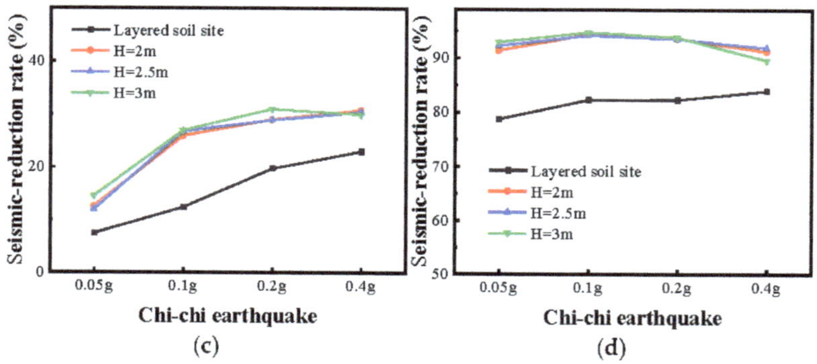

Figure 31. Decreasing amplitude ratio of the force of structure under the Chi-Chi earthquake. (**a**) Bending moment. (**b**) Shearing force. (**c**) Bending moment. (**d**) Shearing force.

4. Conclusions

This study investigates the seismic response of the subway station structure with a double-layer double-span rectangular underground foundation. The influence of factors such as weak interlayer thickness, location, and strength on the seismic response of subway station structures is investigated by establishing the two-dimensional finite element model of the soil-structure interaction. This paper proposes a seismic mitigation structure that involves installing FPB on the top of the central column within the subway station. Then, the seismic mitigation effectiveness of FPB in weak interlayer conditions is discussed by the two-dimensional dynamic time history analysis method of the soil structure systems. The main conclusions are as follows:

(1) When the weak interlayer is situated in the middle of the subway station structure, the seismic effect is most pronounced. If the weak interlayer is located at the top or bottom of the subway station structure, the structural impact is less pronounced. The influence is close to that of a layered soil site condition when the weak interlayer is situated in the upper and lower sections of the station. These patterns become especially evident when the seismic acceleration reaches 0.4 g;

(2) When the weak interlayer is situated in the middle of the subway station structure, the maximum horizontal displacement of the structure follows a trend of initially decreasing and then increasing with an increase in the thickness of the weak interlayer. The greater thickness of the weak interlayer exacerbates seismic tensile damage to the structure, particularly affecting the central columns and side wall. A weak interlayer thickness of 3 m is the most detrimental to structural safety;

(3) With the increase in shear wave velocity, there is a decreasing trend in the maximum displacement of the structural side wall and central column. In other words, an increase in the hardness of the weak interlayer can restrain structural deformations and reduce tensile damage to the structural side wall;

(4) In comparison to the original structure, the installation of FPB on the top of central columns reduces the overall structural lateral stiffness of the station, resulting in a slight increase in the maximum displacement of the side wall and the inter-story displacement angle of the upper and lower levels of the subway station structure. It also increases tensile damage to the side wall;

(5) The installation of FPB effectively reduces the deformation of central columns, especially under strong seismic conditions (0.4 g), the maximum reduction could reach 36 mm. This effectively reduces tensile damage of the central column but simultaneously transfers damage at the connection points of the top and bottom slabs with the central columns to the connection points with the side wall;

(6) The seismic mitigation scheme with FPB on the top of central column of subway station structure significantly reduces the internal forces in the central column. The most

pronounced seismic reduction is observed in shear forces, with a reduction rate of up to nearly 95%. In complex site conditions involving weak interlayers, the seismic reduction effect of the seismic mitigation structure is superior to that of layered ground conditions.

In general, based on the findings of this study, it is evident that a weak interlayer has an adverse impact on the seismic performance of station structures. The installation of FPB on the top of the central column of the subway station structure effectively reduces seismic damage to the structure. The proposed seismic mitigation scheme in this paper is suitable for situations where the subway station structure is located in a weak interlayer site. Additionally, the seismic mitigation effectiveness in weak interlayer sites surpasses that in layered soil sites. Particularly in the case of larger seismic events, the seismic mitigation effect becomes more pronounced. This highlights the importance of using FPB for seismic design in complex site conditions. However, it should be noted that the installation of FPB may increase damage to the connections between the structural side wall and the top and bottom slabs. Therefore, it is necessary to strengthen the seismic performance at these locations, especially when subjected to significant seismic input.

Author Contributions: Conceptualization, Z.X. (Zigang Xu) and C.L.; methodology, Z.X. (Zigang Xu) and R.H.; software, C.L.; validation, Z.X. (Zigang Xu); formal analysis, Z.X. (Zongyao Xia); data curation, Z.X. (Zigang Xu); writing—original draft preparation, C.L.; writing—review and editing, Z.X. (Zongyao Xia); visualization, Z.X. (Zongyao Xia); project administration, Z.X. (Zigang Xu); funding acquisition, Z.X. (Zigang Xu) and R.H. All authors have read and agreed to the published version of the manuscript.

Funding: This research is jointly funded by the National Natural Science Foundation of China (No. 52108453), the Youth Talent Support Project Program of Jiangxi Province (No. 2024QT02), the Natural Science Foundation of Jiangxi Province (No. 20212BAB214014 and No. 20232BAB204084), the China Postdoctoral Science Foundation Project (No. 2023M741159), and the Natural Science Foundation of Inner Mongolia Autonomous Region (No. 2023QN05012).

Institutional Review Board Statement: Not applicable.

Informed Consent Statement: Not applicable.

Data Availability Statement: Data are available in a publicly accessible repository.

Conflicts of Interest: The authors declare no conflicts of interest.

References

1. Amorosi, A.; Boldini, D. Numerical modelling of the transverse dynamic behaviour of circular tunnels in clayey soils. *Soil Dyn. Earthq. Eng.* **2009**, *29*, 1059–1072. [CrossRef]
2. Cilingir, U.; Madabhushi, S.G. Effect of depth on the seismic response of square tunnels. *Soils Found.* **2011**, *51*, 449–457. [CrossRef]
3. Tsinidis, G. Response characteristics of rectangular tunnels in soft soil subjected to transversal ground shaking. *Tunn. Undergr. Space Technol.* **2017**, *62*, 1–22. [CrossRef]
4. Hashash, Y.M.; Hook, J.J.; Schmidt, B.; John, I.; Yao, C. Seismic design and analysis of underground structures. *Tunn. Undergr. Space Technol.* **2001**, *16*, 247–293. [CrossRef]
5. Wang, Z.; Zhang, Z. Seismic damage classification and risk assessment of mountain tunnels with a validation for the 2008 Wenchuan earthquake. *Soil Dyn. Earthq. Eng.* **2013**, *45*, 45–55. [CrossRef]
6. Wang, T.; Kwok, O.L.A.; Jeng, F.S. Seismic response of tunnels revealed in two decades following the 1999 Chi-Chi earthquake (Mw 7.6) in Taiwan: A review. *Eng. Geol.* **2021**, *287*, 106090. [CrossRef]
7. An, X.; Shawky, A.; Maekawa, K. The Collapse Mechanism of a Subway Station during the Great Hanshin Earthquake. *Cem. Concr. Compos.* **1997**, *19*, 241–257. [CrossRef]
8. Iida, H.; Hiroto, T.; Yoshida, N.; Iwafuji, M. Damage to Daikai subway station. *Soils Found.* **1996**, *36*, 283–300. [CrossRef]
9. Seokwon, J.; Jongwoo, K.; Youngho, S.; Changwoo, H. Effect of a fault and weak plane on the stability of a tunnel in rock—A scaled model test and numerical analysis. *Int. J. Rock Mech. Min. Sci.* **2004**, *41*, 658–663.
10. Huang, F.; Zhu, H.; Xu, Q.; Cai, Y.; Zhuang, X. The effect of weak interlayer on the failure pattern of rock mass around tunnel–Scaled model tests and numerical analysis. *Tunn. Undergr. Space Technol.* **2013**, *35*, 207–218. [CrossRef]
11. Zhuang, H.; Wang, X.; Chen, G. Earthquake responses of subway station with different depths of soft soil. *Chin. J. Geotech. Eng.* **2009**, *31*, 1258–1266. (In Chinese)
12. Du, N. The Seismic Response Analysis of the Subway Statio in Soil with Saturated Weak Intercalations. Master's Thesis, Beijing Jiaotong University, Beijing, China, 2010. (In Chinese)

13. Dou, Y.; Feng, F.; Wang, J.; Fan, J.; Zheng, S. The 3D seismic response analysis of subway station built in weak interlayer. *Struct. Eng.* **2019**, *35*, 140–148. (In Chinese)
14. Chen, S.; Tang, B.; Zhao, K.; Li, X.; Zhuang, H. Seismic response of irregular underground structures under adverse soil conditions using shaking table tests. *Tunn. Undergr. Space Technol.* **2020**, *95*, 103145. [CrossRef]
15. Huang, Z.; Pitilakis, K.; Tsinidis, G.; Argyroudis, S.; Zhang, D.-M. Seismic vulnerability of circular tunnels in soft soil deposits: The case of Shanghai metropolitan system. *Tunn. Undergr. Space Technol.* **2020**, *98*, 103341. [CrossRef]
16. Huang, Z.; Pitilakis, K.; Argyroudis, S.; Tsinidis, G.; Zhang, D.-M. Selection of optimal intensity measures for fragility assessment of circular tunnels in soft soil deposits. *Soil Dyn. Earthq. Eng.* **2021**, *145*, 106724. [CrossRef]
17. Shimamura, S.; Kasai, H.; Haruumi, M. Seismic isolation effect for a tunnel with a soft isolation layer. *Struct. Eng. Earthq. Eng.* **1999**, *16*, 143–154. [CrossRef] [PubMed]
18. Kim, D.; Konagai, K. Key parameters governing the performance of soft tunnel coating for seismic isolation. *Earthq. Eng. Struct. Dyn.* **2001**, *30*, 1333–1343. [CrossRef]
19. Kim, D.S.; Konagai, K. Seismic isolation effect of a tunnel covered with coating material. *Tunn. Undergr. Space Technol.* **2000**, *15*, 437–443. [CrossRef]
20. Konagai, K.; Kim, D.S. Simple evaluation of the effect of seismic isolation by covering a tunnel with a thin flexible material. *Soil Dyn. Earthq. Eng.* **2001**, *21*, 287–295. [CrossRef]
21. Chen, Z.; Liang, S.; Shen, H.; He, C. Dynamic centrifuge tests on effects of isolation layer and cross-section dimensions on shield tunnels. *Soil Dyn. Earthq. Eng.* **2018**, *109*, 173–187. [CrossRef]
22. Huo, H.; Bobet, A.; Fernandez, G.; Ramírez, J. Load transfer mechanisms between underground structure and surrounding ground: Evaluation of the failure of the Daikai station. *J. Geotech. Geoenviron. Eng.* **2005**, *131*, 1522–1533. [CrossRef]
23. Du, X.; Ma, C.; Lu, D.; Xu, C.; Xu, Z. Collapse simulation and failure mechanism analysis of the Daikai station under seismic loads. *China Civ. Eng. J.* **2017**, *50*, 53–62. (In Chinese)
24. Du, X.; Li, Y.; Xu, C.; Lu, D.; Xu, Z.; Jin, L. Review on damage causes and disaster mechanism of Daikai subway station during 1995 Osaka-Kobe earthquake. *Chin. J. Geotech. Eng.* **2018**, *40*, 223–236. (In Chinese)
25. Xu, C.; Zhang, Z.; Li, Y.; Du, X. Validation of a numerical model based on dynamic centrifuge tests and studies on the earthquake damage mechanism of underground frame structures. *Tunn. Undergr. Space Technol.* **2020**, *104*, 103538. [CrossRef]
26. Chen, Z.; Liu, Z. Optimal shock absorption layout of lead rubber bearing in a multi-story underground structure. In Proceedings of the GeoShanghai 2018 International Conference: Tunnelling and Underground Construction, Shanghai, China, 27–30 May 2018; Springer: Singapore, 2018; pp. 718–727. (In Chinese)
27. Liu, Z.; Chen, Z.; Liang, S.; Li, C. Isolation mechanism of a subway station structure with flexible devices at column ends obtained in shaking-table tests. *Tunn. Undergr. Space Technol.* **2020**, *98*, 103328. [CrossRef]
28. Zheng, Y.; Yue, C. Shaking table test study on the functionality of rubber isolation bearing used in underground structure subjected to earthquakes. *Tunn. Undergr. Space Technol.* **2020**, *98*, 103153. [CrossRef]
29. Chen, Z.; Chen, W.; Bian, G. Seismic Performance upgrading for underground structures by introducing shear panel dampers. *Adv. Struct. Eng.* **2014**, *17*, 1343–1357. [CrossRef]
30. Zhuang, H.; Zhao, C.; Chen, S.; Fu, J.; Zhao, K.; Chen, G. Seismic performance of underground subway station with sliding between column and longitudinal beam. *Tunn. Undergr. Space Technol.* **2020**, *102*, 103439. [CrossRef]
31. Yang, J.; Zhuang, H.; Wang, W.; Zhou, Z.; Chen, G. Seismic performance and effective isolation of a large multilayered underground subway station. *Soil Dyn. Earthq. Eng.* **2021**, *142*, 106560.
32. Bu, X.; Xu, Z.; Ledesma, A.; López-Almansa, F. Seismic target drifts for underground railway stations. Cases of traditional rigid connections and sliding interior columns. *Tunn. Undergr. Space Technol.* **2024**, *144*, 105555. [CrossRef]
33. Xu, Z.; Xia, Z.; Bu, X.; Han, R. Research on Seismic Performance of a Two-Story, Two-Span Underground Subway Station with Split Columns Based on the Quasi-Static Method. *Appl. Sci.* **2024**, *14*, 4077. [CrossRef]
34. Zayas, V.A.; Low, S.S.; Mahin, S.A. A simple pendulum technique for achieving seismic isolation. *Earthq. Spectra* **1990**, *6*, 317–333. [CrossRef]
35. Tesopelas, P.; Constantinou, M.C.; Kim, Y.S.; Okamoto, S. Experimental study of FPS system in bridge seismic isolation. *Earthq. Eng. Struct. Dyn.* **1996**, *25*, 65–78. [CrossRef]
36. Jangid, R.S. Optimum friction pendulum system for near-fault motions. *Eng. Struct.* **2005**, *27*, 349–359. [CrossRef]
37. Mazza, F.; Mazza, M. Nonlinear seismic analysis of irregular rc framed buildings base-isolated with friction pendulum system under near-fault excitations. *Soil Dyn. Earthq. Eng.* **2016**, *90*, 299–312. [CrossRef]
38. Xu, Z.; Du, X.; Xu, C.; Han, R. Numerical analyses of seismic performance of underground and aboveground structures with friction pendulum bearings. *Soil Dyn. Earthq. Eng.* **2020**, *130*, 105967. [CrossRef]
39. Quaglini, V.; Bocciarelli, M.; Gandelli, E.; Dubini, P. Numerical assessment of frictional heating in sliding bearings for seismic isolation. *J. Earthq. Eng.* **2014**, *18*, 1198–1216. [CrossRef]
40. Ma, C.; Lu, D.C.; Du, X.L. Seismic performance upgrading for underground structures by introducing sliding isolation bearings. *Tunn. Undergr. Space Technol.* **2018**, *74*, 1–9. [CrossRef]
41. Systèmes, D. *Abaqus Analysis User's Manual*; Simulia Corp.: Providence, RI, USA, 2007; p. 40.
42. Chen, Z.; Jia, P. Seismic response of underground stations with friction pendulum bearings under horizontal and vertical ground motions. *Soil Dyn. Earthq. Eng.* **2021**, *151*, 106984. [CrossRef]

43. Lee, J.; Fenves, G.L. Plastic-damage model for cyclic loading of concrete structures. *J. Eng. Mech.* **1998**, *124*, 892–900. [CrossRef]
44. Lubliner, J.; Oliver, J.; Oller, S.; Oñate, E. A plastic-damage model for concrete. *Int. J. Solids Struct.* **1989**, *25*, 299–326. [CrossRef]
45. Kampitsis, A.E.; Sapountzakis, E.J.; Giannakos, S.K.; Gerolymos, N.A. Seismic soil-pile-structure kinematic and inertial interaction-a new beam approach. *Soil Dyn. Earthq. Eng.* **2013**, *55*, 211–224. [CrossRef]

applied sciences

MDPI

Article

Seismic Mitigation Effect and Mechanism Analysis of Split Columns in Underground Structures in Sites with Weak Interlayers

Zigang Xu * and Zongyao Xia

State Key Laboratory of Performance Monitoring and Protecting of Rail Transit Infrastructure, East China Jiaotong University, Nanchang 330013, China; 18171912134@163.com
* Correspondence: xuzigang1027@163.com

Abstract: The seismic damage of underground structures has been extensively investigated, and it has been demonstrated that underground structures located at weak interlayer sites are more prone to damage. In this study, a two-story two-span rectangular frame subway station structure is analyzed. A two-dimensional soil-underground structure model is developed using the large-scale finite element analysis software ABAQUS. The equivalent linear soil-underground structure dynamic time-history analysis method is employed to examine the seismic response of underground structures at weak interlayer sites. Variations in the thickness and shear wave velocity of the weak interlayer soil are analyzed. The seismic mitigation effects of split columns and prototype columns in underground structures at weak interlayer sites are systematically compared. The findings indicate that the relative displacement and internal force of key structural components significantly increase when the weak interlayer intersects the underground structure. Furthermore, as the thickness of the interlayer increases, the displacement and internal force also escalate. When the thickness of the weak interlayer remains constant and the shear wave velocity decreases, the relative displacement and internal force of the key structural components gradually intensify. Replacing ordinary columns with split columns substantially reduces the internal force of the middle column, providing an effective seismic mitigation measure for underground structures.

Keywords: underground structures; weak interlayers; split column; subway station; seismic design

1. Introduction

The rapid urbanization and expansion of underground rail transit have become essential for alleviating surface transportation pressures and promoting urban development. As China is geographically situated between the circum-Pacific seismic belt and the Eurasian seismic belt, many of its cities with subway systems are in seismically active regions. Seismic damage data of underground structures reveal that these structures, like surface structures, are significantly affected during strong earthquakes, with catastrophic consequences in some cases. For instance, the 1995 Osaka Kobe earthquake in Japan completely destroyed over 30 columns at the Dakai subway station, leading to its collapse [1–5]. Similarly, the 2001 Chi-Chi earthquake in Taiwan and the 2008 Wenchuan earthquake in China caused extensive damage to urban subway systems [6,7].

At present, substantial progress has been made by Chinese scholars in the seismic research of underground structures [8–13]. However, most studies have primarily focused on ordinary layered sites. As underground construction scales expand, site conditions become increasingly complex, with weak interlayer sites being a notable example. Weak interlayer sites are characterized by layered or banded weak thin layers formed by soft soil. Due to high moisture content, low strength, and low shear strength, weak interlayer soil is prone to deformation and failure under seismic action, thereby exacerbating damage to underground structures. Consequently, it is imperative to investigate methods for mitigating structural damage at weak interlayer sites. Wu et al. [14] examined the seismic response of underground structures by varying the location and thickness of the weak interlayer. Their results indicated that the presence of weak interlayers significantly impacts seismic response, with the strongest effects occurring when the weak interlayer is located at the center of the structure. Huang et al. [15] analyzed the influence of seismic wave intensity on weak interlayers, demonstrating that the amplification effect is primarily governed by seismic wave velocity. Li et al. [16] investigated the seismic response of subway station structures in weak saturated soil interlayer sites using finite element analysis, finding that the weak interlayer amplifies seismic effects. The amplification is most pronounced when the interlayer is situated in the middle of the station. Dou et al. [17] explored the dynamic response of a two-story three-span subway station structure by considering various seismic waveforms and relative positions of the station and weak interlayer. Their findings underscored the significant influence of the station's location within the weak interlayer on structural stress and deformation, with the dynamic response peaking when the station is entirely within the weak interlayer.

In order to reduce the seismic damage to underground structures, two main methods are used to achieve seismic mitigation control: one method involves adding isolation materials between the structure and the soil, while the other involves applying an auxiliary seismic isolation device to the key components of the underground structure (such as the middle column). In 1987, the friction pendulum bearing (FPB) was introduced by Zayas et al. [18] at the University of California, Berkeley. The bearing operates based on the principle of a pendulum, with the structural natural period extended through the swaying motion of the upper support and slider. This motion effectively dissipates seismic energy. In practical engineering applications, FPBs have become widely employed. The idea of applying an isolation layer to underground structures was proposed by Zhao et al. [19] and Chen et al. [20], who carried out a series of theoretical analyses and experimental studies. The experiments showed that the seismic mitigation effect becomes more significant as the elastic modulus ratio of the seismic mitigation layer material to the surrounding rock decreases. Du et al. [21] suggested using the rocking column structure instead of the ordinary middle column to reduce damage to the middle column. The research demonstrated that the seismic performance of the new underground frame structure with a rocking column is significantly improved, and it possesses good self-centering ability, achieving the seismic design goals for underground structures.

The middle column of the underground structure is similar to the middle column of the ground structure, as it must bear a large vertical load. Due to the large shear span ratio of the column, shear failure is likely to occur. In response to this phenomenon, Li et al. [22,23] systematically introduced the split column theory, addressing design, structural requirements, split column frames, and the seismic performance of joints. Hao et al. [24] conducted performance experiments on split columns under low cyclic horizontal load and proposed a calculation model for the bearing capacity of split columns. When the split column is subjected to the same vertical load as the ordinary column, it effectively reduces the lateral stiffness of the column and significantly improves the column's ductility, with enhanced deformation capacity and energy dissipation capacity [25,26]. Li [27] studied the mechanical properties of split columns, examining hysteresis curves, load capacity, and deformation capacity for different separation ratios. The research showed that split columns with varying separation ratios experience compression-bending failure. As the separation ratio decreases, the peak load capacity of the split column increases, ductility decreases, and energy dissipation capacity weakens. This behavior is also applicable to underground structures, effectively alleviating earthquake damage. Xu et al. [28] used the finite element analysis software ABAQUS 2023 to preliminarily examine the influence of split columns on the seismic mitigation effect of underground structures. Xu et al. [28–30] proposed applying split columns to underground two-story three-span subway station structures. The results indicated that split columns can effectively prevent the middle column from bearing excessive shear force and bending moment, fully utilizing both the vertical support capacity and horizontal deformation capacity of the split column.

In summary, this paper examines a two-story double-span rectangular frame subway station structure. The finite element analysis software ABAQUS is used to establish an overall model of the soil-underground structure. The equivalent linear soil-underground structure overall dynamic time history analysis method is employed to study the effects of weak interlayer thickness and shear wave velocity of weak interlayer soil. The seismic response of the underground structure under varying weak interlayer conditions is investigated, and the seismic mitigation effects of split middle columns and prototype cast-in-place middle columns in underground structures at weak interlayer sites are systematically compared.

2. Introduction of Split Column Technology

2.1. Force Characteristics of Split Columns

The split column technology utilizes partition plate material to divide a complete reinforced concrete column into several relatively independent unit columns, with its structural diagram shown in Figure 1. Studies [22–27] have shown that the use of split columns in ground structures can effectively reduce the flexural bearing capacity of the frame structure, causing its failure mode to shift from shear failure to bending failure. The deformation capacity and ductility of frame columns are significantly improved.

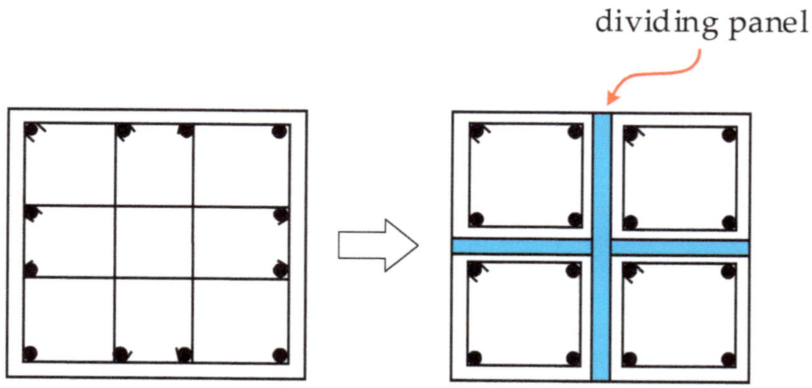

Figure 1. Structure diagram of split column.

The strain distribution pattern of the split column during loading is given in Figure 2. b denotes the overall width of the column structure and b_1 denotes the width of the split unit column.

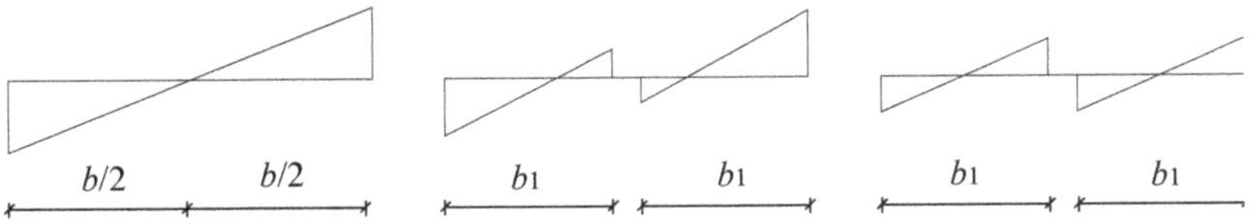

Figure 2. Strain distribution of split column concrete.

When the horizontal displacement is small, the friction effect between the clapboard and the unit column provides a strong bonding effect. At this stage, the clapboard effectively constrains the adjacent unit columns to produce a synergistic effect, and the stress variation in the concrete section resembles that of an ordinary column. As the horizontal displacement gradually increases, the friction effect between the partition and the unit column decreases, though it still provides some effective bonding. The deformation ability of the concrete on both sides of the friction restraint partition weakens, and the strain of the inner edge of the cross-section of the unit column on both sides is smaller than that of the outer edge. When the horizontal displacement continues to increase, the friction effect between the partition plate and the unit column disappears completely, and the partition plate no longer restrains the deformation of the unit column. At this point, each unit column operates independently, and the strain neutral axis of the unit column section shifts toward the center of each section.

2.2. *Comparative Analysis of Skeleton Curves*

Some scholars have designed ordinary reinforced concrete columns and reinforced concrete split columns for pseudo-static experiments based on the actual situation. The failure modes of ordinary reinforced concrete columns and reinforced concrete split columns have been studied. The seismic performance of various components, including ductility, rigidity degeneration, and energy dissipation capacity, is analyzed [25,31]. The reinforcement details of column specimens are shown in Figure 3.

In this paper, finite element analysis software ABAQUS is used to establish models for ordinary concrete columns (PTZ) and reinforced concrete split columns (FTZ), and each module is set up, including components, material properties, assembly, constraints, boundary conditions, analysis steps, and meshing steps. In the model, concrete is modeled using the three-dimensional eight-node reduced integration solid element (C3D8R), the steel bars are modeled using the three-dimensional two-node truss element (T3D2), and the clapboard is also modeled with the three-dimensional eight-node reduced integration solid element (C3D8R). The constitutive model for concrete adopts the plastic damage model, while the ideal double broken line model is used for the steel bars. The concrete constitutive parameters are shown in Figure 4, and additional plastic parameters for concrete are provided in Table 1. To ensure the accuracy of the simulation results, boundary conditions and load settings are kept consistent with the test. The bottom of the column is fixed, and the top of the column is subjected to axial pressure and horizontal displacement constraints. In the interaction module, the steel element is embedded in the concrete element through the built-in area (Embedded) in the module. The simulation model is shown in Figure 5.

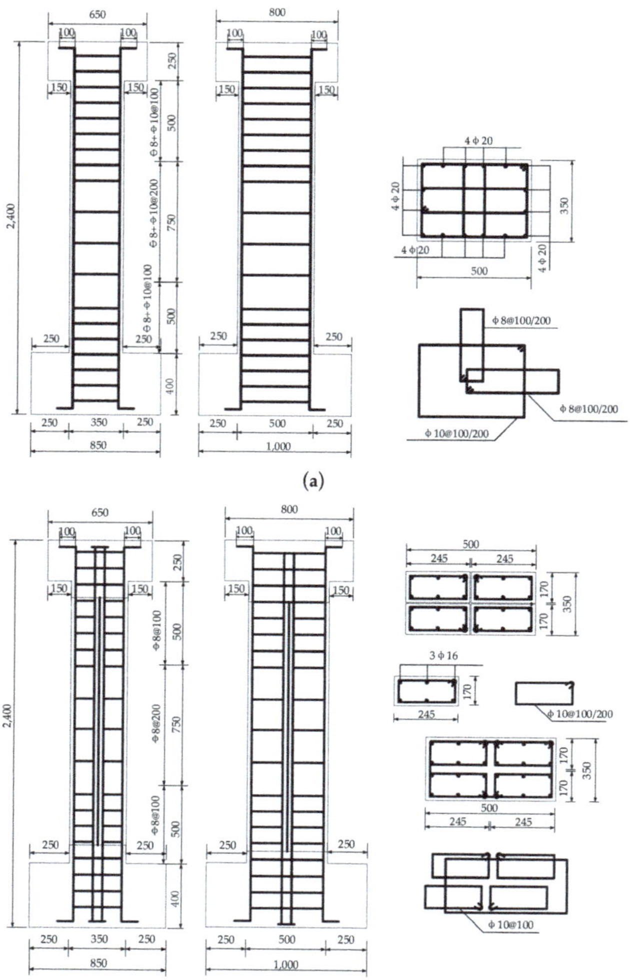

Figure 3. Reinforcement diagram for each specimen. (**a**) PTZ. (**b**) FTZ.

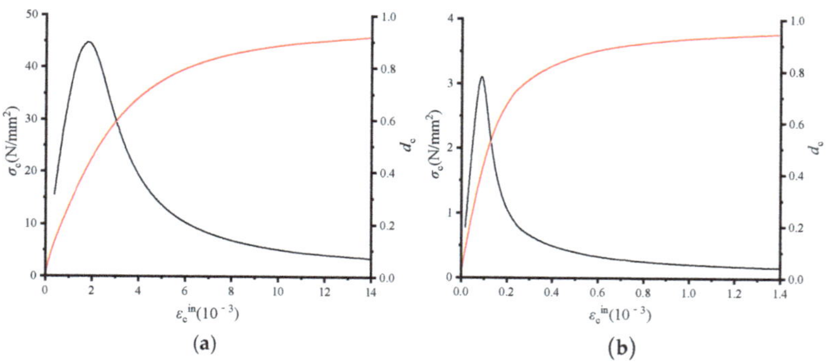

Figure 4. The relationship between stress, damage factor, and inelastic strain. (**a**) Compression behavior. (**b**) Tensile behavior.

Table 1. Other plasticity parameters of concrete.

Expansion Angle	Eccentricity	f_{b0}/f_{c0}	k	Viscosity Parameter
34	0.1	1.16	0.6667	0.0005

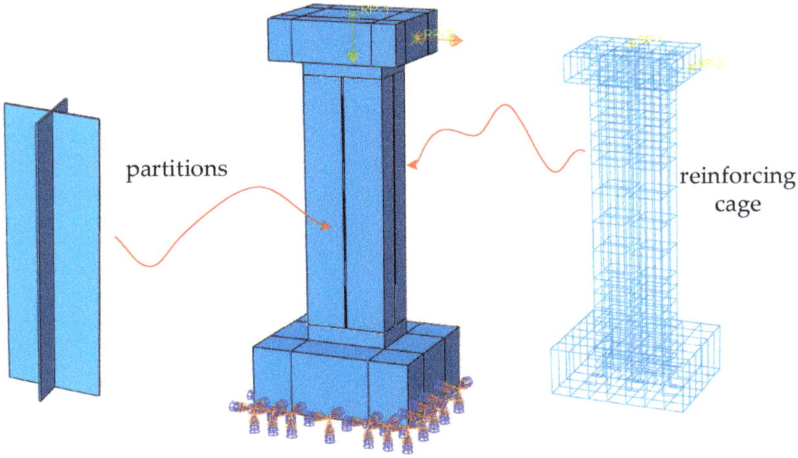

Figure 5. Finite element model of the specimen.

The skeleton curves of the split column and the whole column under both the test and numerical simulation are extracted, as shown in Figure 6. In the elastic stage, the horizontal load of the structure in the numerical simulation and the test is essentially the same. As the displacement increases, for PTZ, when the displacement reaches 55 mm, both the numerical simulation and the test show that the specimen reaches its peak at the same time. The horizontal peak in the test is approximately 245 kN, while the horizontal peak in the numerical simulation is 240 kN. As the displacement continues to increase, the decrease in horizontal load observed in the test becomes more apparent. When the displacement reaches 120 mm, the horizontal load decreases to 199 kN in the test. The numerical simulation shows that when the displacement reaches 120 mm, the horizontal load decreases to 223 kN. For FTZ, the numerical simulation shows that the horizontal load reaches a peak of 154 kN when the displacement reaches 61 mm. The test shows that the horizontal load reaches a peak of 159 kN when the displacement reaches 91 mm. As the displacement increases further, the decrease in horizontal load observed in the test becomes more noticeable. When the displacement reaches 120 mm, the test shows that the horizontal load drops to 131 kN, while the numerical model shows that the horizontal load drops to 143 kN.

As shown in Figure 7, in the elastic stage, the load–displacement skeleton curves of FTZ and PTZ in the test are essentially consistent, and the load–displacement skeleton curve of PTZ in the numerical simulation is slightly larger than that of FTZ. The experiment shows that when the horizontal load reaches the peak, PTZ increases by approximately 54.1% compared to FTZ, while the numerical simulation shows that when the horizontal load reaches the peak, PTZ increases by approximately 55.8% compared to FTZ.

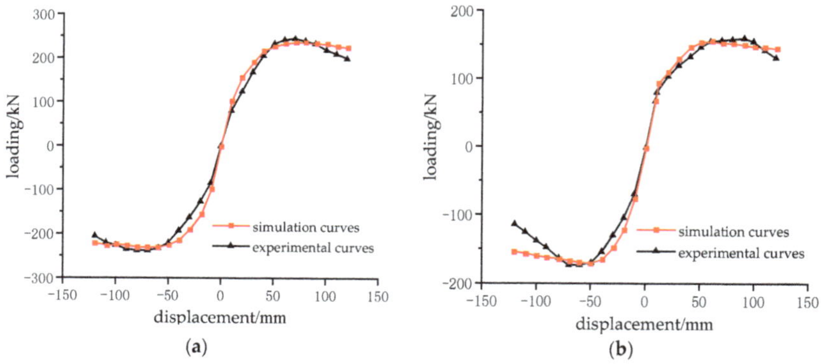

Figure 6. Comparison of the numerical simulation and experimental skeleton curve. (**a**) Ordinary column skeleton curve. (**b**) Split column skeleton curve.

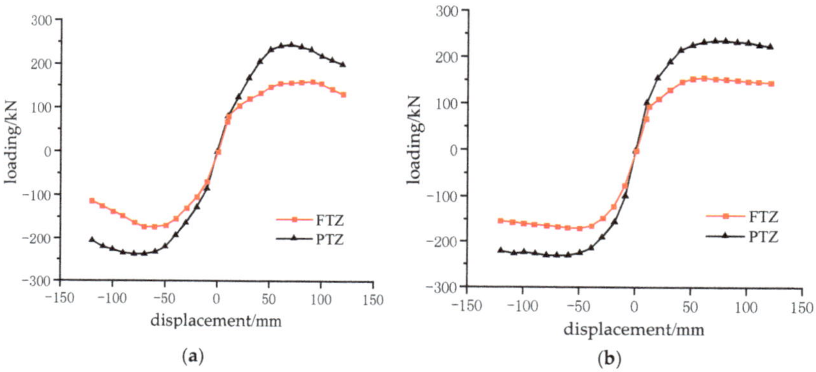

Figure 7. Comparison of skeleton curves for each specimen. (**a**) Test skeleton curve. (**b**) Simulation of skeleton curves.

3. Numerical Model

3.1. Project Profile

The subway station structure model shown in Figure 8 is selected. The model represents a typical double-layer double-span rectangular frame structure. The cross-sectional width of the station is 18 m, and the height of the station is 12 m. The thickness of the left and right walls is 0.8 m, the thickness of the roof is 0.75 m, the thickness of the middle plate is 0.45 m, and the thickness of the bottom plate is 0.8 m. The longitudinal middle column spacing distance is 6 m, and the cross-sectional size of the middle column is 0.8 m × 0.8 m. The station material is C50 strength concrete, with an elastic modulus of 35 GPa, a density of 2500 kg/m^3, and a Poisson's ratio of 0.2. In this model, only the influence of horizontal ground motion is considered. When the middle column of the underground structure is divided, only the direction of the horizontal seismic force is divided, and the longitudinal direction of the station is not divided; that is, the two-column structure. The structural diagram is shown in Figure 9.

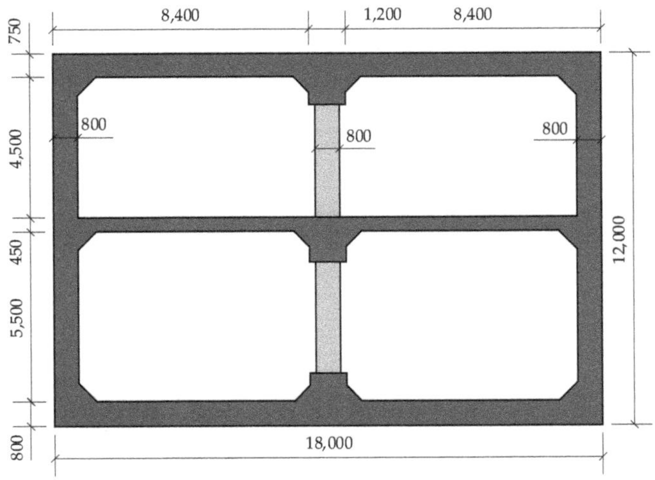

Figure 8. Cross section of prototype structure.

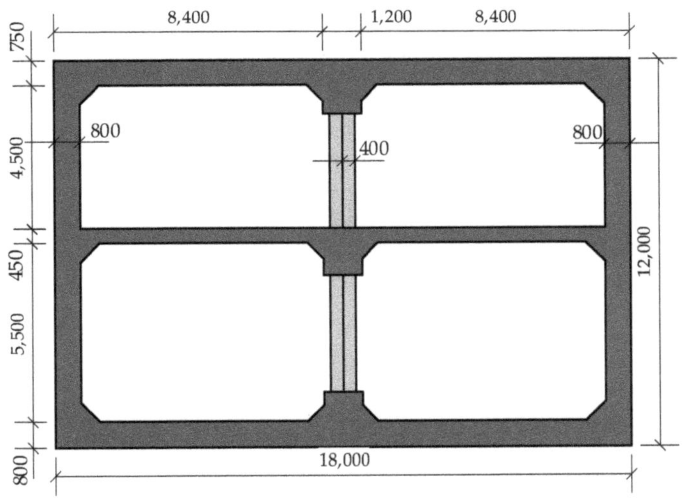

Figure 9. Cross section diagram of seismic mitigation structure.

Figure 10 shows that the buried depth of the station structure is 6 m. Previous studies have shown that when the weak interlayer is located in the middle of the underground structure, the seismic action has the greatest impact on the underground structure [16]. Therefore, in this model, only the case where the weak interlayer site passes through the middle of the underground structure is considered. The soil conditions of the site are shown in Table 2. The density of the homogeneous soil layer and weak interlayer is 2000 kg/m^3, Poisson's ratio is 0.3, the shear wave velocity of the homogeneous soil layer is 380 m/s, and the shear wave velocities of the four weak interlayers are 150 m/s, 125 m/s, 100 m/s, and 75 m/s, respectively.

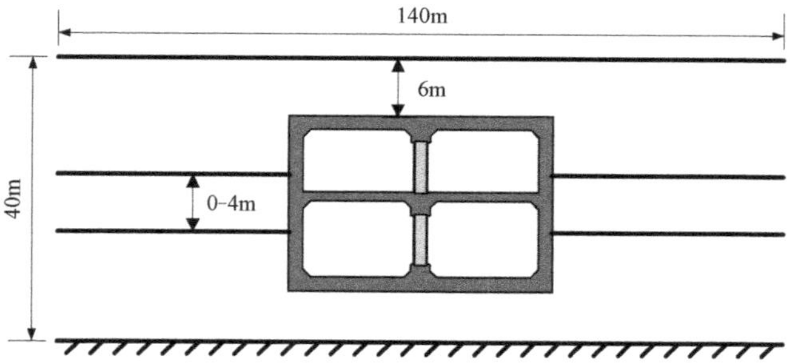

Figure 10. Schematic diagram of the soil-underground structure.

Table 2. Physical properties of soils.

Ground Level	Densities/kg/m^3	Shear Wave Velocity/m/s	Poisson's Ratio
sandy soil layer	2000	380	0.3
clay layer 1	2000	150	0.3
clay layer 2	2000	125	0.3
clay layer 3	2000	100	0.3
clay layer 4	2000	75	0.3

To reflect the influence of the thickness of the weak interlayer on the seismic response of the underground structure, four working conditions with the interlayer thickness increasing from 1 m to 4 m are set in this paper. The shear wave velocity of the weak interlayer is 125 m/s, and a working condition without a weak interlayer is set as the reference working condition for comparison. To reflect the influence of the degree of weakness of the weak interlayer on the seismic response of the underground structure, four additional working conditions are set, where the thickness of the weak interlayer is 3 m, and the shear wave velocities of the weak interlayer are 150 m/s, 125 m/s, 100 m/s, and 75 m/s, respectively. The parameters for each working condition are shown in Table 3.

Table 3. List of operating conditions.

Working Condition	Interlayer Thickness/m	Shear Wave Velocity in Weak Interlayers/m/s	Homogeneous Layer Shear Wave Velocity/m/s
0	interlayer-free	/	380
1	1	125	380
2	2	125	380
3	3	125	380
4	4	125	380
5	3	150	380
6	3	100	380
7	3	75	380

3.2. Finite Element Model

The model was established using the finite element analysis software ABAQUS, as shown in Figure 11 [32]. The calculation and analysis method adopted the overall dynamic time history analysis method of the soil-underground structure, based on equivalent linearization, as proposed by Du et al. [33]. The elastic modulus of the soil layer was calculated using Equations (1) and (2).

$$G = \rho V_S^2 \tag{1}$$

$$E = 2(1 + \nu)G \tag{2}$$

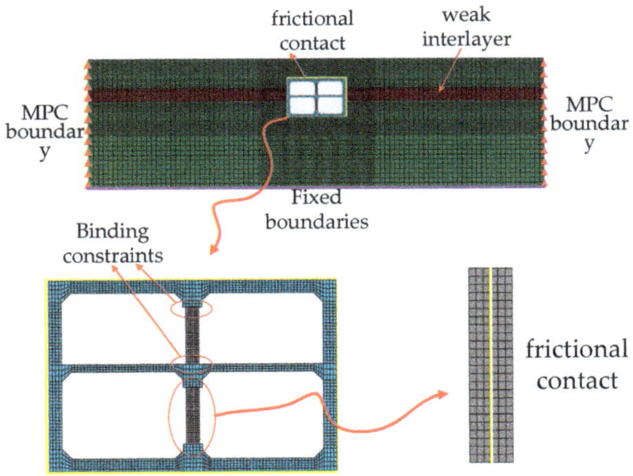

Figure 11. Two-dimensional finite element model.

Here, ρ represents the density of the soil layer, V_s stands for the shear wave velocity of the soil layer, and ν represents the Poisson's ratio of the soil layer.

In defining the material properties of the soil layer, the seismic mitigation of the soil material needed to be specified. In this model, the Rayleigh seismic mitigation model was selected. The frequency analysis module in ABAQUS was used to obtain the natural frequency of the test soil. The first-order and second-order frequencies were selected, and the Rayleigh seismic mitigation coefficient was calculated using Formulas (3) and (4).

$$\alpha = \frac{2D \times w_1 \times w_2}{w_1 + w_2} \tag{3}$$

$$\beta = \frac{2D}{w_1 + w_2} \tag{4}$$

Here, w_1 and w_2 are the first and second order self-oscillation frequencies of the soil layer and D is the seismic mitigation ratio taken as 0.1 [34].

The model materials were all elastic constitutive models. The total width of the soil-structure system model was 140 m, and the height was 40 m, which met the requirement of being more than seven times the width of the structure. The middle columns of the station were distributed equidistantly along the longitudinal direction of the station, with a spacing of 6 m between adjacent middle columns. Due to the two-dimensional model adopted in this paper, the station structure used the plane strain unit to establish the model, and the material of the middle column needed to be reduced according to Formula (5) [35].

$$E' = E\frac{h}{h'} \quad \rho' = \rho\frac{h}{h'} \tag{5}$$

Here, E' is the equivalent elastic modulus, ρ' is the equivalent material density, h is the length of the middle column section along the longitudinal direction of the subway, and h' is the distance between the two adjacent middle columns in the longitudinal direction.

After reduction, the elastic modulus of the middle column material is 4.67 GPa, the density is 333.3 kg/m^3, and the Poisson's ratio is 0.2. The binding boundary utilizes the MPC (Multi-Point Constraint) node degree of freedom coupling feature within ABAQUS, effectively binding the finite element model nodes of the soil to the boundary nodes at the same height, ensuring these nodes share identical displacements throughout the analysis [36]. Previous studies have shown that the boundary form has good calculation accuracy in soil-structure dynamic analysis [37] and pseudo static analysis [38]. The contact type between the station structure and the surrounding soil is surface-to-surface contact, and the contact property is friction contact. The tangential behavior is considered as the friction force to prevent the relative sliding between surfaces, and the friction coefficient is set to be 0.4. A hard contact is established in the normal direction of the interface. In order to simplify the analysis model, the dividing plate element is not set in the middle column of the seismic mitigation structure. The friction contact is set between the two adjacent unit columns, and the friction coefficient is 0.8 [33]. The soil and structure are simulated by the four-node plane strain element (CPE4R) in ABAQUS. On the grid division, the grid size of the station structure is 0.5 m × 0.5 m, and the soil structure is divided into 0.5 m × 0.5 m grids within 2 times the width of the station. Outside the double width, the vertical grid height is 1 m, the width is 0.5 m, the transverse grid width is 1 m, and the height is 0.5 m.

The Chi-chi wave, Duzce wave, and Manjil wave are selected as input ground motions for dynamic time history analysis. The ground motion input only considers the horizontal seismic action. In the calculation, the peak values of the three seismic waves should be adjusted to make the peak value of the seismic wave be adjusted to 0.1 g. The acceleration time history curve of the ground motion is shown in Figure 12.

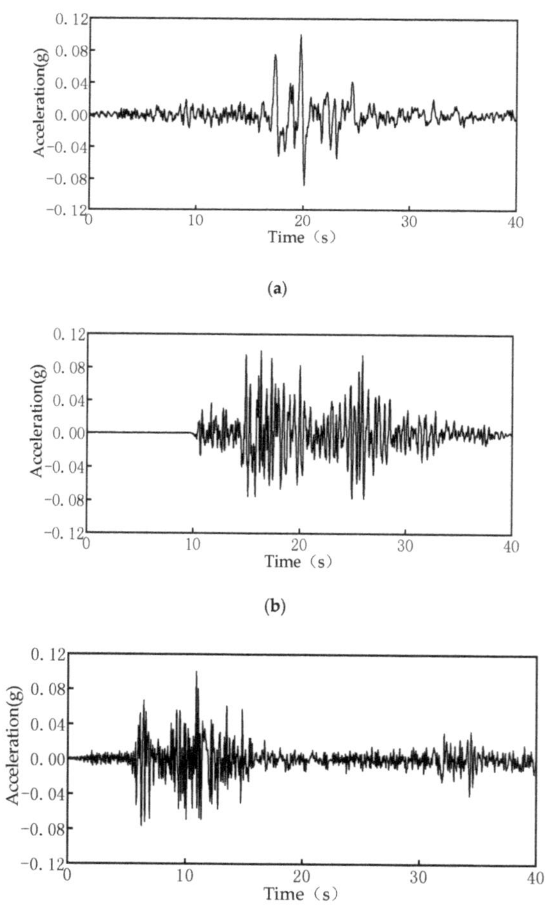

Figure 12. Input ground motion acceleration time history curve. (**a**) Chi-Chi earthquake. (**b**) Dzuce earthquake. (**c**) Manjil earthquake.

4. Calculation Result

4.1. Prototype Structure

This section mainly analyzes the seismic response behavior of the prototype subway station structure by varying the thickness of the weak interlayer and the shear wave velocity of the weak interlayer.

It has been shown that the bottom of the middle column structure will reach the ultimate bearing capacity first due to its small cross-sectional area, making it a more dangerous cross-section. Furthermore, the seismic response characteristics of the underground structure are relatively affected by displacement [2,39]. Therefore, the absolute value of the relative horizontal displacement between points A and B at the top and bottom of the side wall, as shown in Figure 13, is selected as the horizontal displacement value of the side wall. The absolute value of the relative horizontal displacement between points C and D of the upper middle column is selected as the horizontal displacement value of the upper middle column. The absolute value of the relative horizontal displacement between points E and F of the lower middle column is selected as the horizontal displacement value of the lower middle column. The shear force and bending moment of the S1 cross section at the bottom of the upper middle column are selected as the shear force and bending moment at the bottom of the upper middle column, and the shear force and bending moment of the S2 cross section at the bottom of the lower middle column are selected as the shear force and bending moment at the bottom of the lower middle column. The section bending moment is expressed as the bending moment per meter, with the unit kN·m/m, and the shear force of the section is expressed as the shear force per meter, with the unit kN/m. The horizontal displacement is measured in mm.

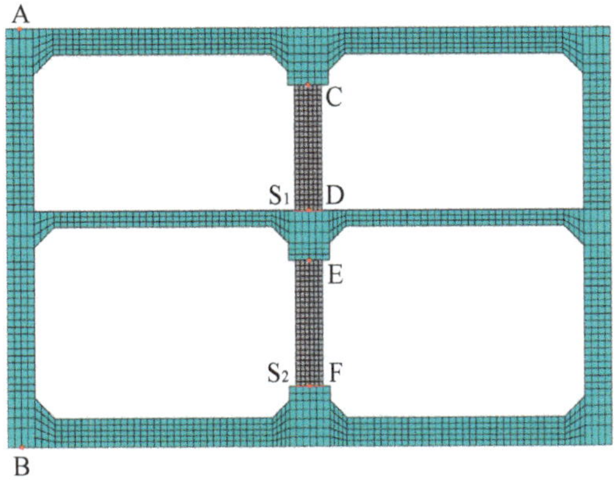

Figure 13. Monitoring points at the structure.

4.1.1. Influence of Weak Interlayer Thickness on the Seismic Response of Underground Structures

It is assumed that the shear wave velocity of the weak interlayer is 125 m/s, and the thickness of the weak interlayer passing through the middle of the underground structure is set to 1 m, 2 m, 3 m, and 4 m, respectively. Four different soil thickness conditions and one condition without a weak interlayer are considered to study the influence of the weak interlayer thickness on the seismic response of the underground structure.

Due to space limitations, only the displacement values of the side wall, the upper middle column, and the lower middle column for the homogeneous site and the weak interlayer with a thickness of 4 m under the Chi-chi earthquake are shown here, as presented in Figure 14. From the figure, it can be observed that when the weak interlayer passes through the middle of the underground structure, the maximum displacement of the side wall of the underground structure is 10.2 mm, which is approximately 171.3% higher than that of the homogeneous site. The maximum displacement of the upper middle column is 4.11 mm, which is increased by approximately 158.5%. The maximum displacement of the lower middle column is 5.1 mm, which is increased by approximately 147.6%. This indicates that the weak interlayer has a negative amplification effect on the deformation of the underground structure.

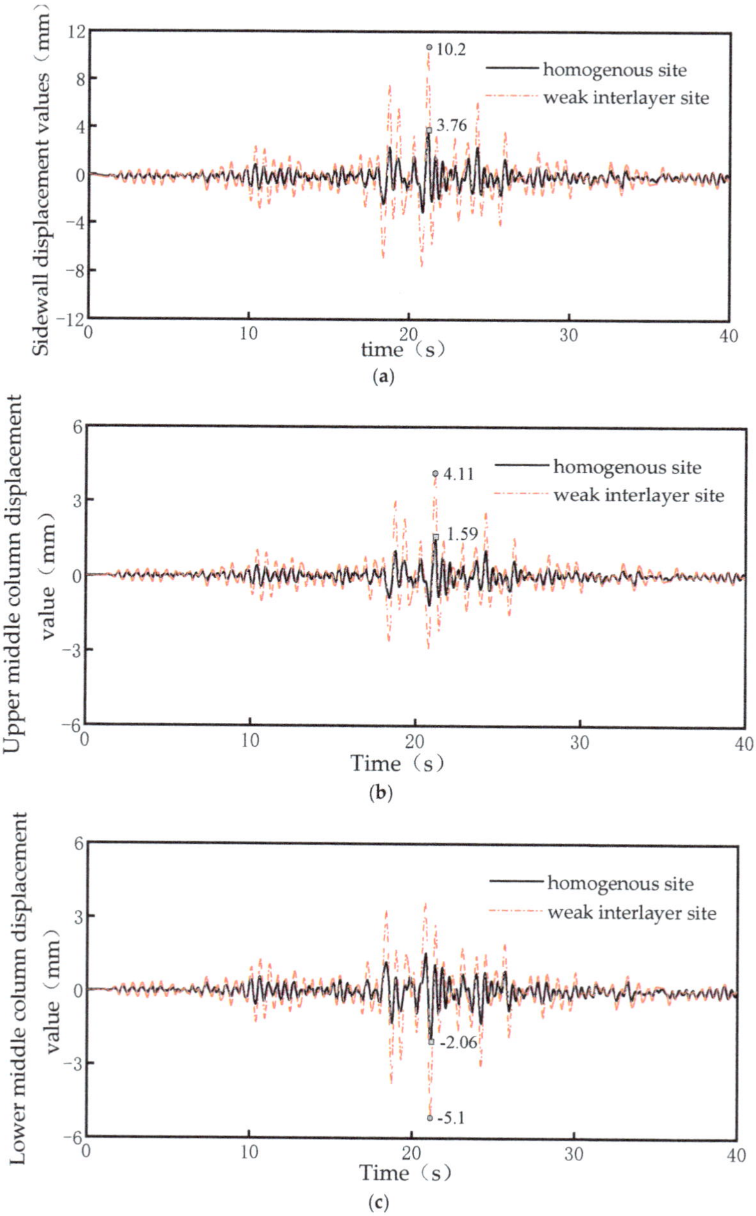

Figure 14. Displacement values of key structural members under the Chi-chi earthquake. (**a**) Displacement value of the side wall. (**b**) Displacement value of the upper middle column. (**c**) Displacement value of the lower middle column.

The displacement values of the side walls, upper middle columns, and lower middle columns under various working conditions for three different ground motions are further extracted, as shown in Figure 15. From the figure, it can be observed that the displacement values under the weak interlayer site are greater than those under the homogeneous site. With the increase in the thickness of the weak interlayer, the displacement values of the side wall, the upper middle column, and the lower middle column gradually increase.

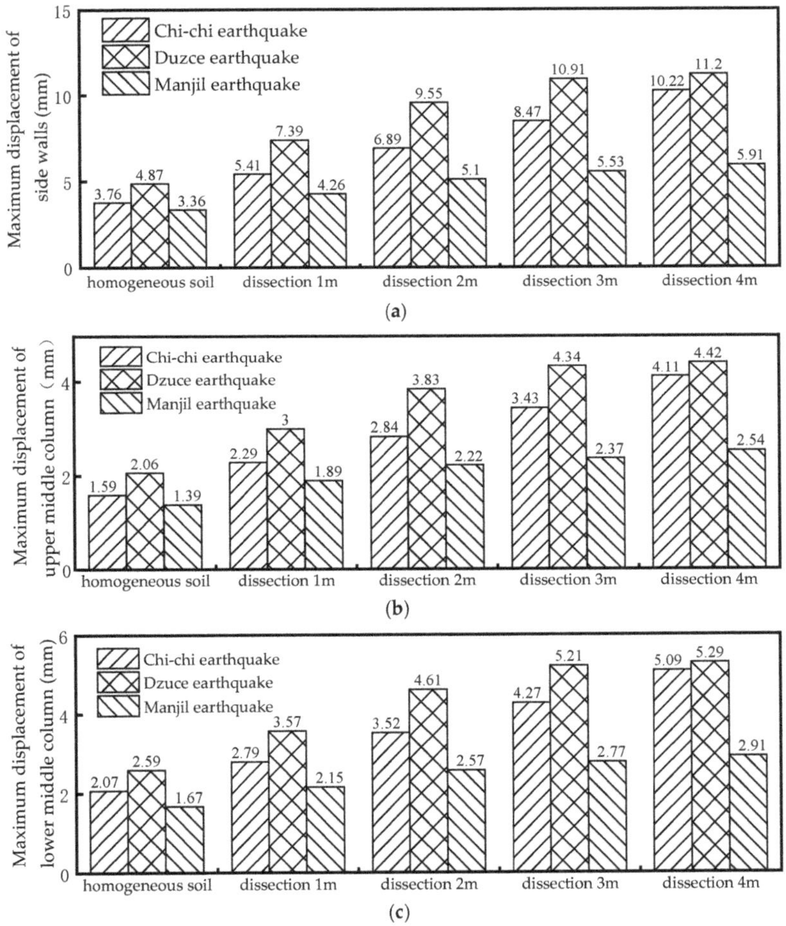

Figure 15. Maximum displacement of structure in different cases. (**a**) Maximum displacement of the side wall. (**b**) Maximum displacement of the upper middle column. (**c**) Maximum displacement of the lower middle column.

In Figure 16, the maximum shear force and bending moment at the bottom of the middle column and the bottom of the lower middle column are extracted. From the calculation results, it can be seen that in the case of weak interlayer crossing, the internal force at the bottom of the middle column is greater than that of the homogeneous site, and with the increase in the thickness of the weak interlayer, the internal force at the bottom of the middle column also increases.

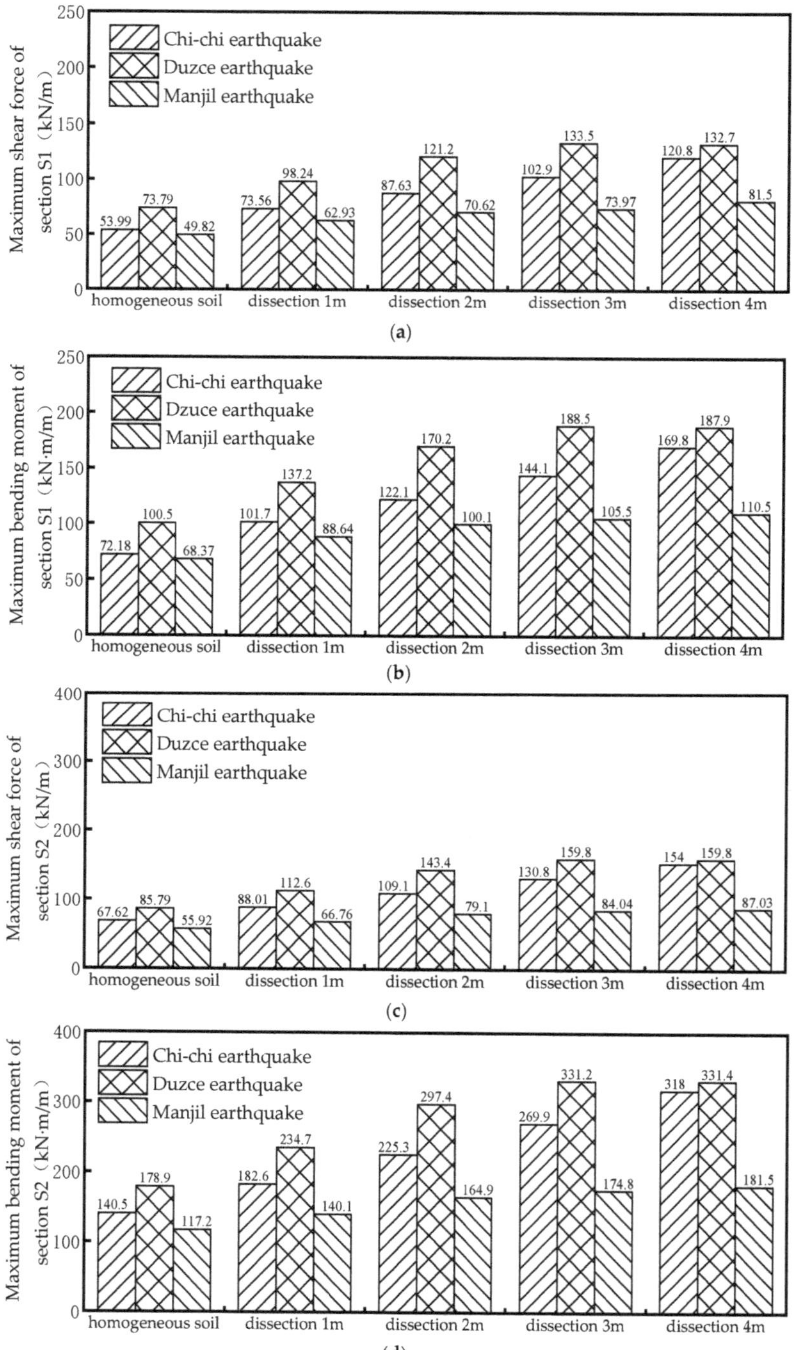

Figure 16. Internal force value at the bottom of the central column in different cases. (**a**) Maximum shear at the base of the upper middle column. (**b**) Maximum bending moment at the bottom of the upper middle column. (**c**) Maximum shear at the base of the lower middle column. (**d**) Maximum bending moment at the bottom of the lower middle column.

4.1.2. Influence of the Strength of a Weak Interlayer on the Seismic Response of the Underground Structure

The thickness of the weak interlayer is 3 m, and the shear wave velocity of the weak interlayer is assumed to be 150 m/s, 125 m/s, 100 m/s, and 75 m/s, respectively. Four working conditions with varying shear wave velocities in different soil layers and one condition without a weak interlayer crossing are set up to study the influence of the weak degree of the weak interlayer on the seismic response of the underground structure.

Figure 17 shows the maximum displacement of the side wall, upper middle column, and lower middle column for the homogeneous site and the weak interlayer with a shear wave velocity of 75 m/s under the Chi-chi earthquake. It can be observed from the figure that when a weak layer crosses the underground structure and in the case of soft soil, the maximum displacement of the side wall reaches 16.5 mm. Compared to the homogeneous site, the displacement value increases by 338.8%. The maximum displacement of the upper middle column reaches 6.35 mm, which is 299.4% higher than that of the homogeneous site. The maximum displacement of the lower middle column reaches 8.24 mm, which is 300% higher than that of the homogeneous site. To further analyze the seismic response behavior of the weak interlayer in the underground structure, as shown in Figure 18, the maximum displacement of the key components of the underground structure under various working conditions is extracted. It can be observed from the calculation results that under different ground motions, with the thickness of the weak interlayer remaining constant, the maximum horizontal displacement of the side wall, the upper middle column, and the lower middle column gradually increases as the shear wave velocity decreases (i.e., the weaker the weak interlayer).

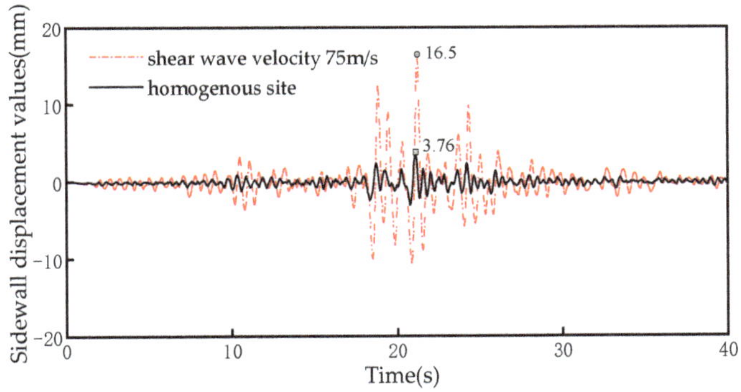

(a)

Figure 17. *Cont.*

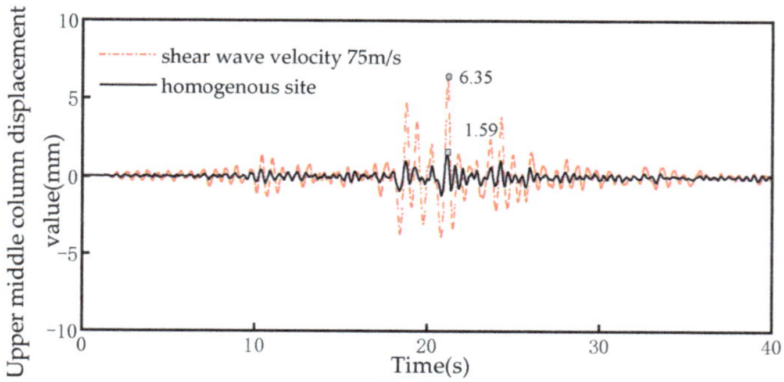

(**b**)

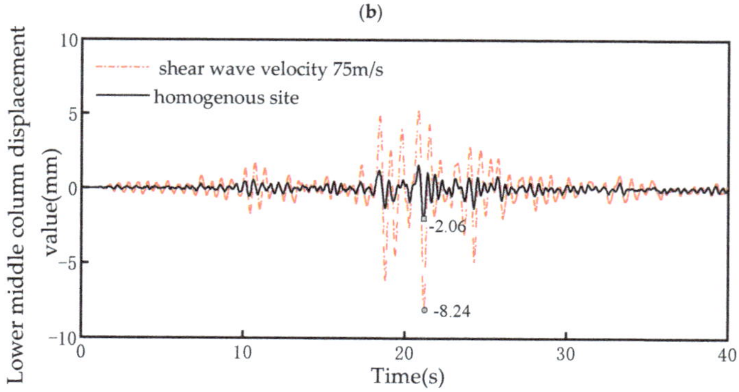

(**c**)

Figure 17. Displacement values of key structural members under the Chi-chi earthquake. (**a**) Displacement value of the side wall. (**b**) Displacement value of the upper middle column. (**c**) Displacement value of the lower middle column.

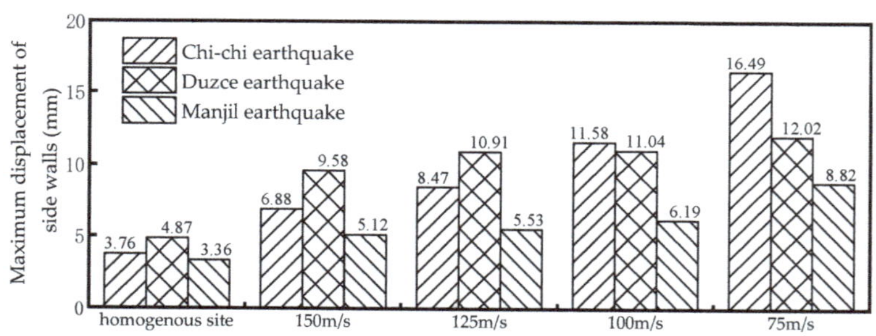

(**a**)

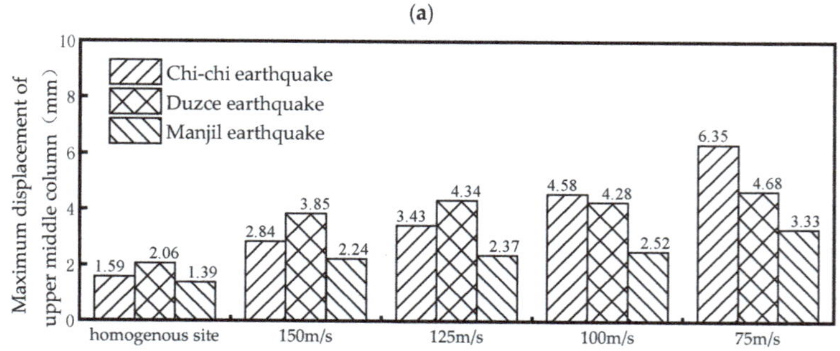

(**b**)

Figure 18. *Cont.*

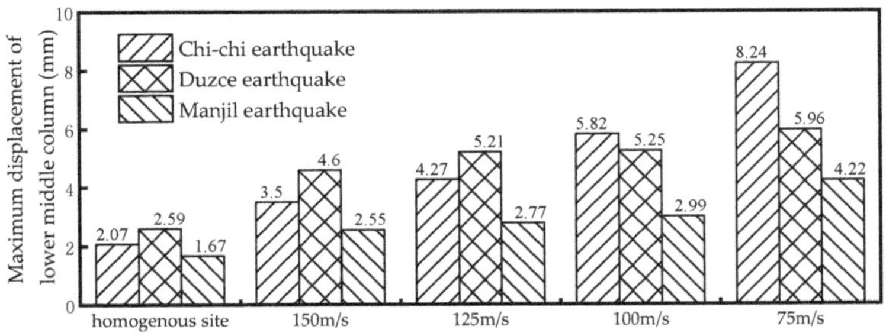

(c)

Figure 18. Maximum displacement of the structure in different cases. (**a**) Maximum displacement of the side wall. (**b**) Maximum displacement of the upper middle column. (**c**) Maximum displacement of the lower middle column.

Figure 19 shows that the maximum shear force and the maximum bending moment at the bottom of the upper and lower columns under each working condition are extracted. From the calculation results, it can be seen that under the action of ground motion, with the decrease in the shear wave velocity of the weak interlayer, the maximum shear force and the maximum bending moment at the bottom of the upper and lower middle columns gradually increase.

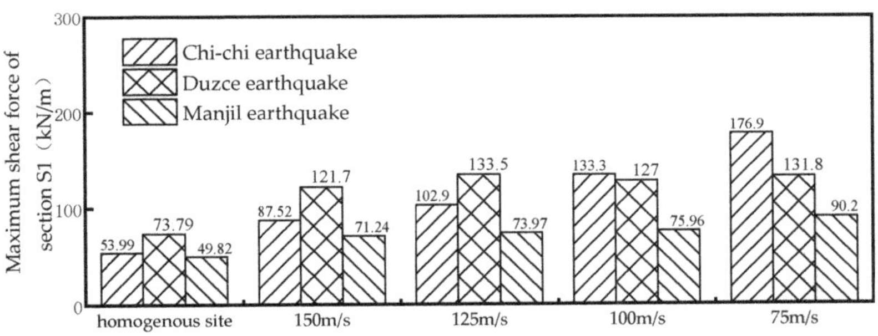

(**a**)

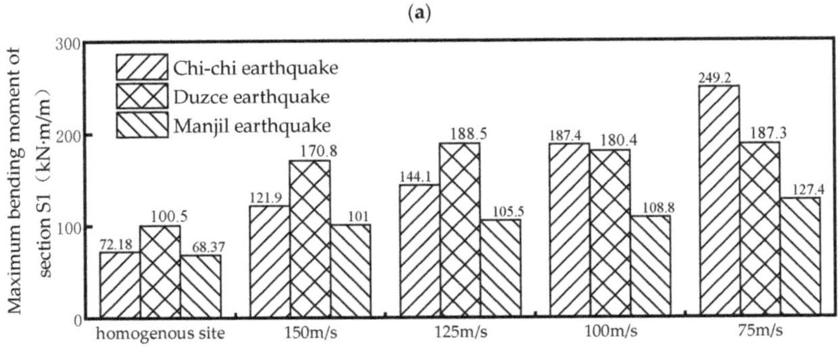

(**b**)

Figure 19. *Cont.*

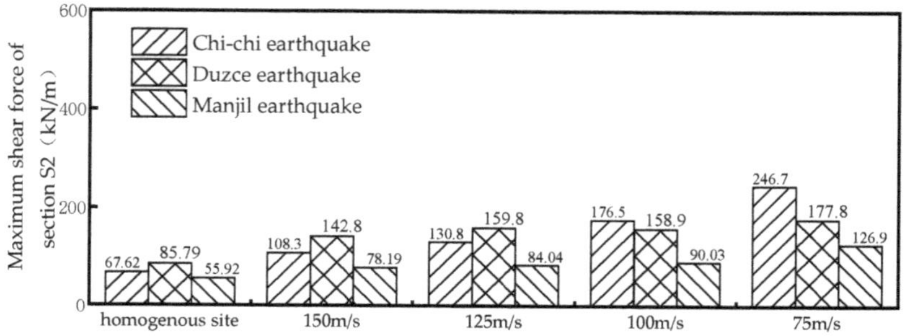

(**c**)

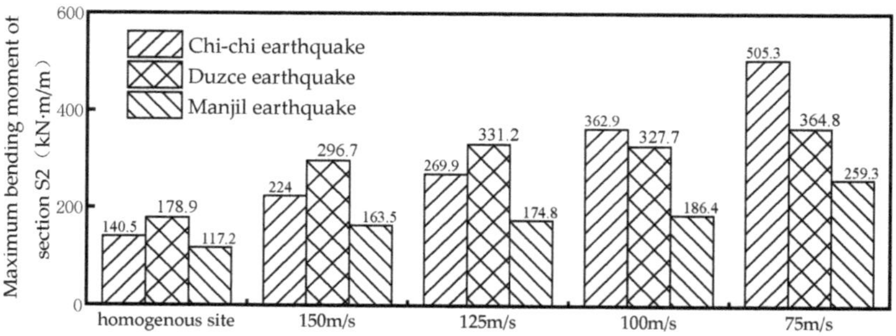

(**d**)

Figure 19. Internal force value at the bottom of the central column in different cases. (**a**) Maximum shear at the base of the upper middle column. (**b**) Maximum bending moment at the bottom of the upper middle column. (**c**) Maximum shear at the base of the lower middle column. (**d**) Maximum bending moment at the bottom of the lower middle column.

4.2. Seismic Mitigation Structure

In this section, a comparison and analysis of the split column structure and the ordinary column structure are performed from the perspectives of structural deformation and cross-sectional internal force, and the seismic mitigation effect of the split column is discussed. In terms of deformation, this study selects the relative horizontal displacement at the top and bottom of the side wall, as well as the relative horizontal displacement at the top and bottom of the upper middle column and the lower middle column. In terms of internal force, the shear force and bending moment at the bottom of the upper middle column and the bottom of the lower middle column are selected to compare the differences between the prototype structure and the seismic mitigation structure. The comparison index is expressed by the seismic mitigation rate in this study. The formula is

$$\delta = \frac{N_0 - N_R}{N_0} \tag{6}$$

In the equation, under the same conditions, N_0 represents the internal force values within the top and bottom cross-sections of the central columns of the prototype structure, while N_R represents the internal force values within the top and bottom cross-sections of the central columns of the seismic mitigation structure.

The comparative values for the prototype structure and the seismic mitigation structure under the homogeneous site are extracted, as shown in Table 4. The table shows that, in terms of structural deformation, the seismic mitigation rate is negative, indicating that the horizontal displacement of the key components of the seismic mitigation structure is greater than that of the prototype structure. The absolute value of the seismic mitigation rate for the lower middle column is the largest, reaching 40%. In terms of internal force, the seismic mitigation effect of the seismic mitigation structure exceeds 60%, and the bending moment at the bottom of the lower middle column reaches 80%. This indicates that the use of the split column instead of the prototype integral column can effectively reduce the internal force of the key components.

Table 4. Comparative index values of prototype structure and seismic reduction structure under homogeneous site conditions.

Comparative Indicators		Prototype Structure	Seismic Mitigation	Seismic Mitigation Rate
Maximum displacement of side walls	Chi-chi earthquake	3.76 mm	3.99 mm	−6.12%
	Duzce earthquake	4.87 mm	5.18 mm	−6.37%
	Manjil earthquake	3.36 mm	3.58 mm	−6.55%
Maximum displacement of upper middle column	Chi-chi earthquake	1.59 mm	1.89 mm	−18.87%
	Duzce earthquake	2.06 mm	2.45 mm	−18.93%
	Manjil earthquake	1.39 mm	1.67 mm	−20.14%
Maximum displacement of lower middle column	Chi-chi earthquake	2.07 mm	2.94 mm	−42.03%
	Duzce earthquake	2.59 mm	3.72 mm	−43.63%
	Manjil earthquake	1.67 mm	2.42 mm	−44.91%
Maximum bending moment at the bottom of the upper middle column	Chi-chi earthquake	72.18 kN·m/m	25.92 kN·m/m	64.09%
	Duzce earthquake	100.5 kN·m/m	35.54 kN·m/m	64.64%
	Manjil earthquake	68.37 kN·m/m	23.94 kN·m/m	64.98%
Maximum shear at the base of the upper middle column	Chi-chi earthquake	53.99 kN/m	13.34 kN/m	75.29%
	Duzce earthquake	73.79 kN/m	18.07 kN/m	75.51%
	Manjil earthquake	49.82 kN/m	12.25 kN/m	75.41%
Maximum bending moment at the bottom of the lower middle column	Chi-chi earthquake	140.5 kN·m/m	24.25 kN·m/m	82.74%
	Duzce earthquake	178.9 kN·m/m	31.26 kN·m/m	82.53%
	Manjil earthquake	117.2 kN·m/m	21.33 kN·m/m	81.80%
Maximum shear at the base of the lower middle column	Chi-chi earthquake	67.62 kN/m	17.98 kN/m	73.41%
	Duzce earthquake	85.79 kN/m	23.27 kN/m	72.88%
	Manjil earthquake	55.92 kN/m	15.24 kN/m	72.75%

4.2.1. Influence of the Thickness of the Weak Interlayer on the Seismic Mitigation Effect of the Underground Structure

The shear wave velocity of the weak interlayer is 125 m/s, and the thickness of the weak interlayer crossing the middle of the underground structure is set to 1 m, 2 m, 3 m, and 4 m, respectively. Four working conditions with different soil thicknesses and one condition without a weak interlayer were established to study the influence of weak interlayer thickness on the seismic mitigation effect of underground structures.

Figure 20 shows the seismic mitigation rate of the displacement of the upper and lower middle columns under different ground motions. It can be observed from the figure that the seismic mitigation rate of the upper and lower middle columns under various working conditions is negative, indicating that the displacement of the seismic mitigation structure under earthquake action is greater than that of the prototype structure.

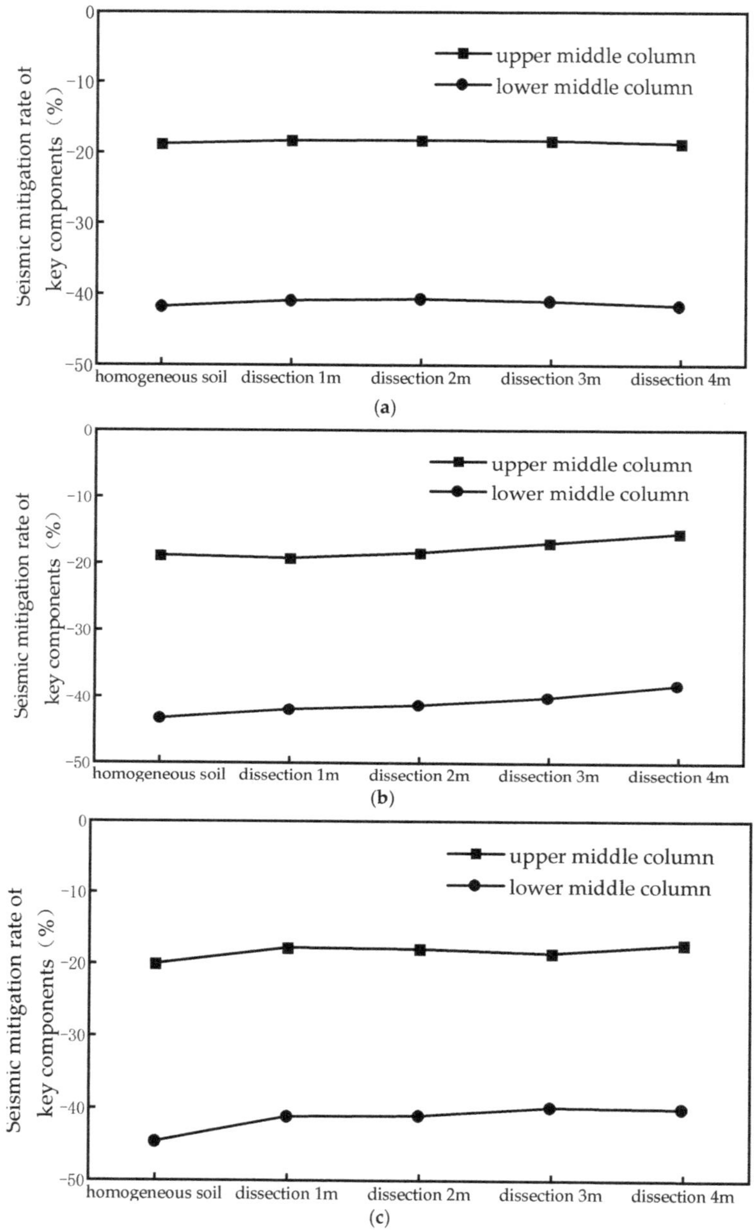

Figure 20. Displacement seismic mitigation rate of middle columns under different ground shocks. (**a**) Displacement seismic mitigation ratio under the Chi-chi earthquake. (**b**) Displacement seismic mitigation ratio under the Duzce earthquake. (**c**) Displacement seismic mitigation ratio under Manjil earthquake.

The internal force seismic mitigation rate at the bottom of the upper middle column and the bottom of the lower middle column under the Chi-chi earthquake is extracted, as shown in Figure 21. It can be seen from the figure that as the thickness of the weak interlayer increases, the seismic mitigation rate of the internal force in the section of the upper middle column tends to decrease, though the minimum value remains close to

60%. For the lower middle column, the seismic mitigation rate of the internal force in the section tends to increase, with the maximum value approaching 85%. This indicates that the application of split columns in underground stations can significantly reduce the internal force in the key sections. As the thickness of the weak interlayer increases, the influence on the seismic mitigation rate of the upper middle column decreases, while the influence on the seismic mitigation rate of the lower middle column increases.

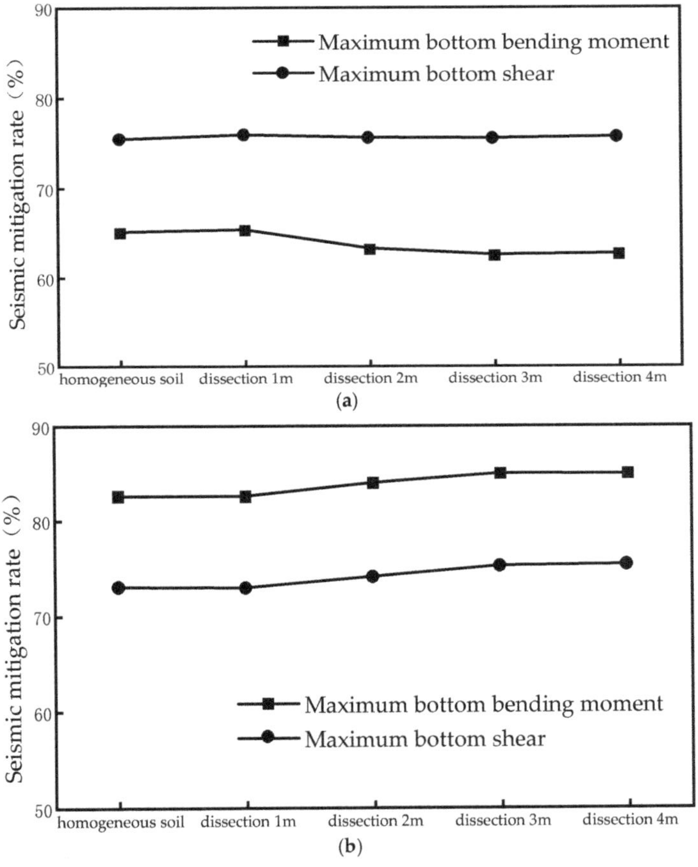

Figure 21. Comparative index values of key components under the Chi-chi earthquake. (**a**) Seismic mitigation ratio of the upper middle column under the Chi-chi earthquake. (**b**) Seismic mitigation ratio of the lower middle column under the Chi-chi earthquake.

4.2.2. Influence of the Weak Degree of the Weak Interlayer on the Seismic Mitigation Effect of the Underground Structure

The thickness of the weak interlayer is set to 3 m, and the shear wave velocity of the weak interlayer is considered to be 150 m/s, 125 m/s, 100 m/s, and 75 m/s, respectively. Four working conditions with varying shear wave velocities in the soil layers and one condition without a weak interlayer were set up to study the influence of the weak interlayer on the seismic reduction effect of underground structures.

Figure 22 shows that the seismic mitigation rate of the displacement of the upper and lower middle columns under different ground motions is negative.

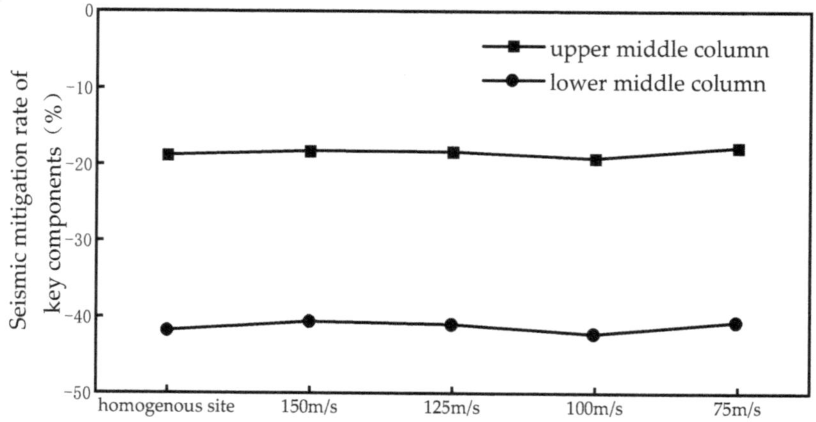

(a)

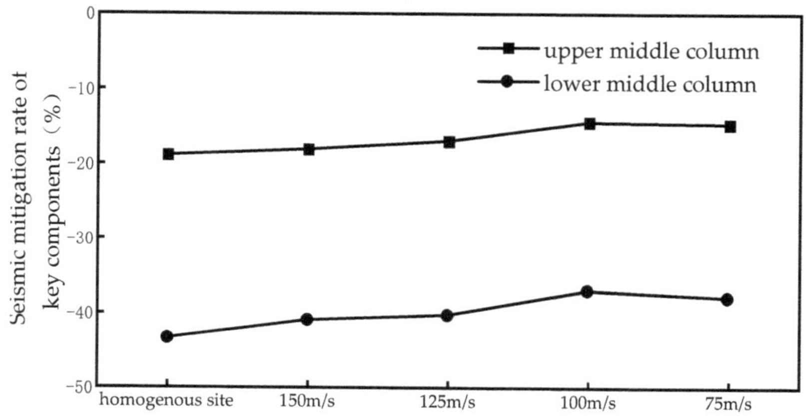

(b)

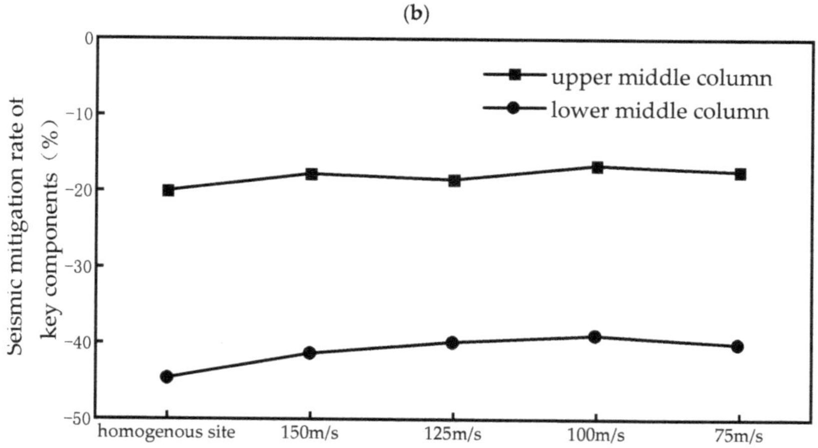

(c)

Figure 22. Displacement seismic mitigation rate of middle columns under different ground shocks. (**a**) Displacement seismic mitigation ratio under the Chi-chi earthquake. (**b**) Displacement seismic mitigation ratio under the Duzce earthquake. (**c**) Displacement seismic mitigation ratio under the Manjil earthquake.

The internal force seismic mitigation rate at the bottom of the upper middle column and the bottom of the lower middle column under the Chi-chi earthquake is extracted, as shown in Figure 23. It can be observed from the figure that, under the action of the earthquake, the internal force seismic mitigation effect of the seismic mitigation structure section is evident. The shear seismic mitigation rate at the bottom section of the upper middle column reaches 80%, and the maximum shear force at the bottom section of the lower middle column reaches 85%. As the shear wave velocity decreases, the seismic mitigation rate of the internal force in the upper middle column tends to decrease, and the bending moment at the bottom decreases significantly. For the lower middle column, the seismic mitigation rate of the internal force in the section tends to increase. The results indicate that as the shear wave velocity decreases (i.e., the weak layer becomes softer), the seismic mitigation effect is weakened for the upper middle column and enhanced for the lower middle column.

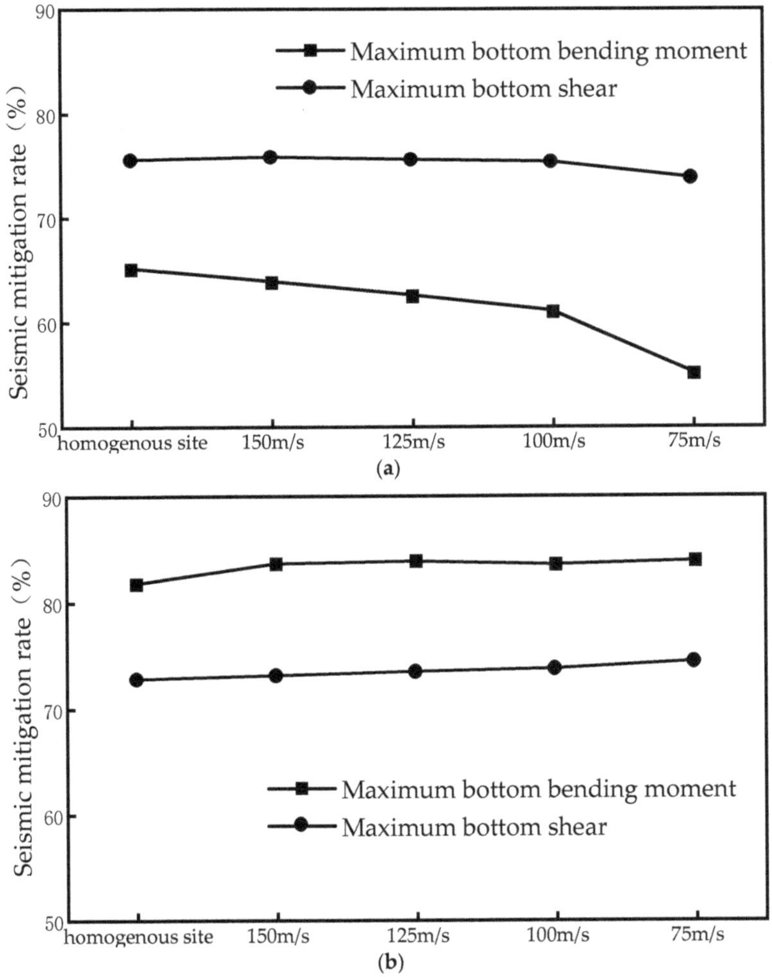

Figure 23. Comparative index values of key components under the Chi-chi earthquake. (**a**) Seismic mitigation ratio of the upper middle column under the Chi-chi earthquake. (**b**) Seismic mitigation ratio of the lower middle column under the Chi-chi earthquake.

4.3. Mechanism Analysis

The soil-structure stiffness ratio [40] defining a rectangular frame structure is shown in Figure 24. In the finite element analysis software ABAQUS, the equivalent soil element and the underground structure are set to be fixed at the bottom, and the unit shear load, τ, is applied at the top and both sides of these elements. The deformations under the respective unit shear loads are then calculated separately. The deformation of the iso-substituted soil unit is recorded as Δ_{soil}, and the deformation of the structure is recorded as Δ_{str}. The soil-structure stiffness ratio, F, is then solved according to Equation (7).

$$F = \frac{\frac{1}{\Delta_{soil}}}{\frac{1}{\Delta_{str}}} = \frac{\Delta_{str}}{\Delta_{soil}} \tag{7}$$

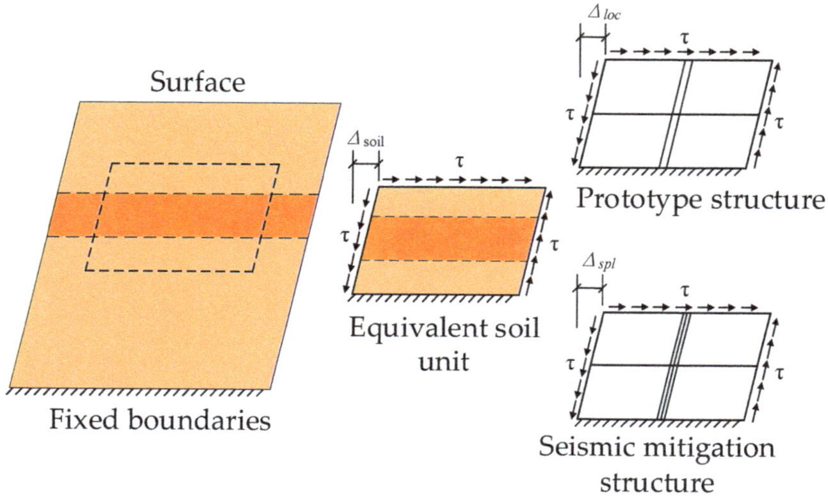

Figure 24. Calculate model of the soil-structure stiffness ratio.

Figure 25 shows the soil-structure stiffness ratios of the prototype structure and the seismic mitigation structure under five conditions of the extracted weak interlayer, with a shear wave velocity of 125 m/s and a thickness of the weak interlayer passing through the middle of the underground structure. These conditions are presented in the following order: 1 m, 2 m, 3 m, 4 m, and no weak interlayer passing through the homogeneous site. The soil-structure stiffness ratios of the prototype structure and the seismic mitigation structure for five conditions, with shear wave velocities of 150 m/s, 125 m/s, 100 m/s, 75 m/s, and no weak interlayer passing through the homogeneous site, are shown in Figure 26 for the extracted weak interlayer with a thickness of 3 m. The weak interlayer traversing the central part of the subsurface structure is also considered in the order of 150 m/s, 125 m/s, 100 m/s, 75 m/s, and no weak interlayer traversing the homogeneous site. From the calculation results, it can be observed that the soil-structure stiffness ratio of the seismic mitigation structure is greater than that of the prototype structure under all working conditions. This indicates that the horizontal displacement of the seismic mitigation structure is greater than that of the prototype structure due to the use of split columns, which leads to a reduction in the overall stiffness of the structure during seismic mitigation action.

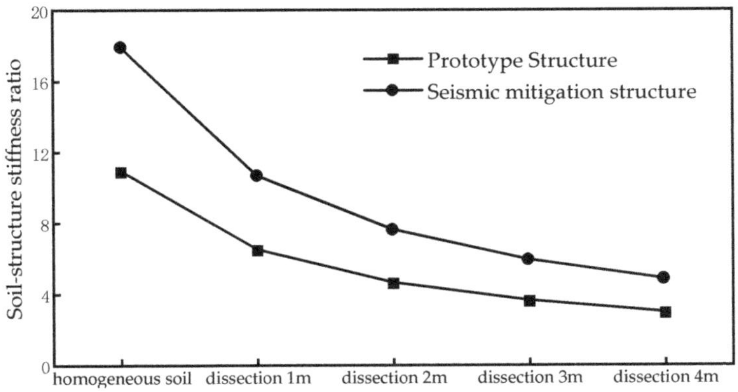

Figure 25. Soil-structure stiffness ratios for different weak interlayer thicknesses.

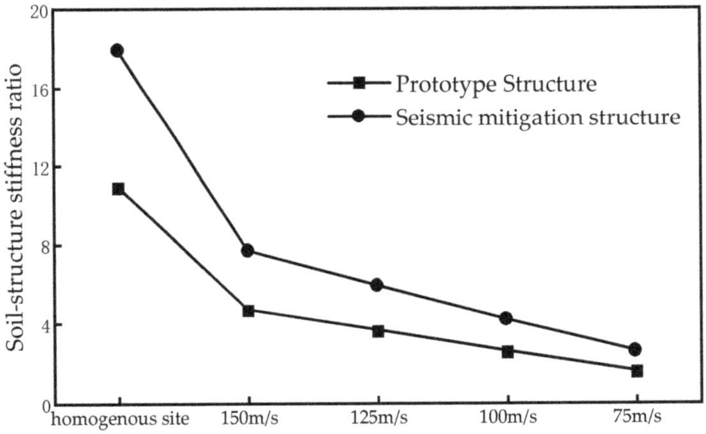

Figure 26. Soil-structure stiffness ratios for different degrees of softness in soft interlayers.

5. Conclusions

This paper considers the two-story double-span rectangular frame underground station as the research object. A two-dimensional numerical model is established using the finite element analysis software ABAQUS. The soil-underground structure overall dynamic time-history analysis is adopted to study the effect of the soft interlayer thickness and soil condition (i.e., shear wave velocity) on the seismic response of the underground station structure. The seismic mitigation effect of the structure formed by the use of split columns is compared with that of the prototype structure in terms of the lateral displacement rate of key components and the seismic mitigation rate. The following conclusions are drawn:

(1) The quasi-static test results from the numerical simulation and model test show good agreement. The same conclusion can be drawn from the skeleton curves of both the numerical simulation and the model test. When the horizontal load reaches its peak, the traditional column is found to be approximately 55% larger than the split column;
(2) Under earthquake action, when the weak interlayer passes through the underground station, the deformation of the key components of the underground structure increases, and the internal force at the bottom of the middle column also increases. As the thickness of the weak interlayer increases, the relative displacement and internal force of the key components also increase. Additionally, with the increase in

weak interlayer thickness, the dynamic response of the interlayer thickness to the underground structure is gradually weakened;

(3) The softer the soil in the weak layer passing through the underground structure (the smaller the shear wave velocity), the greater the relative displacement and internal force of the key components. With the decrease in shear wave velocity, the influence of weakness on the seismic response of underground structures is gradually enhanced;

(4) Compared to the prototype structure, the use of split columns in the underground structure results in a reduction in the overall stiffness of the structure, leading to an increase in the horizontal displacement of the structure;

(5) The application of split columns in underground structures can significantly reduce the internal forces in key sections. The most pronounced seismic mitigation effect is observed in the bending moment at the bottom of the lower middle column, where reductions of up to 85% can be achieved.

In general, at weak interlayer sites, the seismic performance of the station structure can be effectively improved by replacing the ordinary columns with split columns in the subway station structure. The worse the soil quality of the weak layer, the more significant the seismic mitigation effect, making it an effective seismic mitigation measure. This paper only considers the weak interlayer passing through the middle of the subway station. The influence of weak interlayers in other locations and additional factors require further research and analysis.

Author Contributions: Conceptualization, Z.X. (Zigang Xu) and Z.X. (Zongyao Xia); methodology, Z.X. (Zigang Xu); software, Z.X. (Zongyao Xia); validation, Z.X. (Zigang Xu); formal analysis, Z.X. (Zongyao Xia); data curation, Z.X. (Zigang Xu); writing—original draft preparation, Z.X. (Zongyao Xia); writing—review and editing, Z.X. (Zongyao Xia); visualization, Z.X. (Zongyao Xia); project administration, Z.X. (Zigang Xu); funding acquisition, Z.X. (Zigang Xu) and Z.X. (Zongyao Xia). All authors have read and agreed to the published version of the manuscript.

Funding: This research is jointly funded by the National Natural Science Foundation of China (No. 52108453), the Natural Science Foundation of Jiangxi Province (No. 20212BAB214014 and No. 20232BAB204084), the China Postdoctoral Science Foundation Project (No. 2023M741159), and Special Funding Project for Basic Research Operating Costs of the Institute of Engineering Mechanics, China Earthquake Administration (No. 2023D49).

Institutional Review Board Statement: Not applicable.

Informed Consent Statement: Not applicable.

Data Availability Statement: Data are available in a publicly accessible repository.

Conflicts of Interest: The authors declare no conflicts of interest.

References

1. Iida, H.; Hiroto, T.; Yoshida, N.; Iwafuji, M. Damage to Daikai subway station. *Soils Found.* **1996**, *36*, 283–300. [CrossRef]
2. Du, X.L.; Li, Y.; Xu, C.S.; Lu, D.; Xu, Z.; Jin, L. Review on damage causes and disaster mechanism of Daikai subway station during 1995 Osaka-Kobe Earthquake. *Chin. J. Geotech. Eng.* **2018**, *40*, 223–236. (In Chinese)
3. Jiang, L.; Chen, J.; Li, J. Seismic response of underground utility tunnels: Shaking table testing and FEM analysis. *Earthq. Eng. Eng. Vib.* **2010**, *9*, 555–567. [CrossRef]
4. Ulgen, D.; Saglam, S.; Ozkan, M.Y. Dynamic response of a flexible rectangular underground structure in sand: Centrifuge modeling. *Bull. Earthq. Eng.* **2015**, *13*, 2547–2566. [CrossRef]
5. Jiang, X.; Wang, F.; Yang, H.; Sun, G.; Niu, J. Dynamic Response of Shallow-Buried Small Spacing Tunnel with Asymmetrical Pressure: Shaking Table Testing and Numerical Simulation. *Geotech. Geol. Eng.* **2018**, *36*, 2037–2055. [CrossRef]
6. Wang, W.L.; Wang, T.T.; Su, J.J.; Lin, C.H.; Seng, C.R.; Huang, T.H. Assessment of damage in mountain tunnels due to the Taiwan Chi-Chi earthquake. *Tunn. Undergr. Space Technol.* **2001**, *16*, 133–150. [CrossRef]

7. Wang, Z.; Gao, B.; Jiang, Y.; Yuan, S. Investigation and assessment on mountain tunnels and geotechnical damage after the Wenchuan earthquake. *Sci. China Ser. E Technol. Sci.* **2009**, *52*, 546–558. [CrossRef]
8. Zhuang, H.; Tang, B.Z.; Yu, B.Y.; Xu, Z.G.; Zhang, J.; Zhou, Z.W. Review and Prospect of Earthquake Resistance and Seismic Isolationof Underground Structures. *Technol. Earthq. Disaster Prev.* **2023**, *18*, 1–12.
9. Chen, G.X.; Su, C.; Du, X.L.; Lu, D.C. Review of Seismic Damage, Model Test, Available Design and Analysis Methods of Urban Underground Structure: Retrospect and Prospect. *J. Disaster Prev. Mitig. Eng.* **2016**, *36*, 1–23.
10. Yang, W.; Xu, M.; Peng, W.; Liu, T. Optimization Study on Key Technology of Improved Arch Cover Method Construction for Underground Metro Stations Based on Similar Model Test. *Appl. Sci.* **2024**, *14*, 3982. [CrossRef]
11. Maleska, T.; Beben, D.; Vaslestad, J.; Sergei Sukuvara, D. Application of EPS Geofoam below Soil–Steel Composite Bridge Subjected to Seismic Excitations. *J. Geotech. Geoenviron. Eng.* **2024**, *150*, 04024115. [CrossRef]
12. Chen, Z.H.; Ni, P.P.; Cheng, C.; Jia, P.; Hu, D.; Wu, M. Monitoring interaction of shallowly buried, large-section, and long-distance twin box jacking tunnels with small spacing under the Beijing-Hangzhou Grand Canal. *Tunn. Undergr. Space Technol.* **2024**, *153*, 105973. [CrossRef]
13. Mubarak, A.G.; Knappett, J.; Brown, M. Numerical simulation of large tunnel alignments under seismic loading: The Large Hadron Collider as a case study. *Tunn. Undergr. Space Technol.* **2025**, *157*, 106315. [CrossRef]
14. Wu, Y.; Xu, Z.G.; Du, X.L. Analysis of effects of weak interlayer on seismic responses of underground structures. *J. Zhengzhou Univ. (Eng. Sci.)* **2021**, *42*, 91–97.
15. Huang, R.Q.; Yu, J.S. Modelling of the Effects of Properties of a Buried Weak Layer on Seismic Waves. *J. Eng. Geol.* **2003**, *11*, 312–317.
16. Li, W.H.; Zhao, C.G.; Du, N.X. Analysis of effects of sat-urated soft interlayer on seismic responses of metro station. *Rock Soil Mech.* **2010**, *31*, 3958–3963. (In Chinese)
17. Dou, Y.M.; Feng, F.; Wang, J.N.; Fan, J.C.; Zheng, S.Y. The 3D Seismic Response Analysis of Subway Station Built in Weak Interlayer. *Struct. Eng.* **2019**, *35*, 140–148.
18. Zayas, V.A.; Low, S.S.; Mahin, S.A. A simple pendulum technique for achieving seismic isolation. *Earthq. Spectra* **1990**, *6*, 317–333. [CrossRef]
19. Zhao, W.S.; Chen, W.Z.; Yang, D.S. Interaction between strengthening and isolation layers for tunnels in rock subjected to SH waves. *Tunn. Undergr. Space Technol.* **2018**, *79*, 121–133. [CrossRef]
20. Chen, Z.Y.; Liang, S.B.; Shen, H.; He, C. Dynamic centrifuge tests on effects of isolation layer and cross-section dimensions on shield tunnels. *Soil Dyn. Earthq. Eng.* **2018**, *109*, 173–187. [CrossRef]
21. Du, X.L.; Wang, Z.L.; Liu, H.T. Study of a seismic new system of underground frame structure based on toughness design. *Technol. Earthq. Disaster Prev.* **2018**, *13*, 493–501. (In Chinese)
22. Li, Z.X. Theory and technology of split reinforced concrete columns. *Eng. Mech.* **2005**, *22*, 128–140. (In Chinese) [CrossRef]
23. Li, X.Z.; Hao, Y.X.; Zhang, J.Y.; Kang, G.Y.; Hu, Q.C.; Xu, Y.F. Research on seismic behavior of interior beam-column joints of reinforced concrete frames with split columns. *J. Build. Struct.* **2001**, *22*, 55–60. (In Chinese)
24. Hao, Y.C.; Hu, Q.C.; Xu, Y.F.; Kang, G.Y. Experimental study on bearing capacity of separated columns under horizontal cyclic loading. *J. Build. Struct.* **1999**, *20*, 18–25. (In Chinese)
25. Li, Z.X.; Hao, Y.C.; Zhou, B.; Kang, G.; Hu, Q.; Xu, Y. Model experimental study on seismic behavior of reinforced concrete frame with split columns. *J. Build. Struct.* **2003**, *24*, 1–10. (In Chinese)
26. Ma, Q.Y.; Zhao, G.Q.; Jiang, C.Y. Experimental research on seismic performance of central column in metro station. *Build. Struct.* **2021**, *51*, 77–85. (In Chinese)
27. Li, R.W. Experimental Study on the Mechanical Behavior of Split Columns with Different Separation Ratios. Master's Thesis, Yantai University, Yantai, China, 2017. (In Chinese).
28. Xu, C.S.; Wang Yang, X.S.; Du, X.L.; Xu, Z.G.; Zhang, Z.H. Seismic mitigation effects of split columns in underground station structures. *Chin. J. Geotech. Eng.* **2021**, *43*, 624–633. (In Chinese)
29. Xu, Z.G.; Xia, Z.Y.; Zhuang, H.Y.; Zhang, Q.; Zhang, J.; Tang, B.Z. Research on seismic performance of subway station structure based on split column. *Technol. Earthq. Disaster Prev.* **2023**, *18*, 27–36.
30. Xu, Z.; Xia, Z.; Bu, X.; Han, R. Research on Seismic Performance of a Two-Story, Two-Span Underground Subway Station with Split Columns Based on the Quasi-Static Method. *Appl. Sci.* **2024**, *14*, 4077. [CrossRef]
31. Zhao, G.Q. Research on Seismic Performance of Split Central Column of Metro Station. Master's Thesis, Chang'an University, Xi'an, China, 2021.
32. Systèmes, D. Abaqus analysis user's manual. *Simulia Corp. Provid. RI USA* **2007**, *671*, 672.
33. Du, X.L.; Xu, Z.G.; Xu, C.S.; Li, Y.; Jiang, J.W. Time-history analysis method for soil-underground structure system based on equivalent linear method. *Chin. J. Geotech. Eng.* **2018**, *40*, 2155–2163.
34. Xu, Z.G.; Du, X.L.; Xu, C.S.; Zhang, C.Y.; Jiang, J.W. Comparison of determination methods of site Rayleigh seismic mitigation coefficients in seismic responses analysis of underground structures. *Rock Soil Mech.* **2019**, *40*, 4838–4847. (In Chinese)

35. Xue, R.L. Seismic Elasto-plastic Analysis of Subway Station Structure. Master's Thesis, Southeast University, Nanjing, China, 2018. (In Chinese).
36. Kampitsis, A.E.; Sapountzakis, E.J.; Giannakos, S.K.; Gerolymos, N.A. Seismic soil-pile-structure kinematic and inertial interaction-A new beam approach. *Soil Dyn. Earthq. Eng.* **2021**, *55*, 211–224. [CrossRef]
37. Zhong, Z.L.; Shi, Y.B.; Li, J.Q.; Zhao, M.; Du, X.L. Seismic performance assessment of subway station structures considering fuzzy probability of damage states. *Chin. J. Geotech. Eng.* **2022**, *44*, 2196–2205. (In Chinese)
38. Han, R.B.; Xu, C.S.; Xu, Z.G.; Jiang, J.W.; Du, X.L. A boundary forced response displacement method for seismic analysis of symmetrical underground structures. *Eng. Mech.* **2021**, *38*, 50–60. (In Chinese)
39. Du, X.L.; Xu, Z.G.; Xu, C.S.; Li, Y. Inertia force-displacement method for seismic analysis of shallow buried underground structure. *Chin. J. Geotech. Eng.* **2018**, *40*, 583–591. (In Chinese)
40. Du, X.L.; Kang, K.L.; Xu, Z.G.; Li, Y.; Xu, C.S. Main characteristics and rules of seismic response for underground structures. *China Civ. Eng. J.* **2018**, *51*, 11–21.

applied sciences

MDPI

Article

Response of Corroded Steel Pipeline Without and with CFRP Reinforcement to Reverse Fault Movement

Junyan Han [1,2], Yansong Bi [1], Benwei Hou [1,*], Wenle Zhao [1] and Mohamed Hesham El Naggar [2,*]

1 Key Laboratory of Urban Security and Disaster Engineering of Ministry of China Education, Beijing University of Technology, Beijing 100124, China; junyanhan@bjut.edu.cn (J.H.); bys0807@emails.bjut.edu.cn (Y.B.); materialism0301@163.com (W.Z.)

2 Geotechnical Research Centre, Department of Civil and Environmental Engineering, Western University, London, ON N6A 3K7, Canada

* Correspondence: benweihou@bjut.edu.cn (B.H.); naggar@uwo.ca (M.H.E.N.)

Abstract: Corroded steel pipelines are particularly vulnerable to failure due to ground movement, which highlights the need to improve their seismic resistance through reinforcement methods. This paper establishes a three-dimensional finite element model of a corroded steel pipeline subjected to a reverse fault, which considers the effects of the corrosion position and depth, winding thickness, and length of carbon fiber-reinforced polymer (CFRP), to investigate the stress, strain, elliptic deformation, and failure modes of the pipeline before and after CFRP reinforcement. Results indicate that the main failure mode of the intact and corroded pipeline crossing the reverse fault is local buckling. Corrosion intensifies the response of the cross-fault pipeline, accelerates its failure occurrence, and promotes transformation from a single failure mode to multiple failure modes. For CFRP reinforcement, an increase in CFRP winding thickness can effectively inhibit the growth of the pipeline's compressive strain, thus reducing the buckling potential. Each additional CFRP layer can further enhance the overall buckling resistance but at a decreasing rate. Similarly, longer CFRP winding improves buckling resistance though the effectiveness per meter decreases. Therefore, it is recommended that the thickness and length of CFRP winding on the pipeline should be optimized to obtain the best reinforcement at a reasonable cost.

Keywords: corroded pipeline; reverse fault; CFRP reinforcement; mechanical response; failure mode

1. Introduction

Long gas pipelines are crucial national infrastructure and civil engineering projects [1] that span extensive areas and inevitably cross seismically active regions. For example, the China–Myanmar pipeline crosses 18 active faults and multiple seismic zones and is highly susceptible to fault-induced permanent ground displacement (PGD) [2]. Post-earthquake investigations have revealed that PGD will compromise the structural integrity of buried pipelines. Significant damage has been observed in pipelines subjected to fault-induced PGD, including the 1971 San Fernando earthquake [3], the 1999 İzmit earthquake [4], the 1999 Chi-Chi earthquake [5], and, most recently, the 2023 Gaziantep earthquake [6]. Most studies that investigated the performance of pipelines crossing faults primarily focus on their failure modes [7], reinforcement measures [8], and failure criteria of non-corroded pipelines across faults [9].

During the transportation of natural gas through the pipelines, there is a large temperature difference between the transport medium and the environment outside. Corrosive media in the pipeline, such as CO_2, H_2S, and volatile organic acids, will condense into droplets at the top of the pipeline along with the water vapor [10], leading to top-of-the-line corrosion (TLC). This process will lead to the thinning of the pipeline wall and reduction of its strength; in serious cases, leakage will occur causing accidents and economic losses [11]. Most studies on corroded pipelines mainly focus on failure modes under simple loads like internal pressure [12]. However, there is limited research on the damage amplification effects and failure modes of corrosion in hanging wall and footwall pipelines subjected to faults.

Research on the performance analysis methods and model prediction of buried corroded pipelines has gradually become a key point in pipeline engineering research. Researchers have carried out a large amount of work on the modeling, simulation, and prediction of pipeline corrosion characteristics. Zhou et al. [13] proposed a nonuniform corrosion field modeling method based on the random field theory to simulate the complex corrosion morphology on the outer wall of the pipeline and, combined with finite element analysis, predict the pipeline burst bearing capacity. In addition, for the corrosion defect behavior of high-strength steel pipes, Ren et al. [14] used the extended finite element method and artificial neural network to analyze the stress field change and rupture behavior at the crack tip, further improving the adaptability and accuracy of steel pipe corrosion prediction. Long et al. [15] evaluated the seismic resilience of urban water supply networks considering the soil corrosion environment and the service age of pipelines. The results showed that different corrosion environments and service ages would significantly affect the seismic resilience of the pipelines. In summary, the research on the performance analysis and model prediction of buried corroded pipelines has made significant progress in aspects such as nonuniform corrosion field modeling and machine learning-based prediction methods.

There are numerous studies on the reinforcement of buried pipelines subject to fault-induced PGD, such as selecting higher-grade steel, reducing the pipeline diameter-to-thickness ratio or buried depth, and setting flexible joints between pipeline segments. Other solutions have also been considered, including increasing the width of the backfill trench and reducing the pipeline–ground friction. Retrofitting corroded steel pipeline CFRP materials has been proposed, owing to many features including lightweight, high tensile strength, and high flexibility for application in complex terrains and irregular pipeline shapes. Consequently, pipelines can be retrofitted quickly and more economically by employing CFRP compared with conventional methods [16]. In addition, CFRP materials have excellent fatigue strength and superior corrosion resistance. Recent studies have also explored the application of CFRP materials inside pipelines. For instance, Rusch et al. [17] developed a bonded internal reinforcement system that boosts pipeline stiffness. Bruce et al. [18] tested a repair system for X52 pipelines with carbon fiber and glass fiber composites and found that the repaired pipeline had a 17% higher burst strength than the unrepaired ones. However, there are no studies on the effectiveness of CFRP treatment for corroded steel pipelines in enhancing its performance when subject to fault-induced PGD.

To address this identified gap in the literature, this paper investigates the response of corroded steel pipelines without and with CFRP reinforcement to reverse fault movement. It evaluates the effects of corrosion location and depth on the failure modes of buried pipelines and, further, on adopted reinforcement measures. A three-dimensional finite element model of the pipeline with corrosion defects subjected to reverse fault is developed and employed to examine the effectiveness of CFRP treatment considering the influence of CFRP reinforcement thickness and length on the responses and failure modes of pipelines crossing faults.

2. Finite Element Model

2.1. Geometric Dimension and Mesh Elements

A 3D finite element model of a buried pipeline with localized corrosion subjected to reverse fault movement is developed employing ABAQUS 2021. As shown in Figure 1, the buried depth is set at 2 m, outer diameter $D = 0.9144$ m, and wall thickness $t = 0.0119$ m [19]. To eliminate boundary effects, the soil domain size is set as 8 m $\times$ 10 m $\times$ 60 m and is discretized using eight-node reduced solid units (C3D8R) [20], with a total of 13,260 units. The reverse fault features a crossing angle of $\theta = 90^\circ$ and a dip angle of $\beta = 60^\circ$ [21]. The pipeline material is API 5L X65 steel and is discretized using four-node reduced shell elements (S4R).

To account for the small soil deformation at the far end of the pipeline model and its effect on the large deformation of the section subject to PGD, the equivalent boundary condition proposed is adopted; nonlinear springs are attached to both ends of the pipeline to simulate the far-field end. The relationship between the axial external load and the pipeline elongation is given by Liu et al. [22]:

$$F = \sqrt{2f_s AE\Delta L}, \tag{1}$$

where F is the axial force of the equivalent nonlinear springs at both ends of the pipeline; f_s is the axial force per unit length along the pipeline axis; ΔL is the elongation of the equivalent nonlinear springs when subjected to axial force; A and E are the cross-sectional area and elastic modulus of the pipeline.

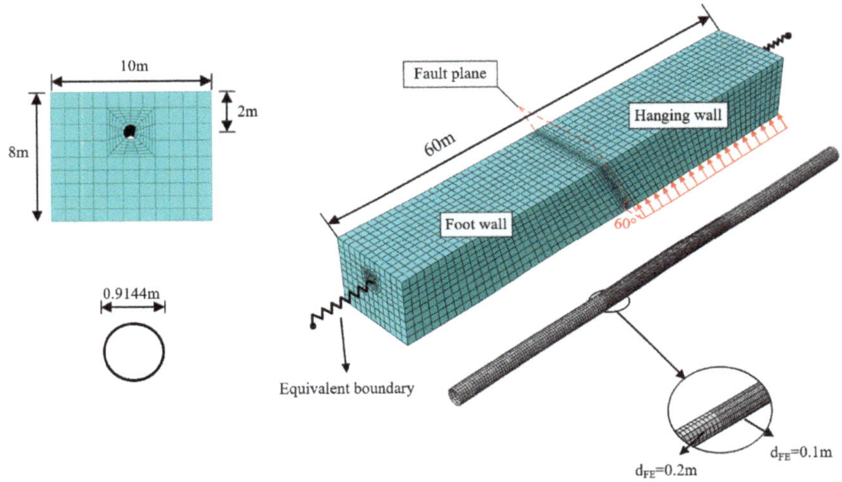

Figure 1. Reverse fault-pipeline 3D finite element model.

2.2. Material Constitutive Model and Related Parameters

The Mohr–Coulomb constitutive model is used to simulate soil behavior, with parameters listed in Table 1. The Ramberg–Osgood stress–strain constitutive model is used to simulate the pipelines' plastic deformation under fault displacement, with the stress–strain relationship given by:

$$\varepsilon = \frac{\sigma_y}{E_p}\left[\frac{\sigma}{\sigma_y} + \lambda\left(\frac{\sigma}{\sigma_y}\right)\right], \tag{2}$$

where σ is stress; ε is total strain; σ_y is yield stress; E_p is initial elastic modulus; λ and N are parameters of the Ramberg–Osgood model. For X65 steel material considered in this study, σ_y = 448.5 MPa, E_p = 206 MPa, λ = 0.533, N = 14 [23]. Figure 2 shows the stress–strain curve of X65 steel. The range of the elastic and elastoplastic stages of X65 steel can be obtained from Figure 2, and the material parameters required for numerical simulation are also provided.

Table 1. Mechanical parameters of soil.

Type	Density (kg/m^3)	Elastic Modulus (MPa)	Poisson's Ratio ν	Friction Angle φ (°)	Cohesion c (kPa)	Dilatation Angle ψ (°)
Clay	1960	20	0.3	33.5	26	3

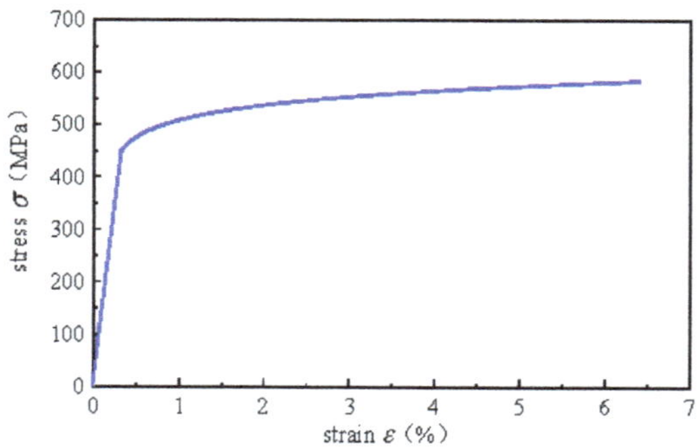

Figure 2. The Ramberg–Osgood constitutive model of X65 steel.

2.3. Constitutive Model and Reinforcement Form of CFRP

To accurately simulate the mechanical response of CFRP-reinforced steel pipelines subject to fault displacement, the orthogonal elastic plane stress model (Lamina model) is applied to simulate the constitutive behavior of CFRP, which can simulate the mechanical properties of composite laminates [24]. The Hashin failure criterion is used to predict the anisotropic damage degree of CFRP under fault displacement, which considers four damage initiation mechanisms: fiber tension, fiber compression, substrate tension, and substrate compression.

The CFRP material considered in the analysis refers to that used in the testing program of Teng et al. [25], with its mechanical properties shown in Table 2. The failure parameters of CFRP are defined according to the Hashin failure criteria. The thickness of single-layer CFRP is 0.18 mm, and each single-layer is assumed perfectly combined. Finally, shell elements are used to discretize the CFRP material.

Table 2. Mechanical properties of CFRP material.

E_1/MPa	E_2/MPa	N_u	G_{12}/MPa	G_{13}/MPa	G_{23}/MPa	t_{CFRP}/mm
230,000	1900	0.3	3387	3387	3387	0.18

2.4. Simulation of Fault Displacement and Pipe-Soil Interaction

The frictional contact model, face-to-face discretization, penalty function, and hard contact are used to simulate fault displacement [26]. The friction coefficient at soil interfaces is set to 0.7, while the coefficient at the pipe–soil interface is set to 0.3 [27]. The pipeline mesh is refined within 10 m near the fault plane, while the mesh of the farther part is relatively coarse.

The loading is applied in two stages. First stage: the soil domain and bottom surface are fixed, and the gravity load is applied to the whole model to simulate the initial geostatic stress conditions and the deformation of the whole model. Second stage: the footwall is fixed, and the hanging wall represents the displacement applied by the moving block 60° along the fault direction to simulate the reverse fault movement, and both are fixed in the normal direction. The fault displacement caused by an earthquake is predicted based on an empirical statistical equation [28], i.e.,:

$$\begin{aligned} \log_{10} d_{max} = \mathrm{M} \quad & - 2.247\log_{10}\left(\frac{k}{L_R}\right) + 0.6489M + 0.0581s - 0.3407v - 2.985 \\ & - 0.1369M^2 - 0.0306S_L + 0.2302S_L^2 + 0.5792S_L^3 \\ & + [-0.3898R - 0.2749(1-R)]R/100, \end{aligned} \quad (3)$$

where M is earthquake magnitude, R is epicentral distance (km), H is depth of seismic source, and LR is rupture length.

2.5. Simulation of Local Corrosion

Table 3 shows the data measured in the total distance of 62.8 km from the first station to the first middle station of the Qing–Harbin Pipeline where 9625 volumetric defects in steel pipelines were identified [29]. Five working conditions of residual wall thickness of 0.6 t, 0.7 t, 0.8 t, 0.9 t, and t (t = 0.0119 m) are designed according to the proportion of different corrosion extents. With the deepening of corrosion extent, the remaining wall thickness decreases accordingly.

Table 3. Interval proportions of different residual wall thicknesses.

Corrosion Depth	Number of Defective Points	Percentage
90~100%	1295	13.4%
80~90%	3277	34.5%
70~80%	2672	27.8%
60~40%	2261	23.5%

Considering the most unfavorable condition of the pipeline stress based on the results of seismic response analysis for non-corroded pipelines [30], corrosion defects are set at the maximum strain of the hanging wall and footwall pipelines (L = 3.2 m away from the fault plane). Figure 3 displays a schematic diagram of the corroded pipeline model, Figure 3a shows the front profile of the pipeline, and Figure 3b presents the longitudinal profile of the pipeline. In Figure 3, d is the corrosion depth, S is the side length of the corroded area, and t is the wall thickness. Studies have revealed that the size of the corrosion area is mostly 50% of the annular length of the upper half of the pipeline [31]. Therefore, the proposed annular length of the corrosion area is 0.28 m, and it is assumed that the corrosion area is a square with $S^2 = 0.28\ m^2$.

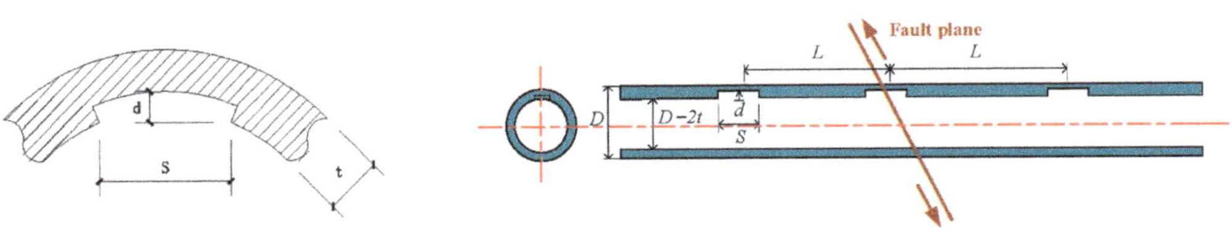

(a) (b)

Figure 3. Schematic diagram of pipeline mode with local corrosion. (**a**) Front cross-sectional view of the corroded pipeline. (**b**) Longitudinal section diagram of the corroded pipeline.

3. Failure Criteria for Pipeline Crossing Fault

The strain-based pipeline design criteria are chosen according to the code provision GB/T 50470-2017 [32]. When the pipeline is subjected to fault-induced displacement, the following three different limit states are considered for analysis and quantification and are used as the failure criteria of the pipeline.

3.1. Tensile Fracture

Due to the fault displacement, the tensile strain of the pipe wall exceeds the ultimate tensile strain of the pipeline material, resulting in rupture. Considering the ductility of the steel material itself, the allowable tensile strain $[\varepsilon_T]$ used in this study is 3% according to the recommended value in IRRK-GSDMA [33].

3.2. Local Buckling

Under fault displacement, the pipeline will produce compressive folding deformation at a certain position away from the fault plane. When the compressive strain exceeds a certain limit, the pipe wall will sag and bulge due to instability, and when buckling further progresses, more serious collapse will occur. Therefore, according to the relevant provisions of the CSA Z662 [34], the allowable compressive strain $[\varepsilon_C]$ in this study is expressed as:

$$[\varepsilon_c] = 0.5\frac{t}{D} - 0.0025 + 3000\left(\frac{\sigma_h}{E}\right)^2, \quad (4)$$

$$\sigma_h = \begin{cases} \frac{PD}{2t}, \frac{PD}{2t\sigma_y} \leq 0.4 \\ 0.4\sigma_y, \frac{PD}{2t\sigma_y} > 0.4' \end{cases} \quad (5)$$

where $[\varepsilon_C]$ is an allowable compressive strain (taken here as −0.016); t is the thickness of pipe wall; D is the pipe outer diameter; σ_y is the pipe circumferential stress; P is the pipe internal pressure; and σ_h is the yield stress of pipe material.

3.3. Elliptic Deformation

The loss of pipe wall thickness caused by corrosion will lead to a change in the pipeline cross-section shape under fault displacement. Correspondingly, serious bending will cause the circular section of the pipeline to flatten, thus reducing the pipeline transportation capacity. The deformation of the pipeline cross-section is usually presented by an ellipse. In this study, the dimensionless ellipticity parameter f is used to measure the extent of pipeline ellipticity, which is expressed as follows:

$$f = \frac{\Delta D}{D}, \quad (6)$$

where D is the change in the pipe's outer diameter. When the ellipticity reaches 0.15, the pipeline will be damaged [35]. The final adopted pipeline failure limits in this study are shown in Table 4.

Table 4. Identification of failure modes in this study.

Failure Modes	Corresponding Indicators	Selected Limits
Tensile fracture	Allowable tensile strain $[\varepsilon_T]$	0.03
Local buckling	Allowable compressive strain $[\varepsilon_C]$	−0.016
Elliptic deformation	Allowable ovality $[f]$	0.15

4. Results and Analyses

4.1. Validation of Numerical Model

To validate the finite element model, its predictions are compared with the results of the full-scale physical model test conducted by Jalali et al. [36]. The developed model adopts the same material structure and size as in the physical test, and according to the test loading scheme, 0.2 m and 0.6 m fault displacements are applied to this model. Figure 4a shows the local buckling predicted by the finite element model in this study. Compared with Jalali's test, it can be seen that the model successfully predicts the buckling position of the pipeline. Furthermore, Figure 4b compares the measured and calculated longitudinal strains after fault displacements of 0.2 m and 0.6 m. Figure 4b demonstrates that the finite element model successfully predicted the maximum strain and the position of the buckling pipeline section. The measured and calculated peak strains at the top of the pipeline at the hanging wall are 0.02 and 0.018, and the peak strains at the footwall are −0.095 and −0.081, respectively (i.e., discrepancy < 20%). This difference is attributed to the possible soil inhomogeneity and porosity in the test soil caused by fault displacement. In conclusion, the developed finite element model is suitable for conducting the response and failure mode analysis of buried pipeline crossing a reverse fault.

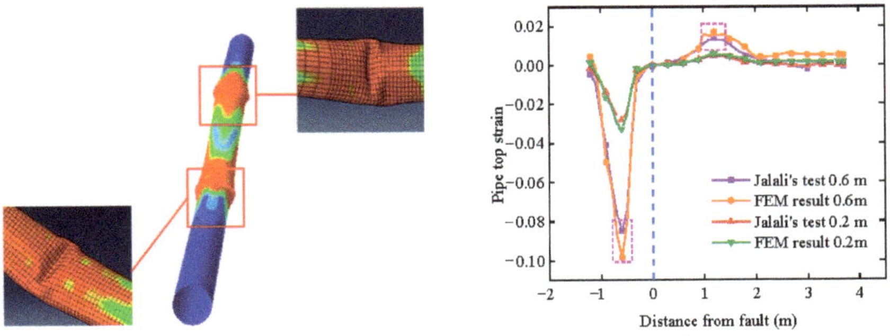

(**a**) (**b**)

Figure 4. A comparison of numerical calculation and Jalali's test. (**a**) Local buckling of the steel pipeline. (**b**) The axial strain of the pipe's top.

4.2. Mechanical Response of Corroded Pipeline Crossing Reverse Fault

4.2.1. Mechanical Response of Pipeline with Corrosion at Hanging Wall

Figure 5 shows the axial strain diagrams at the top of the pipeline with different residual wall thicknesses when the corrosion occurs at the hanging wall at a distance L = 3.2 m from the fault plane and 0.9 m fault displacement. Figure 5 shows that as the corrosion deepens, its influence on the pipeline strain increases within a certain range x = [17 m, 19.3 m] outside the corrosion area due to the reduced remaining wall thickness. As the pipe wall becomes thinner, the strain at the corrosion location exhibits an approximately linear increase (approximately 0.01 per every 10% reduction in thickness). When the residual wall thickness is 0.8 t, the tensile strain in the etched region exceeds 0.03 (i.e., failure).

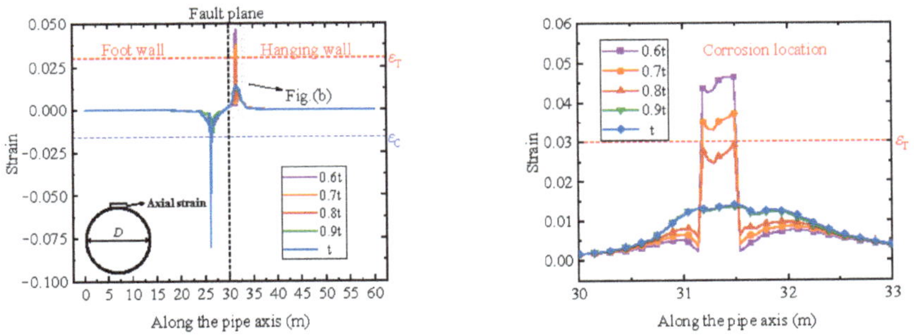

(**a**) (**b**)

Figure 5. The axial strain of the pipeline's top with corrosion at the hanging wall. (**a**) The axial strain of the pipe's top. (**b**) Enlarged image.

Figure 6 shows the von Mises stress diagram of the pipeline with different residual wall thicknesses when corrosion occurs at the hanging wall at a distance L = 3.2 m from the fault plane and 0.9 m fault displacement. Figure 6 demonstrates that the whole pipeline exhibits S-shape deformation and is obviously bent at two locations 3.2 m from both sides of the fault plane. With the stress concentration, the pipe wall is folded due to the pressure. Combined with analyses of Figure 5, the results reveal that local buckling is still the main failure mode of the pipeline crossing a reverse fault when the residual wall thickness is more than 0.8 t. With the deepening corrosion, the main failure mode will change from a single local buckling to a double form controlled by local buckling and tensile fracture when the residual wall thickness is less than 0.8 t. In other words, the increased corrosion weakens the seismic performance of the pipeline and promotes the development of multiple failure modes.

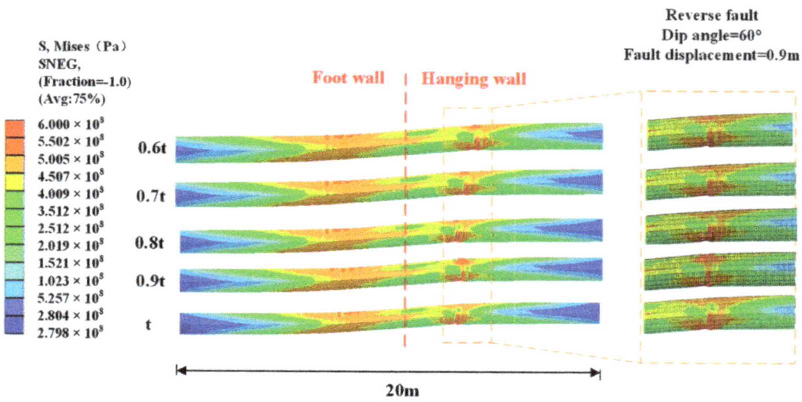

Figure 6. Stress diagram of the pipeline with corrosion at the hanging wall.

4.2.2. Mechanical Response of Pipeline with Corrosion at Footwall

Figure 7 shows the axial strain diagrams at the top of the pipeline with different residual wall thicknesses when the corrosion occurs at the footwall at a distance L = 3.2 m from the fault plane and 0.9 m fault displacement. It is seen from Figure 7 that compared with the peak strain of a non-corroded pipeline, the peak strain is increased by 24%, 70%, and 99% for 0.9 t, 0.8 t, and 0.7 t pipelines, respectively. Meanwhile, the peak strain of the pipeline at the footwall increases by 126% from −0.0801 for the intact pipeline to −0.1817 for 0.6 t. When the pipe wall thickness is reduced by 10%, the strain in the corroded area changes dramatically, showing an irregular nonlinear growth trend.

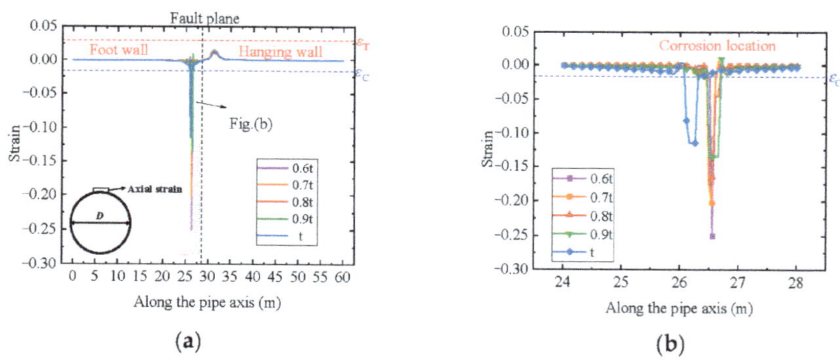

Figure 7. The axial strain of the pipeline's top with corrosion at the footwall. (**a**) The axial strain of the pipe's top. (**b**) Enlarged image.

Figure 8 shows the von Mises stress diagram of the pipeline with different residual wall thicknesses when corrosion occurs at the footwall at a distance L = 3.2 m from the fault plane and 0.9 m fault displacement. It reveals that the deformation mode of the whole pipeline is consistent with the corrosion of the hanging wall pipeline; however, the difference is that the corrosion of the footwall pipeline has a significant effect on its section deformation. As the corrosion extent increases, the stress level of the corrosion area increases gradually compared with the surrounding area, and the deformation becomes more concentrated, resulting in large section deformation. Though the main failure mode of the pipeline, in this case, is still local buckling, the axial compressive strain makes it easier to reach the failure limit with the deepening corrosion.

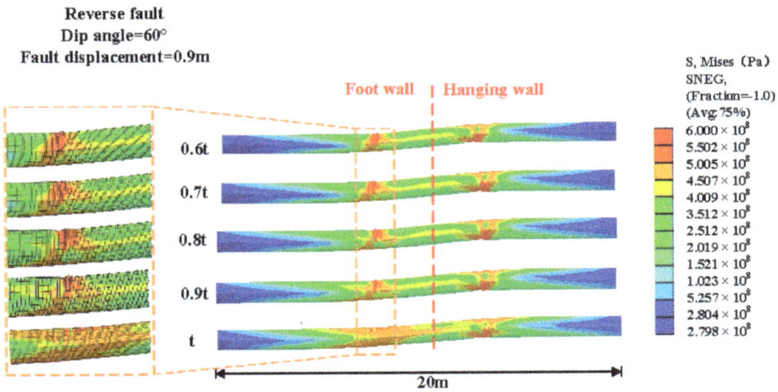

Figure 8. The stress diagram of the pipeline with corrosion at the footwall.

For corrosion at the hanging wall, as the fault displacement increases, the failure mode of the pipeline changes from a single failure mode of local buckling to a double failure mode controlled by local buckling and tensile fracture. On the other hand, the corrosion at the footwall does not change the main failure mode of local buckling; however, the corrosion effect intensifies the local buckling. Thus, different corrosion positions have a significant amplification effect on the local strain of the pipeline. The strain response caused by corrosion at the footwall is larger than that caused by corrosion at the hanging wall.

4.3. Failure Modes of Corroded Pipeline Crossing Reverse Fault

4.3.1. Failure Modes of Pipeline with Corrosion at Hanging Wall

Figure 9 shows the peak strain of the hanging and footwall pipelines with different corrosion extents at the hanging wall at a distance L = 3.2 m from the fault plane (C is compressive strain, T is tensile strain). It is observed in Figure 9a that there is no significant difference in the compressive strain curve at the bottom of the hanging wall pipeline for different levels of corrosion, while the difference in the tensile strain curve is evident. When the fault displacement exceeds 0.6 m, a single failure mode of local buckling occurs at the bottom of the hanging wall pipeline almost simultaneously for different corrosion extents, and the difference in tensile strain at the top of the pipeline is obvious. Double failure modes of local buckling and tensile fracture occur when the fault displacement exceeds 0.8 m. Figure 9b shows that the corrosion at the hanging wall has a minor influence on the strain of the pipeline at the footwall: it does not affect its main failure mode.

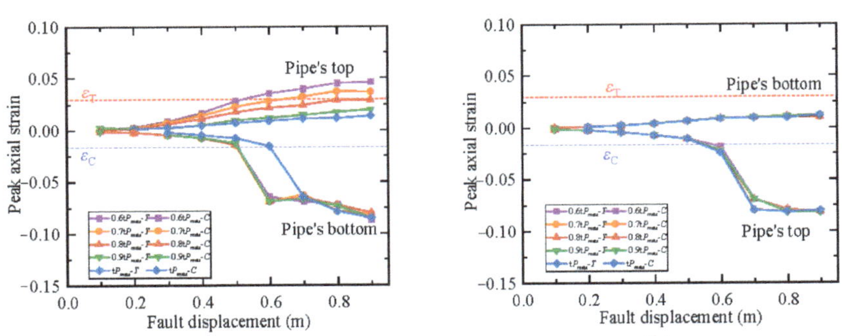

Figure 9. The peak axial strain of the pipeline with corrosion at the hanging wall. (**a**) Hanging wall section. (**b**) Footwall section.

Figure 10 shows the variation trend of pipeline peak ellipticity under different fault displacements. Figure 10 shows that the ellipticity of the pipeline increases as the fault displacement increases. When the fault displacement exceeds 0.9 m, the pipeline with different corrosion extents gradually presents an elliptic deformation failure.

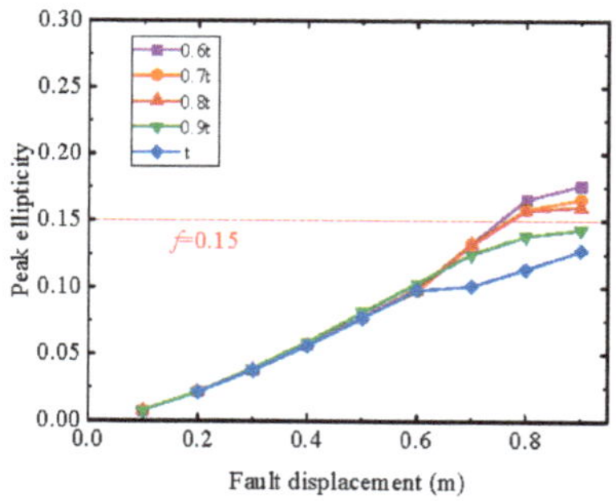

Figure 10. Peak ellipticity of the pipeline with corrosion at the hanging wall.

Table 5 shows the critical fault displacement of the pipeline with corrosion at the hanging wall. Taking 0.8 t as an example, when fault displacement exceeds 0.9 m, the hanging wall pipeline simultaneously has three failure modes: local buckling at the bottom, tensile fracture at the top, and elliptic deformation. The results demonstrate that when corrosion occurs at the hanging wall, the increasing corrosion extent will weaken the pipeline's resistance to tensile fracture and elliptic deformation. With the increase in fault displacement and corrosion extent, the failure mode of the hanging wall pipeline changes from single to multiple forms.

Table 5. Critical fault displacement of the pipeline with corrosion at the hanging wall (m).

Failure Mode	Remaining Wall Thickness				
	0.6 t	**0.7 t**	**0.8 t**	**0.9 t**	**t**
Local buckling	0.51	0.51	0.51	0.51	0.6
Tensile fracture	0.52	0.62	0.8	1.1	1.21
Elliptic deformation	0.78	0.79	0.79	0.92	0.99

4.3.2. Failure Modes of Pipeline with Corrosion at Footwall

Figure 11 shows the peak strain of the hanging and footwall pipelines with different corrosion extents at the footwall at a distance L = 3.2 m from the fault plane. Figure 11a indicates that the corrosion at the footwall has a minor influence on the strain of the hanging wall pipeline. Although the top compression and bottom tensile strain fluctuate to a certain extent, this does not affect its main failure mode. Figure 11b reveals that as the fault displacement increases, the compressive strain of the pipe's top with different corrosion extents varies significantly. As the corrosion extent increases, the critical fault displacement of local buckling decreases gradually. If the residual wall thickness is only 0.6 t, local buckling failure occurs at the pipe's top when the fault displacement is only 0.33 m.

Figure 12 shows the variation trend of pipeline peak ellipticity under different fault displacements. From Figure 12, it is noted that the ellipticity increases with the increase in fault displacement. However, when the fault displacement exceeds 0.4 m, the influence of corrosion extent becomes evident. When the fault displacement exceeds 0.9 m, the pipeline with different corrosion extents gradually exhibits an elliptic deformation failure.

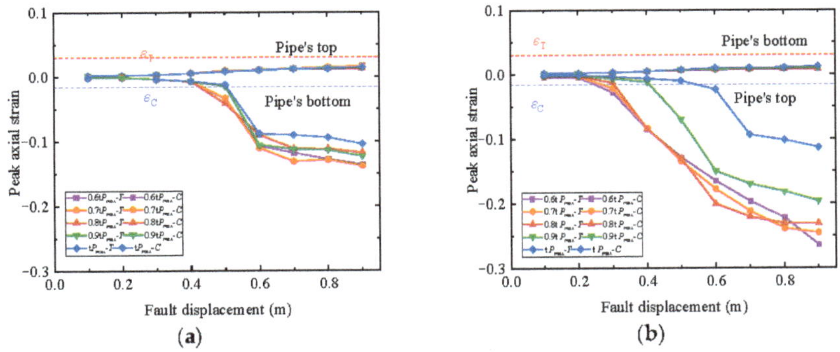

Figure 11. The peak axial strain of the pipeline with corrosion at the footwall. (a) Hanging wall section. (b) Footwall section.

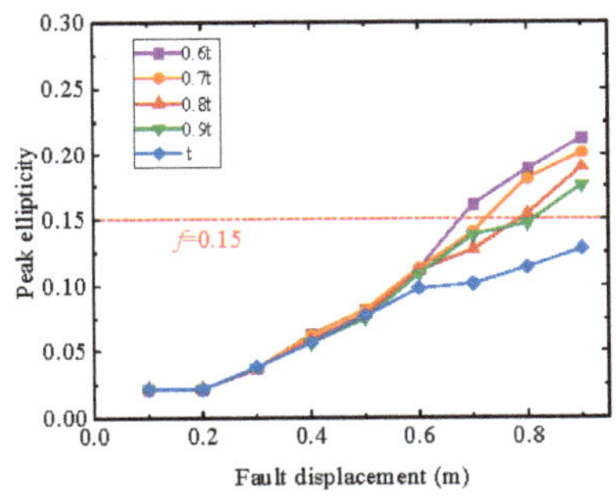

Figure 12. Peak ellipticity of the pipeline with corrosion at the footwall.

Table 6 shows a critical fault displacement of the pipeline with corrosion at the footwall. It is evident from Table 6 that as the corrosion extends, the critical fault displacement of local buckling and elliptic deformation of the pipeline continues to decrease, especially the local buckling, which decreases sharply, while the tensile fracture is the largest and unaffected by corrosion. When the fault displacement exceeds 0.87 m but does not reach 1.15 m, a double failure mode of local buckling and elliptic deformation occurs at the footwall pipeline section. When the fault displacement exceeds 1.15 m, there are multiple failure modes of local buckling, tensile fracture, and elliptic deformation at the footwall pipeline section. The results indicate that when corrosion occurs at the footwall, an increasing corrosion extent will weaken the resistance of the pipeline to local buckling and elliptic deformation. With the increase of fault displacement and corrosion extent, the failure mode of the footwall pipeline will change from single one to multiple forms.

Table 6. Critical fault displacement of the pipeline with corrosion at the footwall (m).

Failure Mode	Remaining Wall Thickness				
	0.6 t	0.7 t	0.8 t	0.9 t	t
Local buckling	0.26	0.28	0.3	0.41	0.55
Tensile fracture	1.3	1.3	1.3	1.3	1.3
Elliptic deformation	0.68	0.72	0.8	0.81	1.09

4.4. Effectiveness of CFRP Reinforcement

When the top corrosion occurs in a certain position of the footwall pipeline, the original failure mode of the pipeline changes from single to multiple forms, and the failure damage will be more serious than that in other positions. Therefore, it is necessary to undertake reinforcement measures for the corroded footwall pipeline.

4.4.1. Design Conditions

A 3D numerical model is established for a corroded pipeline crossing a reverse fault with dip angle β = 60°, buried depth h = 2 m, and corrosion depth of 0.6 t (t = 0.0119 m) to investigate the effectiveness of different CFRP external winding reinforcement schemes on the mechanical response and the failure mode of the pipeline. The number of reinforced layers is either 4, 8, or 12, and the reinforcement lengths considered are 5 m, 10 m, 15 m, and 20 m.

4.4.2. Response and Failure of Corroded Pipeline with CFRP Reinforcement

Figure 13 shows the peak strain and peak ellipticity of the footwall pipeline reinforced with CFRP of a certain 20 m length and different thicknesses. It is observed in Figure 13a that compared with the unreinforced pipeline, CFRP can significantly reduce the strain response and delay the occurrence of local buckling failure. The buckling resistance increased by 56.7%, 82.4%, and 86.9% for the 4-, 8- and 12-layer cases, respectively. The strain of the severely corroded pipeline (0.6 t) can be restored to 90% of the intact pipeline by using 4-layer CFRP. It is further noted that in the interval from 0 to 4 layers, the average buckling resistance increased by 14.17% for each layer compared with 0 layers; in the interval from 4 to 8 layers, the average buckling resistance increased by 6.43% for each layer compared with 4 layers; and in the interval from 8 to 12 layers, the average buckling resistance increased by 1.2% for each layer compared with 8 layers. Therefore, as the CFRP thickness increases, the buckling resistance increases but at a decreasing rate.

Figure 13b shows that, compared with the unreinforced pipeline, the peak ellipticity of the 4-layer reinforced pipeline is increased by 27.8%, that of the 8-layer reinforced pipeline is increased by 36.79%, and that of the 12-layer reinforced pipeline is increased by 44.8%, indicating that more CFRP reinforced layers mean stronger resistance of the pipeline to cross-section deformation. Moreover, when the fault displacement is less than 0.3 m, the section deformation of the pipeline reinforced with CFRP is basically the same as that without the reinforcement, and the effect and difference of CFRP reinforcement with different thicknesses cannot be reflected temporarily.

Table 7 shows the critical fault displacement of the pipeline with different CFRP thicknesses. It is noted from Table 7 that the critical fault displacement of local buckling and elliptic deformation increases significantly after the reinforcement, and the reinforcement effect is the most significant in the 4-layer case. The failure sequence of the pipeline is first local buckling, then elliptic deformation, and finally tensile fracture.

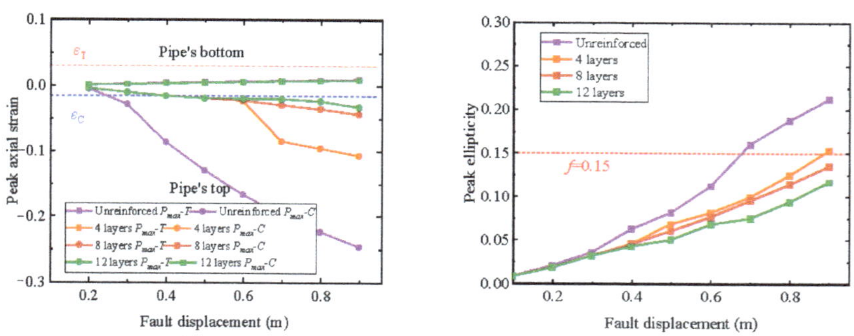

(a) (b)

Figure 13. A mechanical response of the pipeline with different CFRP thicknesses. (**a**) Peak axial strain. (**b**) Peak ellipticity.

Figure 14 shows the peak strain and peak ellipticity of the footwall pipeline reinforced with CFRP of 8 layers and different lengths. Figure 14a shows that compared with the

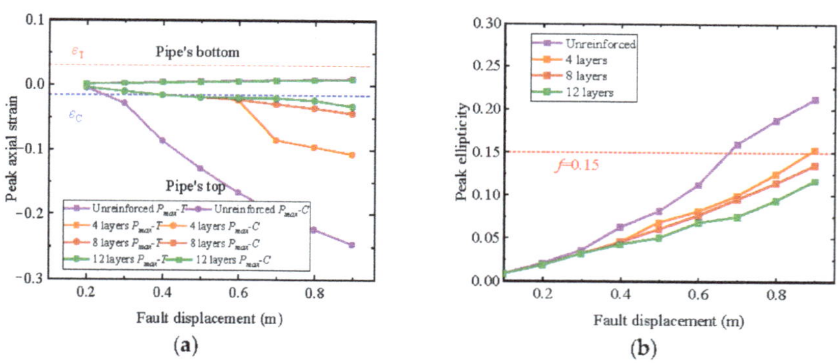

MC-ConvTasNet can not only effectively suppress the majority of direct waves but also maintain most of the reflected wave signals.

3.6. Comparison of the Wave Separation Capabilities of MC-ConvTasNet and Wave-U-Net

We compare the wave separation performance of MC-ConvTasNet and Wave-U-Net. Wave-U-Net is constructed based on the encoder–decoder architecture proposed in [21], with the encoder comprising four layers. The number of filters doubles in each subsequent layer, specifically totaling 32, 64, 128 and 256. In the decoder, we utilize one-dimensional transposed convolutional layers to replace the upsampling method originally used in the paper. All other details of the Wave-U-Net remain consistent with the description in [21]. After training, the Wave-U-Net achieves a scale-invariant signal-to-noise ratio of 8.7 dB on the test set. Among the three models, namely, the hard-to-hard single-interface model, the soft-to-hard single-interface model and the double-interface model, Wave-U-Net exhibits a lower R_{std} than MC-ConvTasNet. The R_{std} of Wave-U-Net is 23.7 dB, 21.7 dB and 23.7 dB, respectively. In field logging data, the amplitude of the direct wave after suppression using Wave-U-Net is greater than that of MC-ConvTasNet (see Figure 13). Despite having a simpler structure compared to MC-ConvTasNet, Wave-U-Net exhibits an inferior performance and generalization capability when it comes to extracting reflected waves.

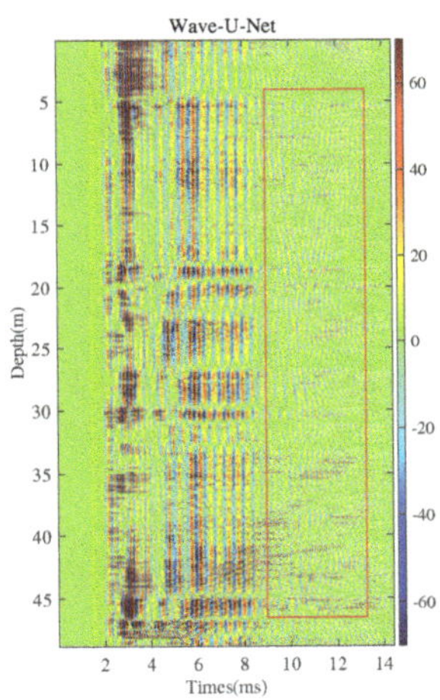

Figure 13. Reflection waves extracted from field logging data by Wave-U-Net. The area containing the reflected waves is highlighted with a solid red box. The values displayed in the variable density map are clipped to a range of +/−70.

4. Discussion

Both the median filter and parameter estimation offer the advantage of simplicity in operation, often allowing for direct use without frequent parameter adjustments in practical applications. However, both methods have their respective limitations. Specifically, the median filter may cause signal distortion and is not applicable to formations parallel to the wellbore axis, while parameter estimation only suppresses the amplitude of direct waves to a certain extent. In contrast, the F-K filter, high-resolution Radon transform and shear wave transform excel in extracting reflected waves from reflectors close to the wellbore. However, they rely on expert experience to adjust method parameters during use, resulting in uncertain required time, and are similarly inapplicable to formations parallel to the wellbore axis. In field logging data, the types of reflectors are often unknown, and each method has its specific limitations. Therefore, when extracting reflected waves, experts often need to judge the possible types of reflectors in each wellbore section based on their experience and select appropriate signal processing methods for extraction. This process is not only time consuming but also inefficient. MC-ConvTasNet overcomes the dependence on expert experience and achieves automatic extraction of reflected waves across the entire wellbore section. Although the performance of MC-ConvTasNet may decline when reflectors are very close to the wellbore, its applicable range still covers the typical distances involved in single-well imaging, ranging from a few meters to several tens of meters.

Although our current research focuses on synthetic data, studies [42,43] have shown that integrating synthetic and real data within datasets can significantly enhance model performance. Therefore, we plan to incorporate partially labeled actual field data into our future work to further improve the robustness of our models. It is recommended that readers first generate synthetic datasets using analytical algorithms to train the neural network. The trained model can then be effectively applied to extract clear reflection waves from field logging data. Subsequently, the reflection waves extracted from each well can be used as known labels to fine-tune the model with field logging data. Through fine-tuning, the model's generalization capability will be significantly enhanced, thereby further improving its performance in reflection wave extraction.

This paper only focuses on the separation of flexural and reflected waves for dipole S-wave imaging. Further study is being conducted to apply this algorithm to the wave separation of monopole P-wave imaging.

5. Conclusions

We apply MC-ConvTasNet to wave separation to achieve effective extraction of reflected waves. Due to the differences in propagation distance and velocity between the direct and reflected waves, their arrival time differences between channels in the CSG are different. The Conv2d module in MC-ConvTasNet is used to obtain spatial cues between channels to help separate the direct wave and the reflected wave. Because reflectors are rarely known in field logging, synthetic data are used as a dataset in this study. A large number of theoretical waveforms of the CSG are obtained by analytical algorithms.

After training, MC-ConvTasNet exhibits an excellent ability to extract reflected waves. Using three models, we compare the wave separation performance of parameter estimation, MC-ConvTasNet, a median filter and an F-K filter. In the hard-to-hard single-interface model and soft-to-hard single-interface model, the average R_{std} of parameter estimation and the median filter is less than 12.4 dB, less than 31 dB for the F-K filter and more than 33 dB for MC-ConvTasNet. The reflected waves extracted by MC-ConvTasNet are closest to the theoretical reflected waves in both phase and shape, with the largest average R_{std}. In the double-interface model, the average R_{std} of the other three methods is less than 0, while the average R_{std} of MC-ConvTasNet is more than 32 dB. Despite the absence of double-interface samples in the dataset, MC-ConvTasNet exhibits excellent performance in dealing with double-interface cases, and MC-ConvTasNet trained on waveforms in the CSG domain is proven to be suitable for interfaces parallel to the wellbore axis. In the hard-to-hard single-interface model, we compare the wave separation capability of MC-ConvTasNet on waveforms without noise and waveforms with Gaussian white noise. Compared with noise-free data, the increase in the root mean square error is less than 0.06 when the signal-to-noise ratio is 20 dB. MC-ConvTasNet is applicable to data containing noise and maintains good performance.

In the field logging data, MC-ConvTasNet is compared with parameter estimation, a median filter and an F-K filter. Surprisingly, MC-ConvTasNet extracts clearer and more complete reflected waves. The reflector obtained from the imaging of reflected waves extracted by MC-ConvTasNet is the clearest one, with the fewest artifacts, and

MC-ConvTasNet achieves both an R_{dsr} and an R_{snr} exceeding 55 dB, demonstrating a superior direct wave suppression capability. Despite the dataset not including field logging data, MC-ConvTasNet is effective for field logging data analysis. When MC-ConvTasNet is used to process the waveforms of the CSG, end-to-end wave separation can be achieved. By inputting the full wave of the CSG containing the currently processed channel, MC-ConvTasNet outputs the time-domain waveforms of direct and reflected waves. It is expected to extract reflected waves in real time when the acoustic logging tool is continuously lifted.

Author Contributions: Conceptualization, data curation, investigation, methodology, software, validation, visualization, writing—original draft, writing—review and editing, W.L.; conceptualization, funding acquisition, resources, writing—review and editing, J.X.; funding acquisition, project administration, supervision, writing—review and editing, H.H. All authors have read and agreed to the published version of the manuscript.

Funding: This work was supported by the National Natural Science Foundation of China (grant no. 12472087, grant no. 42204117, grant no. 12272107).

Institutional Review Board Statement: Not applicable.

Informed Consent Statement: Not applicable.

Data Availability Statement: The original contributions presented in this study are included in the article. Further inquiries can be directed to the corresponding author.

Acknowledgments: We would like to express our sincere gratitude to Xiugang Ma for providing the field logging data, which were essential for the completion of this study. We are also deeply grateful to Yuan Gao for her valuable suggestions during the revision stage of the manuscript, which significantly improved the quality of this paper.

Conflicts of Interest: The authors declare no conflicts of interest.

Abbreviations

The following abbreviations are used in this manuscript:

MC-ConvTasNet	Multi-channel convolutional time-domain audio separation network
CSG	Common-source gather
CRG	Common-receiver gather
COG	Linear dichroism
Conv-TasNet	Convolutional time-domain audio separation network
ICD	Inter-channel convolution difference
Conv2d	Two-dimensional convolution

Appendix A. Expressions for Direct and Reflected Waves

Excited by a dipole source along the x-axis, the displacement expression in the fluid is [26]

$$u_1 = \nabla(\phi_d + \phi_r), \tag{A1}$$

where ϕ_d and ϕ_r are the displacement potentials associated with the source and reflected from the borehole wall, respectively; and ∇ is the Laplacian operator in the cylindrical coordinate system. The displacement potential function in the borehole can be expressed as

$$\begin{aligned} \phi_d(r,\varphi,z,\omega) &= \tfrac{-V_0T(\omega)}{4\pi^2}\int\limits_{-\infty}^{+\infty} r_0 f K_1(fr)\mathrm{e}^{\mathrm{j}kz}\cos\varphi \mathrm{d}k, \\ \phi_r(r,\varphi,z,\omega) &= \tfrac{-V_0T(\omega)}{4\pi^2}\int\limits_{-\infty}^{+\infty} \left(\tfrac{r_0 f}{2}\right) A(k,\omega) I_1(fr)\mathrm{e}^{\mathrm{j}kz}\cos\varphi \mathrm{d}k, \end{aligned} \tag{A2}$$

where $f^2 = k^2 - \frac{\omega^2}{\alpha_f^2}$; f and k are the radial wavenumber and axial wavenumber of fluid; w is the angular frequency; α_f is the fluid velocity in the borehole; I_1 and K_1 are the first-order modified Bessel function of the first and second kinds, respectively; r_0 is the radius of the dipole source; $T(\omega)$ represents the spectrum of the time function of the acoustic source; V_0 represents the peak amplitude of the source volume change; φ indicates the azimuth angle; and the coefficient A can be obtained from borehole boundary conditions. The direct wave excited by a dipole source is a flexural wave whose waveform can be calculated by the real axis integration method [27] according to Formula (A1).

Excited by a dipole source along the x-axis, the reflected displacement along the x-axis received in the borehole is [28]

$$u_2 = u_{\mathrm{SH}} + u_{\mathrm{SV}} + u_{\mathrm{P}} + u_{\mathrm{SP}} + u_{\mathrm{PS}}, \tag{A3}$$

where

$$\begin{aligned} u_{SH} &= T(\omega) R_{SH} \sin^2\varphi, \\ u_{SV} &= T(\omega) R_{SV} \cos^2\varphi, \\ u_P &= T(\omega) R_P \cos^2\varphi, \\ u_{PS} &= T(\omega) R_{PS} \cos^2\varphi, \\ u_{SP} &= T(\omega) R_{SP} \cos^2\varphi, \end{aligned} \tag{A4}$$

$$\begin{aligned} R_{SH} &= \tfrac{P_0 R_0 r_0}{\bar{L}\rho\beta^2} e^{j\omega\bar{L}/\beta} G_{SH} F_{SH}[k_S(\theta_1),\omega] F_{SH}[k_S(\theta_2),\omega], \\ R_{SV} &= \tfrac{P_0 R_0 r_0}{\bar{L}\rho\beta^2} e^{j\omega\bar{L}/\beta} G_{SV} F_{SV}[k_S(\theta_1),\omega] F_{SV}[k_S(\theta_2),\omega], \\ R_P &= \tfrac{P_0 R_0 r_0}{\bar{L}\rho\alpha^2} e^{j\omega\bar{L}/\alpha} G_P F_P[k_P(\theta_1),\omega] F_P[k_P(\theta_2),\omega], \\ R_{PS} &= \tfrac{P_0 R_0 r_0}{\tilde{L}_S\rho\alpha^2} e^{j\omega\bar{L}_S/\beta} G_{PS} F_P\left[k_P\left(\theta_1^P\right),\omega\right] F_{SV}\left[k_S\left(\theta_2^S\right),\omega\right], \\ R_{SP} &= \tfrac{P_0 R_0 r_0}{\tilde{L}_P\rho\beta^2} e^{j\omega\bar{L}_P/\alpha} G_{SP} F_{SV}\left[k_S\left(\theta_1^S\right),\omega\right] F_P\left[k_P\left(\theta_2^P\right),\omega\right], \end{aligned} \tag{A5}$$

$$\begin{aligned} F_{\mathrm{P}}(k_P,\omega) &= \tfrac{fC(k_P,\omega)}{2\rho_f\omega} j\rho\alpha, \\ F_{\mathrm{SH}}(k_S,\omega) &= -\tfrac{fD(k_S,\omega)}{2\rho_f\omega} j\rho\beta\sin\theta, \\ F_{\mathrm{SV}}(k_S,\omega) &= -\tfrac{fE(k_S,\omega)}{2\rho_f} \rho\sin\theta. \end{aligned} \tag{A6}$$

Here, u_{SH}, u_{SV}, u_{P}, u_{SP} and u_{PS} are the borehole displacement stimulated by SH-SH, SV-SV, P-P, P-SV and SV-P waves, respectively; $R = \sqrt{r^2 + z^2}$, $k_P = \omega/\alpha\cos\theta$, $k_S = \omega/\beta\cos\theta$, $\cos\theta = z/R$; ρ_f, ρ, α and β are the fluid density in the borehole, the formation density outside the borehole and the P-wave and S-wave velocities of the formation, respectively; G_{SH}, G_{SV}, G_P, G_{PS} and G_{SP} are the plane reflection coefficients of SH-SH, SV-SV, P-P, P-SV and SV-P waves, respectively [31]; $\bar{L}$ is the total propagation distance of the reflected wave; $\bar{L}_P$ and $\bar{L}_S$ are the distances between the receiver in the borehole and the virtual source of the SV-P and P-SV waves; $\tilde{L}_P$ and $\tilde{L}_S$ are distances associated with the geometric spreading of SV-P and P-SV waves; expressions for $\bar{L}$, $\bar{L}_P$, $\bar{L}_S$, $\tilde{L}_P$ and $\tilde{L}_S$ can be found in the study of Xu and Hu [28]; θ_1 is the radiation take-off angle of SH-SH, SV-SV and P-P waves from the borehole of the reflected wave; θ_2 is the reception angle of SH-SH, SV-SV and P-P waves upon the borehole of the reflected wave; θ_1^P and θ_1^S represent the incident angle and reflection angle of the P-SV wave relative to the geologic interface, respectively; θ_1^S and θ_2^P represent the incident angle and reflection angle of the SV-P wave relative to the geologic interface, respectively; and the amplitude coefficients C, D and E are obtained by the borehole boundary conditions.

References

1. Tang, X. Imaging near-borehole structure using directional acoustic-wave measurement. *Geophysics* **2004**, *69*, 1378–1386. [CrossRef]
2. Esmersoy, C.; Chang, C.; Kane, M.; Coates, R.; Tichelaar, B.; Quint, E. Acoustic imaging of reservoir structure from a horizontal well. *Lead. Edge* **1998**, *17*, 940–946. [CrossRef]
3. Hu, H.; Xu, J.; Wang, Z. Analytical Simulation of the Waves Reflected From a Distant Slanted Interface Outside a Borehole. In Proceedings of the 2019 13th Symposium on Piezoelectrcity, Acoustic Waves and Device Applications (SPAWDA), Shijiazhuang, China, 1–4 November 2019; pp. 1–5. [CrossRef]
4. Xu, J.; Hu, H.; Liu, Q.H. Combination of FDTD with Analytical Methods for Simulating Elastic Scattering of 3-D Objects Outside a Fluid-Filled Borehole. *IEEE Trans. Geosci. Remote Sens.* **2021**, *59*, 5325–5334. [CrossRef]
5. Gao, Y.; Hu, H.; Xu, J. Elastic Reverse Time Migration Method for Single-Well Imaging and Elimination of Azimuthal Ambiguity With a Combined Receiver System. *IEEE Trans. Geosci. Remote Sens.* **2023**, *61*, 5921215. [CrossRef]
6. Borland, W.; Holderson, J.; Rougha, H.; Sultan, A.; Chakravorty, S.; Al-Raisi, M. Integration of Microelectrical and Sonic Reflection Imaging Around the Borehole-Offshore UAE. In Proceedings of the IPTC International Petroleum Technology Conference, Doha, Qatar, 11 November 2005; p. IPTC-11021. [CrossRef]
7. Hirabayashi, N.; Leaney, W.S. Wavefield separation for borehole acoustic reflection survey using parametric decomposition and waveform inversion. *Geophysics* **2019**, *84*, D151–D159. [CrossRef]
8. Tang, X.M.; Zheng, Y.; Patterson, D. Processing array acoustic-logging data to image near-borehole geologic structures. *Geophysics* **2007**, *72*, E87–E97. [CrossRef]
9. Hornby, B.E. Imaging of near-borehole structure using full-waveform sonic data. *Geophysics* **1989**, *54*, 747–757. [CrossRef]
10. Li, C.; Yue, W. High-resolution Radon transforms for improved dipole acoustic imaging. *Geophys. Prospect.* **2017**, *65*, 467–484. [CrossRef]
11. Li, J.; Innanen, K.A.; Tao, G. Extraction of reflected events from sonic-log waveforms using the Karhunen-Loève transform. *Geophysics* **2017**, *82*, D265–D277. [CrossRef]
12. Chen, J.; Yue, W.; Li, C.; Zeng, F. Extracting Reflected Waves From Acoustic Logging Data Based on the Shearlet Transform. *IEEE Geosci. Remote Sens. Lett.* **2019**, *16*, 1688–1692. [CrossRef]
13. Wang, Q.; Wang, H.; Fang, Z.; Xu, D.; Qi, X.; Yu, Y. Weak Reflection Extraction in Borehole Acoustic Reflection Imaging Using an Unsupervised Machine Learning Method. In Proceedings of the SPWLA Annual Logging Symposium, Rio de Janeiro, Brazil, 18–22 May 2024; SPWLA: Houston, TX, USA, 2024.
14. Li, Y.; Zhou, R.; Tang, X.M.; Jackson, J.; Patterson, D. Single-Well Imaging with Acoustic Reflection Survey at Mounds, Oklahoma, USA. In Proceedings of the 64th EAGE Conference & Exhibition, Florence, Italy, 27–30 May 2002. [CrossRef]
15. Gannot, S.; Vincent, E.; Golan, S.M.; Ozerov, A. A Consolidated Perspective on Multimicrophone Speech Enhancement and Source Separation. *IEEE/ACM Trans. Audio Speech Lang. Process.* **2017**, *25*, 692–730. [CrossRef]
16. Hershey, J.R.; Chen, Z.; Le Roux, J.; Watanabe, S. Deep clustering: Discriminative embeddings for segmentation and separation. In Proceedings of the 2016 IEEE International Conference on Acoustics, Speech and Signal Processing (ICASSP), Shanghai, China, 20–25 March 2016; pp. 31–35. [CrossRef]
17. Jansson, A.; Humphrey, E.J.; Montecchio, N.; Bittner, R.M.; Kumar, A.; Weyde, T. Singing Voice Separation with Deep U-Net Convolutional Networks. In Proceedings of the International Society for Music Information Retrieval Conference, Suzhou, China, 23–27 October 2017.
18. Kolbaek, M.; Yu, D.; Tan, Z.; Jensen, J.H. Multitalker Speech Separation With Utterance-Level Permutation Invariant Training of Deep Recurrent Neural Networks. *IEEE/ACM Trans. Audio Speech Lang. Process.* **2017**, *25*, 1901–1913. [CrossRef]
19. Wang, D.; Chen, J. Supervised Speech Separation Based on Deep Learning: An Overview. *IEEE/ACM Trans. Audio Speech Lang. Process.* **2018**, *26*, 1702–1726. [CrossRef] [PubMed]
20. Luo, Y.; Mesgarani, N. TasNet: Time-Domain Audio Separation Network for Real-Time, Single-Channel Speech Separation. In Proceedings of the 2018 IEEE International Conference on Acoustics, Speech and Signal Processing (ICASSP), Calgary, AB, Canada, 15–20 April 2018; pp. 696–700.
21. Stoller, D.; Ewert, S.; Dixon, S. Wave-U-Net: A Multi-Scale Neural Network for End-to-End Audio Source Separation. *arXiv* **2018**, arXiv:1806.03185.
22. Luo, Y.; Mesgarani, N. Conv-TasNet: Surpassing Ideal Time–Frequency Magnitude Masking for Speech Separation. *IEEE/ACM Trans. Audio Speech Lang. Process.* **2019**, *27*, 1256–1266. [CrossRef] [PubMed]
23. Wang, Q.; Wang, H.; Shi, S. A deep learning workflow for weak reflection extraction in pitch-catch measurements in the cased hole. *Geophysics* **2023**, *88*, D147–D157. [CrossRef]
24. Gu, R.; Zhang, S.X.; Chen, L.; Xu, Y.; Yu, M.; Su, D.; Zou, Y.; Yu, D. Enhancing End-to-End Multi-Channel Speech Separation Via Spatial Feature Learning. In Proceedings of the ICASSP 2020–2020 IEEE International Conference on Acoustics, Speech and Signal Processing (ICASSP), Barcelona, Spain, 4–8 May 2020; pp. 7319–7323. [CrossRef]

25. Grumiaux, P.A.; Kitić, S.; Girin, L.; Guerin, A. A survey of sound source localization with deep learning methods. *J. Acoust. Soc. Am.* **2022**, *152*, 107–151. [CrossRef]
26. Kurkjian, A.L.; Chang, S.K. Acoustic multipole sources in fluid-filled boreholes. *Geophysics* **1986**, *51*, 148–163. [CrossRef]
27. Tsang, L.; Rader, D. Numerical evaluation of the transient acoustic waveform due to a point source in a fluid-filled borehole. *Geophysics* **1979**, *44*, 1706–1720. [CrossRef]
28. Xu, J.; Hu, H. Solutions of P-SV and SV-P waves in single-well imaging through reciprocity relations. *Geophysics* **2020**, *85*, D245–D259. [CrossRef]
29. Luo, Y.; Chen, Z.; Yoshioka, T. Dual-Path RNN: Efficient Long Sequence Modeling for Time-Domain Single-Channel Speech Separation. In Proceedings of the ICASSP 2020–2020 IEEE International Conference on Acoustics, Speech and Signal Processing (ICASSP), Barcelona, Spain, 4–8 May 2020; pp. 46–50. [CrossRef]
30. Tang, X.M.; Cao, J.J.; Wei, Z.T. Shear-wave radiation, reception, and reciprocity of a borehole dipole source: With application to modeling of shear-wave reflection survey. *Geophysics* **2014**, *79*, T43–T50. [CrossRef]
31. Aki, K.; Richards, P.G. *Quantitative Seismology: Theory and Methods*, 2nd ed.; University Science Books: Melville, NY, USA, 2002; pp. 128–149,209–215.
32. Wang, Z.; Hu, H.; Yang, Y. Reciprocity relations for the elastodynamic fields generated by multipole sources in a fluid–solid configuration. *Geophys. J. Int.* **2015**, *203*, 883–892. [CrossRef]
33. Xu, J.; Hu, H.; Wang, Z. Asymptotic solution to a 3D dipole single-well imaging system with combined monopole and dipole receivers with an application in elimination of azimuth ambiguity. *Geophysics* **2019**, *84*, D191–D207. [CrossRef]
34. Castagna, J.P.; Batzle, M.L.; Eastwood, R.L. Relationships between compressional-wave and shear-wave velocities in clastic silicate rocks. *Geophysics* **1985**, *50*, 571–581.
35. Li, Q. Velocity regularities of p and s-waves in formations. *Oil Geophys. Prospect. (China)* **1992**, *27*, 5519369.
36. Ma, Z.; Xie, J. Relationship among conpressional wave, shear wave velocities and density in rocks. *Prog. Geophys.* **2005**, *20*, 905–910.
37. Kolbæk, M.; Tan, Z.; Jensen, S.H.; Jensen, J.H. On Loss Functions for Supervised Monaural Time-Domain Speech Enhancement. *IEEE/ACM Trans. Audio Speech Lang. Process.* **2019**, *28*, 825–838. [CrossRef]
38. Roux, J.L.; Wisdom, S.; Erdogan, H.; Hershey, J.R. SDR—Half-baked or Well Done? In Proceedings of the ICASSP 2019—2019 IEEE International Conference on Acoustics, Speech and Signal Processing (ICASSP), Brighton, UK, 12–17 May 2019; pp. 626–630. [CrossRef]
39. Kingma, D.P.; Ba, J. Adam: A Method for Stochastic Optimization. *arXiv* **2014**, arXiv:1412.6980.
40. Gazdag, J.; Sguazzero, P. Migration of seismic data by phase shift plus interpolation. *Geophysics* **1984**, *49*, 124–131. [CrossRef]
41. Sridhar, K.; Cutler, R.; Saabas, A.; Parnamaa, T.; Loide, M.; Gamper, H.; Braun, S.; Aichner, R.; Srinivasan, S. ICASSP 2021 Acoustic Echo Cancellation Challenge: Datasets, Testing Framework, and Results. In Proceedings of the ICASSP 2021—2021 IEEE International Conference on Acoustics, Speech and Signal Processing (ICASSP), Toronto, ON, Canada, 6–11 June 2021; pp. 151–155. [CrossRef]
42. Zhu, W.; Mousavi, S.M.; Beroza, G.C. Chapter Four—Seismic signal augmentation to improve generalization of deep neural networks. In *Machine Learning in Geosciences*; Advances in Geophysics; Moseley, B., Krischer, L., Eds.; Elsevier: Amsterdam, The Netherlands, 2020; Volume 61, pp. 151–177. [CrossRef]
43. Yan, Z.; Zhang, Z.; Liu, S. Improving Performance of Seismic Fault Detection by Fine-Tuning the Convolutional Neural Network Pre-Trained with Synthetic Samples. *Energies* **2021**, *14*, 3650. [CrossRef]

MDPI AG
Grosspeteranlage 5
4052 Basel
Switzerland
Tel.: +41 61 683 77 34

Applied Sciences Editorial Office
E-mail: applsci@mdpi.com
www.mdpi.com/journal/applsci

www.ingramcontent.com/pod-product-compliance
Lightning Source LLC
Chambersburg PA
CBHW061442210926
51850CB00008B/1586

unreinforced pipeline, the increase in buckling resistance is 31.8%, 48.5%, 66.9%, and 82.4% for the 5 m-, 10 m-, 15 m-, and 20 m-long reinforcement, respectively. However, the rate of increase in resistance also decreases as the reinforcement length increases, similar to the case of different CFRP thicknesses.

Table 7. Critical fault displacement of the pipeline with different CFRP thicknesses.

Failure Modes	Reinforcement Thicknesses			
	0 Layer	4 Layers	8 Layers	12 Layers
Local buckling	0.25	0.46	0.46	0.46
Tensile fracture	1.3	1.3	1.3	1.3
Elliptic deformation	0.68	0.98	0.95	1.2

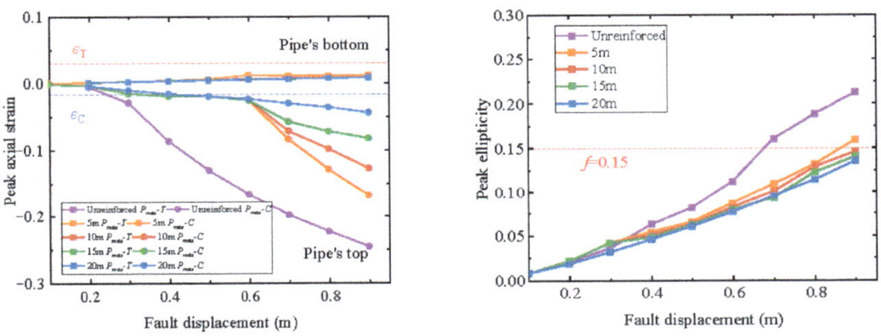

(a) (b)

Figure 14. A mechanical response of the pipeline with different reinforcement lengths. (**a**) Peak axial strain. (**b**) Peak ellipticity.

Furthermore, Figure 14b indicates that compared with the unreinforced pipeline, the resistance to section deformation increases by 25%, 31%, 33.9%, and 36.3% for the 5 m-, 10 m-, 15 m-, and 20 m-long reinforcements. This shows that the resistance of the pipeline to cross-section deformation increases as the CFRP reinforcement length increases but at a decreasing rate. Same as the case of different CFRP thicknesses, when the fault displacement is less than 0.3 m, the CFRP reinforcement has no effect on the pipeline behavior.

Table 8 shows the critical fault displacement of the pipeline with different CFRP lengths. Table 8 indicates that the critical fault displacement of local buckling and elliptic deformation increases significantly after the reinforcement, and the reinforcement effect is the most significant in the 5 m case. The failure sequence of the pipeline is still local buckling, then elliptic deformation, and finally tensile fracture.

Table 8. Critical fault displacement of the pipeline with different reinforcement lengths.

Failure Mode	Reinforcement Lengths				
	0 m	5 m	10 m	15 m	20 m
Local buckling	0.25	0.47	0.47	0.47	0.47
Tensile fracture	1.3	1.3	1.3	1.3	1.3
Elliptic deformation	0.68	0.87	0.91	0.93	0.97

4.4.3. Comparison of Corroded Pipeline Before and After CFRP Reinforcement

Figure 15 shows the Mises stress diagram of the pipeline reinforced with CFRP when the top corrosion occurs at the footwall with 0.9 m fault displacement and residual wall thickness of 0.6 t. Figure 15 demonstrates that the overall stress of the pipeline before and with reinforcement remains unchanged; the top of the footwall pipeline is under pressure and the bottom is under tension, while the top of the hanging wall pipeline is under tension and the bottom is under pressure. Furthermore, the overall pipeline exhibits an S-shaped deformation. However, the pipeline presents different buckling patterns at the corrosion position without and with reinforcement, which is manifested as local buckling without the reinforcement, while there is no obvious stress concentration with the reinforcement. The Mises stress at the corrosion position is significantly reduced after the CFRP reinforcement.

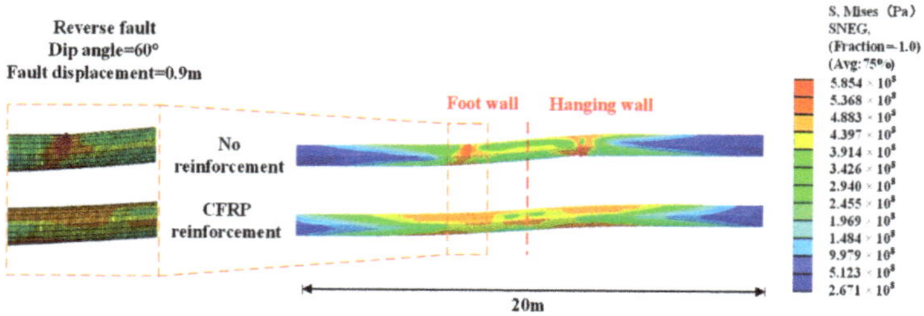

Figure 15. The stress diagram of the corroded pipeline before and after CFRP reinforcement.

Figure 16 shows the cross-section deformation diagram of the corrosion position of the pipeline before and after CFRP reinforcement. Figure 16 shows that without the reinforcement, the corrosion position has a large extent of depression and the shape is greatly distorted. However, after applying the reinforcement, the section deformation changes moderately. The irregular deformation of the pipeline is improved, and the maximum deformation occurs at 0.84 m displacement instead of 0.78 m. The reinforcement at the corrosion position improves the deformation resistance by 7% and the cross-section becomes more circular.

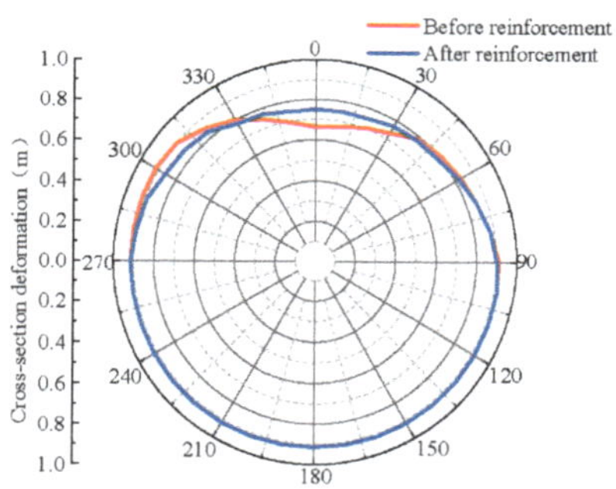

Figure 16. Cross-section deformation of the pipeline before and after CFRP reinforcement.

5. Conclusions

It is necessary to reinforce the cross-fault pipeline with corrosion defects. In this study, a 3D finite element model of a pipeline with corrosion is developed to analyze the impact of corrosion on the response and failure mode of the pipeline crossing a reverse fault. The effectiveness of CFRP reinforcement and its thickness and length are further evaluated to investigate the positive effect on corroded pipelines. The following conclusions are drawn:

(1) Corrosion defect amplifies the axial strain response and weakens the seismic performance of the pipeline crossing a reverse fault. The strain of the pipeline in the corroded area increases linearly with corrosion extent, at about 0.01 increase for every 10% reduction in thickness. Corrosion at different positions of the pipeline will significantly amplify the mechanical response and promote the pipeline's failure process.

(2) Corrosion defect promotes the transition of cross-fault pipeline from a single failure mode to multiple failure modes, increasing the risk of pipeline failure. When the fault displacement exceeds 0.8 m, the pipeline with corrosion at the hanging wall already

experienced failure of local bulking and tensile fracture compared with the intact pipeline only experiencing local bulking.

(3) CFRP reinforcement can alleviate strain concentration, eliminate local buckling, and improve irregular deformation of the pipeline crossing a reverse fault. The compressive resistance of the corroded pipeline is moderately enhanced by the CFRP reinforcement. The section deformation at the reinforced position is like that of a non-corroded pipeline. In addition, it is found that the increase in thickness and length of CFRP reinforcement can enhance the pipeline buckling resistance, but at a decreasing rate, as the reinforced thickness/length increases. Considering the economy of the actual project, combined with the relevant calculation results in this study, it is suggested to wrap 4 layers of CFRP with a length of 5 m on the pipeline wall at the corrosion location to obtain the most cost-efficient reinforcement effect.

Author Contributions: Conceptualization, J.H.; methodology, B.H.; software, Y.B. and W.Z.; validation, Y.B., W.Z. and M.H.E.N.; formal analysis, Y.B. and W.Z.; investigation, J.H.; resources, J.H. and B.H.; data curation, J.H. and Y.B.; writing—original draft preparation, J.H.; writing—review and editing, M.H.E.N.; visualization, Y.B. and W.Z.; supervision, B.H. and M.H.E.N.; project administration, J.H. and B.H.; funding acquisition, J.H. and B.H. All authors have read and agreed to the published version of the manuscript.

Funding: This research is funded by the National Key Research and Development Program of China (2023YFC3007404) and the National Natural Science Foundation of China (No. 52478487).

Institutional Review Board Statement: Not applicable.

Informed Consent Statement: Not applicable.

Data Availability Statement: The raw data supporting the conclusions of this article will be made available by the authors upon request.

Conflicts of Interest: The authors declare no conflicts of interest.

References

1. Zhao, C. Some fundamental problems in lifeline earthquake engineering. *Earthq. Eng. Eng. Vib.* **1993**, *13*, 54–61. (In Chinese)
2. Ballantyne, D. Seismic vulnerability assessment and design of pipelines. *J.-Am. Water Work. Assoc.* **2010**, *102*, 88–97. [CrossRef]
3. MaCaffrey, M.A. Buried pipeline response to reverse faulting during the 1971 San Fernando Earthquake. *ASME, PVP* **1983**, *77*, 151–159.
4. Eidinger, J. Performance of Thames Water 2.2 meter diameter steel pipe at north Anatolian fault crossing. *G&E Rep.* **2001**, *48*, 1.
5. Tsai, J.S.; Jou, L.D.; Lin, S.H. Damage to buried water supply pipelines in the chichi (Taiwan) earthquake and a preliminary evaluation of seismic resistance of pipe joints. *J. China Inst. Eng.* **2000**, *23*, 395–408. [CrossRef]
6. Görüm, T.; Tanyas, H.; Karabacak, F.; Yılmaz, A.; Girgin, S.; Allstadt, K.E.; Süzen, M.L.; Burgi, P. Preliminary documentation of coseismic ground failure triggered by the February 6, 2023 Türkiye earthquake sequence. *Eng. Geo* **2023**, *327*, 107315. [CrossRef]
7. Liu, X.; Sun, S. Strain Based Design of Buried Pipelines Crossing Faults. *Spec. Struct.* **2005**, *22*, 81–85. (In Chinese)
8. Zhang, S.; Wang, R.; Xu, L. Analysis on mechanical properties of the tubular textile composite material. *Acta Mater. Compos. Sin.* **2009**, *26*, 178–185. (In Chinese)
9. Fan, S.; Yan, X. Strain Design of Long Pipeline Crossing Fault under Seismic Loading. *Chin. Pet. Mach.* **2015**, *43*, 114–118. (In Chinese)
10. Cheng, X.; Ma, C.; Huang, R.; Huang, S.; Yang, W. Failure mode analysis of X80 buried steel pipeline under oblique-reverse fault. *Soil. Dyn. Earthq. Eng.* **2019**, *125*, 105723. [CrossRef]
11. Zheng, S.; Xie, X.; Dong, S. Study on Seismic response law of multi-age buried steel pipes considering random corrosion. *J. Disaster Prev. Mitig. Eng.* **2021**, *41*, 255–264. (In Chinese)
12. Benjamin, A.C.; de Andrade, E.Q.; Jacob, B.P.; Pereira, L.C.; Machado, P.R., Jr. Failure behavior of colonies of corrosion defects composed of symmetrically arranged defects. In Proceedings of the International Pipeline Conference, Calgary, AB, Canada, 1 January 2006.
13. Zhou, W.; Bao, J.; Cui, X.; Hong, H.P. Modeling and simulating nonhomogeneous non-Gaussian corrosion fields on buried pipelines and its use in predicting burst capacities of corroded pipelines. *Eng. Struct.* **2021**, *245*, 112957. [CrossRef]
14. Ren, M.; Zhang, Y.; Fan, M.; Xiao, Z. Numerical simulation and ANN prediction of crack problems within corrosion defects. *Materials* **2024**, *17*, 3237. [CrossRef] [PubMed]
15. Long, L.; Yang, H.; Zheng, S.; Cai, Y. Seismic resilience evaluation of urban multi-age water distribution systems considering soil corrosive environments. *Sustainability* **2024**, *16*, 5126. [CrossRef]

16. Wilson, J.M.; Kessler, M.R.; Duell, J.M. Rupture testing of A-106, Grade B steel pipes repaired with carbon/epoxy composites. In Proceedings of the ASME Pressure Vessels and Piping Conference, San Diego, CA, USA, 25–29 July 2004.
17. Rusch, D.W.; Romano, M. Internal repair of pipeline leaks using pressure-activated sealant. In Proceedings of the SPE Eastern Regional Meeting, Charleston, SC, USA, 15 September 2004.
18. Vazouras, P.; Dakoulas, P.; Karamanos, S.A. Pipe–soil interaction and pipeline performance under strike–slip fault movements. *Soil. Dyn. Earthq. Eng.* **2015**, *72*, 48–65. [CrossRef]
19. Mohitpour, M.; Golshan, H.; Murray, M.A.; Mohitpour, M. *Pipeline Design & Construction: A Practical Approach*, 3rd ed.; ASCE Press: New York, NY, USA, 2007.
20. Trifonov, O.V. Numerical stress-strain analysis of buried steel pipelines crossing active strike-slip faults with an emphasis on fault modeling aspects. *J. Pipeline Syst. Eng.* **2015**, *6*, 04014008. [CrossRef]
21. Anastasopoulos, I.; Gazetas, G.; Bransby, M.F.; Davies, M.C.R.; El Nahas, A. Fault rupture propagation through sand: Finite-element analysis and validation through centrifuge experiments. *J. Geotech. Geoenviron. Eng.* **2007**, *133*, 943–958. [CrossRef]
22. Liu, A.; Hu, Z.; Zhao, F.; Li, X.; Takada, S.; Zhao, L. An equivalent-boundary method for the shell analysis of buried pipelines under fault movement. *Acta Seismol. Sin.* **2004**, *17*, 150–156. (In Chinese) [CrossRef]
23. Xu, Z. Study on constitutive relation and failure criteria of X65 pipeline steel. *Pet. Eng. Constr.* **2014**, *40*, 23–27. (In Chinese)
24. Lapczyk, I.; Hurtado, J.A. Progressive damage modeling in fiber-reinforced materials. *Compos. Part. A Appl. Sci. Manuf.* **2007**, *38*, 2333–2341. [CrossRef]
25. Teng, J.; Hu, Y. Behaviour of FRP-jacketed circular steel tubes and cylindrical shells under axial compression. *Constr. Build. Mater.* **2007**, *21*, 827–838. [CrossRef]
26. Ren, Y.; Liu, Y.; Gu, X. The problem of contact surface in pipe-soil interaction solved by ABAQUS software. *Mech. Eng.* **2004**, *26*, 43–45. (In Chinese)
27. *QSY GJX 0114*; Enterprise Crossing Project Construction Technical Specification for Second Westeast Natural Gas Transmission Pipeline Project. Standard of China Petroleum Pipeline Construction Project Management Department: Beijing, China, 2008. (In Chinese)
28. Liu, Y. Study on Permanent Seismic Surface Displacement of Active Fault. Ph.D. Thesis, Institute of Engineering Mechanics, Harbin, China, 2013. (In Chinese).
29. Ma, M. Study on Corrosion Defect Evaluation and Reinforcing Technology of Qinghai-Harbin Pipeline. Ph.D. Thesis, Northeast Petroleum University, Daqing, China, 2011. (In Chinese).
30. Zhao, W. Study on Failure Mode and Performance Improvement Measures of Buried Corrosion Pipeline Across Faults. Master's Thesis, Beijing University of Technology, Beijing, China, 2024. (In Chinese).
31. Zhang, Z.; Hinkson, D.; Singer, M.; Wang, H.; Nešic, S. A mechanistic model of top-of-the-line corrosion. *Corrosion* **2007**, *63*, 1051–1062. [CrossRef]
32. *GB/T 50470-2017*; Seismic Technical Code for Oil and Gas Transmission Pipeline Engineering. Ministry of Housing and Urban-Rural Development of China: Beijing, China, 2017. (In Chinese)
33. *IITK-GSDMA*; Guidelines for Seismic design of BURIED Pipelines. Indian Institute of Technology Kanpur, and Gujarat State Disaster Management Authority: Kanpur, India, 2007.
34. *CSA-Z662*; Standard Association Canadian. Oil and Gas Pipeline Systems: Mississauga, ON, Canada, 2007.
35. Gresnigt, A.M. Plastic Design of Buried Steel Pipelines in Settlement Area. *Heron.* **1986**, *31*, 1–113.
36. Jalali, H.H.; Rofooei, F.R.; Attari, N.K.; Samadian, M. Experimental and finite element study of the reverse faulting effects on buried continuous steel gas pipelines. *Soil. Dyn. Earthq. Eng.* **2016**, *86*, 1–4. [CrossRef]